T0293707

Handbook of Ocean Optics

Handbook of Ocean Optics

Editor: Harold Mason

www.callistoreference.com

Callisto Reference,
118-35 Queens Blvd., Suite 400,
Forest Hills, NY 11375, USA

Visit us on the World Wide Web at:
www.callistoreference.com

ISBN: 978-1-64116-807-6 (Hardback)

Cataloging-in-Publication Data

Handbook of ocean optics / edited by Harold Mason.
p. cm.
Includes bibliographical references and index.
ISBN 978-1-64116-807-6
1. Optical oceanography. 2. Seawater--Optical properties.
3. Physical oceanography. I. Mason, Harold.

GC178.2 .H36 2023
551.460 1--dc23

Table of Contents

Preface

Ocean optics refers to the study of the interaction of sunlight with water and the substances found in water. The research in this area includes the study of sea, lakes, coastal waters, rivers, large ocean basins, and inland waters. Knowledge of ocean optics is required in aquatic remote sensing research for comprehending the information which can be derived from the color of the water, as seen by satellite sensors in space. Ocean optics has been used to investigate topics such as zooplanktons, warming of the upper ocean, and shallow-water habitats (such as coral reefs and seagrass beds). Furthermore, it is applied to study the ocean biological pump as well as marine biogeochemistry. It also uses optical tools for measuring the size spectrum of particles in the ocean. This book contains some path-breaking studies on various topics related to ocean optics. It consists of contributions made by international experts. Students, researchers, experts, and all associated with the study of this field will benefit alike from this book.

After months of intensive research and writing, this book is the end result of all who devoted their time and efforts in the initiation and progress of this book. It will surely be a source of reference in enhancing the required knowledge of the new developments in the area. During the course of developing this book, certain measures such as accuracy, authenticity and research focused analytical studies were given preference in order to produce a comprehensive book in the area of study.

This book would not have been possible without the efforts of the authors and the publisher. I extend my sincere thanks to them. Secondly, I express my gratitude to my family and well-wishers. And most importantly, I thank my students for constantly expressing their willingness and curiosity in enhancing their knowledge in the field, which encourages me to take up further research projects for the advancement of the area.

Editor

Evaluation of Ocean Color Remote Sensing Algorithms for Diffuse Attenuation Coefficients and Optical Depths with Data Collected on BGC-Argo Floats

Xiaogang Xing [1,2,*], Emmanuel Boss [3], Jie Zhang [4] and Fei Chai [1,3]

[1] State Key Laboratory of Satellite Ocean Environment Dynamics, Second Institute of Oceanography, Ministry of Natural Resources, Hangzhou 310012, China; fchai@sio.org.cn
[2] Pilot National Laboratory for Marine Science and Technology (Qingdao), Qingdao 266000, China
[3] School of Marine Sciences, University of Maine, Orono, ME 04469, USA; emmanuel.boss@maine.edu
[4] Laboratoire d'Océanographie de Villefranche, Sorbonne Université, 06238 Villefranche-sur-Mer, France; haijijie@stu.xmu.edu.cn
* Correspondence: xing@sio.org.cn

Abstract: The vertical distribution of irradiance in the ocean is a key input to quantify processes spanning from radiative warming, photosynthesis to photo-oxidation. Here we use a novel dataset of thousands local-noon downwelling irradiance at 490 nm ($E_d(490)$) and photosynthetically available radiation (PAR) profiles captured by 103 BGC-Argo floats spanning three years (from October 2012 to January 2016) in the world's ocean, to evaluate several published algorithms and satellite products related to diffuse attenuation coefficient (K_d). Our results show: (1) MODIS-Aqua $K_d(490)$ products derived from a blue-to-green algorithm and two semi-analytical algorithms show good consistency with the float-observed values, but the Chla-based one has overestimation in oligotrophic waters; (2) The $K_d(PAR)$ model based on the Inherent Optical Properties (IOPs) performs well not only at sea-surface but also at depth, except for the oligotrophic waters where $K_d(PAR)$ is underestimated below two penetration depth ($2z_{pd}$), due to the model's assumption of a homogeneous distribution of IOPs in the water column which is not true in most oligotrophic waters with deep chlorophyll-a maxima; (3) In addition, published algorithms for the 1% euphotic-layer depth and the depth of 0.415 mol photons m^{-2} d^{-1} isolume are evaluated. Algorithms based on Chla generally work well while IOPs-based ones exhibit an overestimation issue in stratified and oligotrophic waters, due to the underestimation of $K_d(PAR)$ at depth.

Keywords: diffuse attenuation coefficient; satellite product assessment; euphotic layer depth; isolume depth

---

1. Introduction

Light from the sun fuels oceanic primary production, heats the upper ocean, and oxidizes chemical compounds such as organic molecules. In order to accurately model these processes, the subsurface light distribution is needed. As light attenuates near-exponentially in water, to describe the subsurface light field, the exponent describing this attenuation needs to be known (referred to as the diffuse attenuation coefficient [1]). This exponent is a function of in-water components themselves and, to a lesser degree, a function of the illumination conditions [2]. Additionally, in order to constrain the layer in which certain processes take place, specific light horizons are of interest. For example, for photosynthesis, a euphotic or isolume depth is defined based on a relative light level (e.g., 1% [3]) or an absolute intensity (e.g., 0.415 mol photons m^{-2} d^{-1} [4]) below which photosynthesis is assumed as zero. Moreover, accurately quantifying the attenuation of sunlight within the upper ocean is essential for physical and biogeochemical models, affecting the modeled upper-ocean temperature (e.g., [5,6]) and ecosystem dynamics (e.g., [7]). Sea-surface sunlight is globally available from space agencies ([8]), and a variety of models to describe its attenuation through the water have been devised (e.g., [9,10]). However, lack of data has limited their validation on global scales.

In the past decade, with the rapid progress of the Biogeochemical-Argo (BGC-Argo) float technology, the BGC-Argo dataset has become the largest data source for optical and biogeochemical observations in the global ocean [11,12]. Downwelling radiometry is one of the six core BGC variables in the International Argo program [13,14], and has been used to study global bio-optical relationships and anomalies [15,16], used to determine the depths of the euphotic zone and a specific isolume [17,18], to improve the parameterization scheme of sunlight attenuation in a biological model [19], as well as to correct and derive other bio-optical measurements performed on the same float [20–22]. An additional important application of BGC-Argo is to validate ocean color remote sensing products (e.g., [23–25]) over the whole year and on the global scale.

In this study, based on a recently compiled global BGC-Argo dataset (described in [15,26,27]), first, we evaluate several ocean color remote sensing algorithms for the diffuse attenuation coefficient at 490 nm ($K_d(490)$), including two empirical [28,29] and two semi-analytical algorithms [9,30]; then, we evaluate the performance of the algorithm for the diffuse attenuation coefficient ($K_d(PAR)$) of photosynthetically available radiation (PAR) [10], at different optical depths; finally, we evaluate several empirical [29,31] and semi-analytical algorithms [32] for the euphotic depth (z_{eu}) and the 0.415 mol photons m^{-2} d^{-1} isolume depth ($z_{0.415}$). Given our dataset, we focus only on open ocean algorithms. The need for validation stems from the fact that these algorithms have not been validated globally and over the whole year.

2. Data and Methods

2.1. BGC-Argo Data

The globally distributed BGC-Argo dataset used here was published in the SEA scieNtific Open data Edition (SEANOE, www.seanoe.org/data/00383/49388/) and is comprised of more than 5000 local-noon profiles of downwelling irradiance (different numbers for different wavelengths due to quality control filtering [26]) obtained using 103 Satlantic OCR504 radiometers deployed on Argo floats, spanning 3 years from 20 October 2012 to 26 January 2016. Each radiometry data depth profile includes downward irradiance (E_d) at three wavelengths (380, 412, 490 nm, [μW cm^{-2} nm^{-1}]), and the instantaneous photosynthetically available radiation (iPAR, [μmol photons m^{-2} s^{-1}]), which is the downwelling photon flux integrated over a wavelength range spanning from 400 to 700 nm. All radiometry data has been quality controlled [26], removing the points and profiles that are significantly affected by clouds and wave focusing. The ocean basin and trophic environment of each profile have been identified in the dataset (www.seanoe.org/data/00383/49388/data/49825.pdf), however, in this study, we reorganized them into 10 regions (Table 1 and Figure 1). All symbols used in this study are listed in Table 2. We note that no correction for dark offset has been applied in this dataset beyond that obtained from the manufacturer. Small (O(0.03 μW cm^{-2} nm^{-1})) nonzero dark values are observed in data from profiling floats but these values are variable in sign and hence will not introduce a bias in our current work.

Table 1. Regions and Basins Classified in This Dtudy and Corresponding Profile Numbers of $E_d(490)$ and iPAR.

Region Name in This Study	Basin Name Classified in the SEANOE-BGC-Argo Dataset	$E_d(490)$ Num *	iPAR Num *
Southern Ocean	ATL (Southern Ocean Atlantic sector) ATOI (Southern Ocean Atlantic to Indian sector) IND (Southern Ocean Indian sector)	974 (894)	1097 (471)
Subpolar Gyre	LAS (Labrador Sea) IRM (Irminger Sea) ICB (Iceland Basin) SLAS (South Labrador Sea)	1618 (1465)	1610 (885)
Transition Zone	NASTZ (North Atlantic Transition Zone) EQNASTZ (North Atlantic South Transition Zone) SASTZ (South Atlantic Transition Zone)	197 (137)	205 (83)
Red Sea	RED (Red Sea)	62 (39)	56 (8)
Black Sea	BLACK (Black Sea)	128 (118)	134 (84)
West Mediterranean (Med.) Sea	NW (Northwestern) SW (Southwestern) TYR (Tyrrhenian Sea)	1121 (939)	1141 (472)
East Med. Sea	ION (Ionian Sea) LEV (Levantine Sea)	886 (683)	801 (206)
Subtropical Gyre	NASTG (North Atlantic Subtropical Gyre) ENASTG (Eastern North Atlantic Subtropical Gyre) WNASTG (Western North Atlantic Subtropical Gyre) SASTG (South Atlantic Subtropical Gyre) SSASTG (South Atlantic South Subtropical Gyre) SPSTG (South Pacific Subtropical Gyre)	549 (505)	630 (282)
Arctic Sea	NOR (Norwegian Sea)	61 (57)	83 (39)
New Caledonia	NC (New Caledonia)	52 (45)	58 (18)
TOTAL		5648 (4882)	5815 (2548)

* The numbers in the parentheses represent the numbers of profiles used in this study, with valid surface $E_d(490,0^-)$ and $iPAR(0^-)$.

Table 2. Symbols used in this study.

Symbol	Definition	Algorithm	Reference
α	Air-sea interface transmission factor	Modeled	[33]
$a(490)$	Absorption coefficient at 490 nm	Quasi-analytical	[34]
$b_b(490)$	Backscattering coefficient at 490 nm	Quasi-analytical	[34]
$Chla_{sat}$	Downloaded MODIS-Aqua chlorophyll-a concentration	Empirical	[35]
$E_d(490)$	Downwelling irradiance at 490 nm	Observed	[26]
iPAR	Instantaneous photosynthetically available radiation	Observed	[26]
$K_d(490)_{M\text{-}L05a}$	Derived MODIS-Aqua $K_d(490)$ product	Semi-analytical	[9]
$K_d(490)_{M\text{-}L13}$	Derived MODIS-Aqua $K_d(490)$ product	Semi-analytical	[30]
$K_d(490)_{M\text{-}KD2M}$	Downloaded MODIS-Aqua $K_d(490)$ product	Empirical	[28]
$K_d(490)_{M\text{-}M07}$	Derived MODIS-Aqua $K_d(490)$ product	Equation (5)	[29]
$K_d(490)_{GC\text{-}L05a}$	Downloaded GlobColour $K_d(490)$ product	Semi-analytical	[9]
$K_d(490)_{GC\text{-}M07}$	Downloaded GlobColour $K_d(490)$ product	Equation (5)	[29]
$K_d(PAR)_z$	Layer-averaged $K_d(490)$ from surface to any depth	Equation (1)	/
$K_d(490)_{zpd}$	Float-observed near surface $K_d(490)$	Equation (3)	[36]
$K_d(490)_{zpd\text{-}O17}$	Float-observed near surface $K_d(490)$	Appendix A	[15]
$K_d(PAR)_{z\text{-}L05b}$	Derived MODIS-Aqua layer-averaged $K_d(PAR)$ from surface to any depth	IOPs-based	[10]
$K_d(PAR)_z$	Layer-averaged $K_d(PAR)$ from surface to any depth	Equation (2)	/
PAR_{sat}	Downloaded MODIS-Aqua daily PAR	/	[8]
$z_{0.415}$	Isolume depth at 0.415 mol photons m^{-2} d^{-1}	Equation (4)	[4]
$z_{0.415\text{-}B10}$	Derived MODIS-Aqua $z_{0.415}$ product	Equations (6) and (7)	[31]
$z_{0.415\text{-}L07}$	Derived MODIS-Aqua $z_{0.415}$ product	IOPs-based	Modified from [32]
$z_{1\%490}$	Float-observed 1% light depth for $E_d(490)$	Equation (1)	/
z_{eu}	Float-observed 1% light depth for iPAR	Equation (2)	[3]
$z_{eu\text{-}L07}$	Downloaded MODIS-Aqua z_{eu} product	IOPs-based	[32]
$z_{eu\text{-}M07}$	Derived MODIS-Aqua z_{eu} product	Equation (6)	[29]
z_{pd}	Penetration depth at 490 nm	Equation (3)	[36]

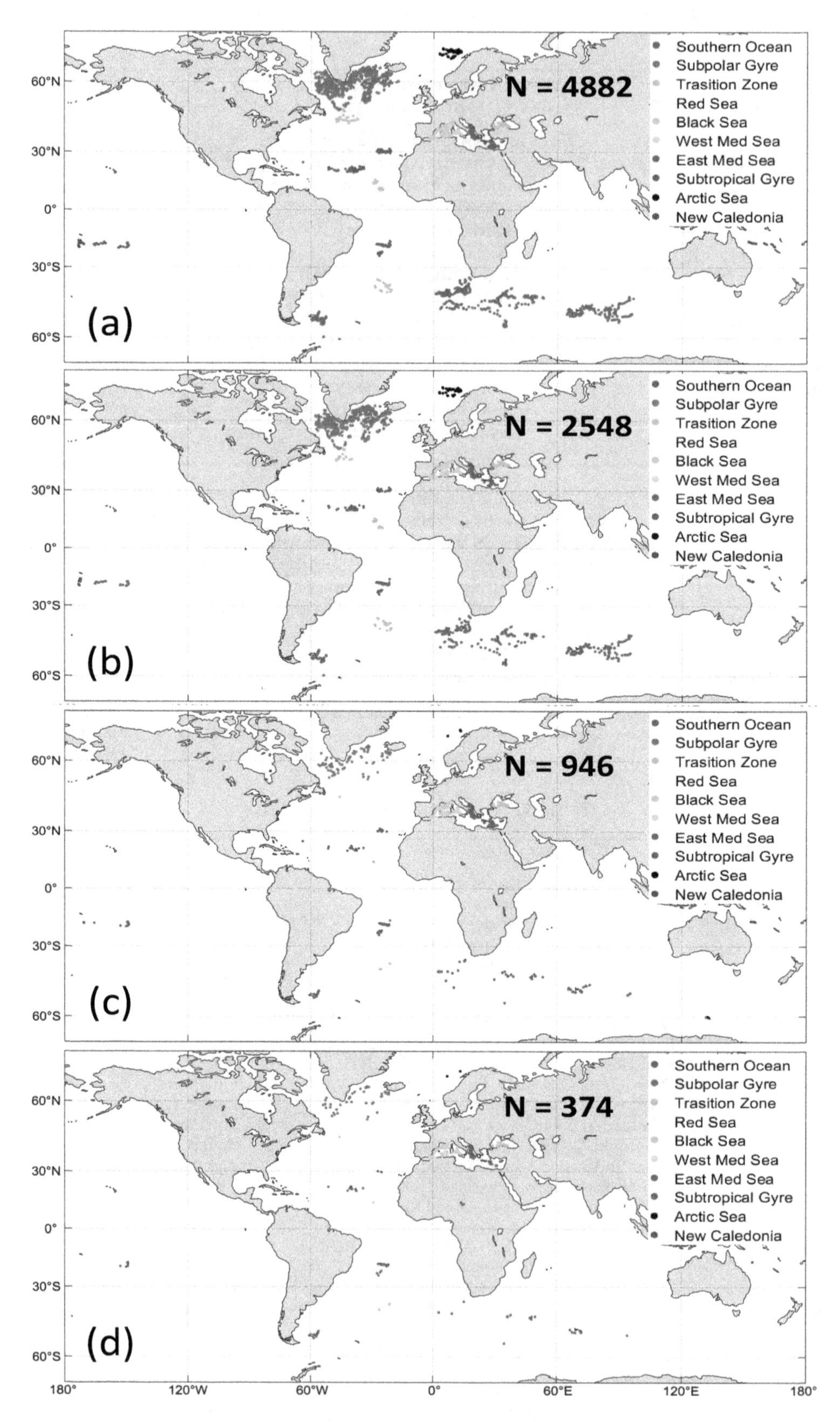

Figure 1. BGC-Argo profile data distribution maps used in this study, for all points (**a**) with valid surface $E_d(490,0^-)$ (N = 4882), (**b**) with valid $iPAR(0^-)$ (N = 2548), (**c**) with matchup with all satellite $K_d(490)$ products (N = 946), and (**d**) with matchup with all satellite $K_d(PAR)$ products (N = 374, same for z_{eu} and $z_{0.415}$ products), respectively.

2.1.1. Layer-Averaged Diffuse Attenuation Coefficients, Penetration, and Euphotic Depths

The attenuation of $E_d(490)$ and of PAR with depth is approximated as an exponential decrease, and the layer-averaged diffuse attenuation coefficients, $K_d(490)_z$ and $K_d(PAR)_z$, from surface (0^-) to any given depth, z, are defined as:

$$K_d(490)_z = \frac{1}{z}\ln\left(\frac{E_d(490,0^-)}{E_d(490,z)}\right) \quad (1)$$

$$K_d(PAR)_z = \frac{1}{z}\ln\left(\frac{iPAR(0^-)}{iPAR(z)}\right) \quad (2)$$

The BGC-Argo radiometry data processing flow chart is shown in Figure 2, where the surface $E_d(490,0^-)$ and $iPAR(0^-)$ are determined first. Since BGC-Argo has no radiometry exactly at $z = 0^-$, they are computed by extrapolating a linear regression for $\ln(E_d(490,z))$ and a second-degree polynomial regression for $\ln(iPAR(z))$ with depth, z, respectively, at the top 10 m of each profile. To ensure the accuracy of extrapolation, we only process the profiles with at least 5 valid values within the top 10 m (which include 4882 $E_d(490)$ and 2548 iPAR profiles, as shown in Table 1). This extrapolation method is validated with in-situ cruise data in Appendix A.

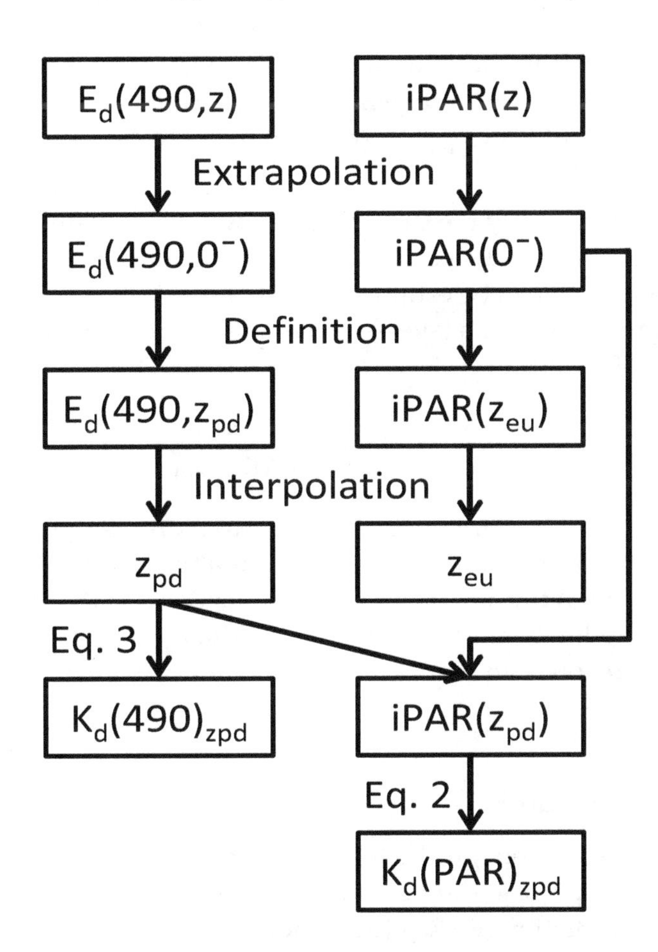

Figure 2. Flow chart of BGC-Argo radiometry data processing. Note that all derived K_d values are layer-averaged ones (the layer from sea surface to depth z, thus, z as the subscript in the symbols, e.g., $K_d(490)_z$), rather than K_d at a specific depth. The same procedure could be used to derive the layer-averaged attenuation coefficient to any depth of interest.

We then determine the penetration depth at 490 nm (z_{pd}) and euphotic depth (z_{eu}), as the depth where $E_d(490,z)$ reaches $E_d(490,0^-) \times e^{-1} \sim 0.37 \times E_d(490,0^-)$ [36] and the depth (denoted by z_{eu})

where iPAR(z) reaches iPAR(0^-) × 1% [3], respectively. z_{pd} and z_{eu}, respectively, are determined through an interpolation of $E_d(490)$ and iPAR profiles to depths where $E_d(490)$ and iPAR reach the calculated $E_d(490, z_{pd}) = E_d(490, 0^-)/e$ and $iPAR(z_{eu}) = 0.01\ iPAR(0^-)$ (Figure 2). Note that, the definition of penetration depth is spectrally dependent (i.e., z_{pd} varies with wavelength), in this study, the wavelength is omitted for simplicity, and all z_{pd} without designated wavelength represents $z_{pd}(490)$. By definition, the remotely-sensed diffuse attenuation coefficient $K_d(490)$ is the layer-averaged one from surface to z_{pd}, and it is determined in situ as:

$$K_d(490)_{zpd} = \frac{1}{z_{pd}} \tag{3}$$

Note that Organelli et al. [15] used an approximation to estimate the penetration depth, to derive $K_d(490)_{zpd}$, namely $z_{pd} = z_{eu}/4.6$. In Appendix B we show that this approximation results in a significant bias in $K_d(490)_{zpd}$ in oligotrophic regions.

2.1.2. The 0.415 mol Photons m^{-2} d^{-1} Isolume Depth ($z_{0.415}$)

Since the work of Letelier et al. [4], the isolume depth $z_{0.415}$, defined as the depth where the daily PAR reaches 0.415 mol photons m^{-2} d^{-1}, has been used as a threshold depth below which light is insufficient to support photosynthesis, and has been regarded as a useful estimate of the euphotic depth [31,37–39]. Phytoplankton living at depth are expected to react to absolute light level rather than a relative level [40], and thus, the isolume is more applicable for phytoplankton growth than the depth of 1% of sea surface light intensity (z_{eu}). Note that recently, Behrenfeld and Boss [41] argued for a lower threshold irradiance value.

Generally, it is difficult to directly measure $z_{0.415}$, as well as any other daily-isolume depth, as it needs the radiometry at both continuous vertical scale (at least a resolution of meters) and continuous temporal scale (at least hourly). It can be estimated using a single observation during a day (on Argo and ships), with the assumption that the layer-averaged $K_d(PAR)$ at $z_{0.415}$ is constant during the day (not a constant $K_d(PAR)$ profile), i.e.,:

$$\frac{iPAR(z_{0.415})}{iPAR(0^-)} = \frac{0.415}{PAR_{sat} \times \alpha} \tag{4}$$

Equation (4) allows us to determine iPAR($z_{0.415}$) based on a single profile of radiometry and PAR_{sat}, the surface daily photosynthetically available radiation product (see Section 2.2.1). α is the transmission of sun light through the air-sea interface, dependent on the latitude and day of year [33]. $z_{0.415}$ is determined as the depth where iPAR(z) reaches iPAR($z_{0.415}$) which is calculated from Equation (4). Using a similar procedure, one can derive the depth of any other isolume.

2.2. Satellite Data

2.2.1. Auxiliary Satellite Data Products

We use several satellite-derived products (Level-3, daily, 4 km-resolution) provided by the Moderate Resolution Imaging Spectroradiometer (MODIS) on the satellite "EOS PM (Aqua)": (1) Daily photosynthetically available radiation (PAR_{sat} [mol photons m^{-2} d^{-1}]) (https://oceandata.sci.gsfc.nasa.gov/MODIS-Aqua/Mapped/Daily/4km/par/) for calculating $z_{0.415}$. This product is derived from the radiative transfer model [8] and the atmospheric properties inferred from MODIS-Aqua; (2) Surface chlorophyll-a concentration ($Chla_{sat}$) (https://oceandata.sci.gsfc.nasa.gov/MODIS-Aqua/Mapped/Daily/4km/chlor_a/). This product combines two algorithms, band-ratio OCx algorithm [35], and the Hu's color index (CI) algorithm [42]; (3) Remote sensing reflectance (R_{rs} [sr^{-1}]) at four wavelengths (443, 488, 555, and 667 nm) (https://oceandata.sci.gsfc.nasa.gov/MODIS-Aqua/Mapped/Daily/4km/Rrs_443/) are used as input to the quasi-analytical algorithm (QAA) [34] to retrieve the absorption and backscattering coefficients at 490 nm (a(490) and $b_b(490)$), and in turn, to calculate $K_d(490)$ following [9] and [30], and to calculate $K_d(PAR)$ following [10]. Note that the Raman effect on R_{rs} is corrected following [30].

2.2.2. Satellite-Based $K_d(490)$

We download three available satellite $K_d(490)$ products and derive three other literature-based $K_d(490)$ for evaluation using the floats' data. They encompass different approaches (empirical vs. semi-analytical) and different sources (MODIS-Aqua vs. GlobColour):

(i) $K_d(490)_{M\text{-}KD2M}$ is an operational MODIS product derived from an empirical algorithm "KD2M" that uses the blue-green reflectance ratio ("B/G") of satellite-measured remote sensing reflectance [28] and is available from NASA (https://oceandata.sci.gsfc.nasa.gov/MODIS-Aqua/Mapped/Daily/4km/Kd_490/);

(ii) $K_d(490)_{GC\text{-}M07}$ is an operational product retrieved from an empirical relationship between $K_d(490)$ and Chla (Equation (5), [29]), provided by the GlobColour project [43], using products merged between MODIS and a visible infrared imaging radiometer (VIIRS) for our data period (ftp://ftp.hermes.acri.fr/GLOB/merged/day/);

$$K_d(490) = 0.0166 + 0.077298 \times [\text{Chla}]^{0.67155} \quad (5)$$

(iii) $K_d(490)_{M\text{-}M07}$ is derived from MODIS-retrieved Chla_{sat} and Equation (5) (the same empirical algorithm as $K_d(490)_{GC\text{-}M07}$);

(iv) $K_d(490)_{GC\text{-}L05a}$ is an operational product retrieved from a semi-analytical IOPs-$K_d(490)$ algorithm [9] available from GlobColour (ftp://ftp.hermes.acri.fr/GLOB/merged/day/);

(v) $K_d(490)_{M\text{-}L05a}$ is derived from MODIS-Aqua R_{rs}, based on the same algorithm as $K_d(490)_{GC\text{-}L05a}$; and

(vi) $K_d(490)_{M\text{-}L13}$ is derived from MODIS-Aqua R_{rs}, based on a similar but updated semi-analytical algorithm [30]. Compared to L05a, the L13 algorithm added a new input: the backscattering coefficient of pure sea water ($b_{bw}(490)$). Here we take it as a constant, 0.001387 m^{-1}, which corresponds to the sea-surface water with temperature at 20 °C and salinity at 35 psu, based on [44]. We have tried to use the Argo-observed temperature and salinity to compute a more accurate $b_{bw}(490)$ following [44], but found no obvious improvement for $K_d(490)$ retrieval, when compared to using a constant value.

2.2.3. Satellite-Based $K_d(PAR)_z$

Based on radiative-transfer computations with Hydrolight [2], Lee et al. [10] proposed an IOPs-$K_d(PAR)$ algorithm, which can be used to estimate the layer-averaged $K_d(PAR)$ at any depth, with the inputs of remotely-sensed surface a(490) and $b_b(490)$. In this study, we evaluate $K_d(PAR)_{M\text{-}L05b}$ at six optical depths (from z_{pd} to $6z_{pd}$).

2.2.4. Satellite-Based z_{eu} and $z_{0.415}$

Two euphotic layer depth algorithms are evaluated here:

(i) $z_{eu\text{-}L07}$, is an operational MODIS product provided by NASA (https://oceandata.sci.gsfc.nasa.gov/MODIS-Aqua/Mapped/Daily/4km/Zeu_lee/), based on [32], which is an implementation of the IOPs-$K_d(PAR)$ model [10];

(ii) $z_{eu\text{-}M07}$, is derived from MODIS-retrieved Chla_{sat}, based on an empirical Chla-z_{eu} relationship proposed by Morel et al. [29]:

$$\log_{10}(z_{eu-M07}) = 1.524 - 0.436X - 0.0145X^2 + 0.0186X^3 \quad (6)$$

Here, $X = \log_{10}(\text{Chla}_{sat})$.

For satellite-based $z_{0.415}$ products, two products are computed:

(i) $z_{0.415\text{-}B10}$, is derived following [31], based on PAR_{sat} and Morel's euphotic layer depth ($z_{eu\text{-}M07}$):

$$z_{0.415-B10} = \ln(0.415 / PAR_{sat} / \alpha) \times (z_{eu-M07} / \ln(0.01)) \quad (7)$$

While $\alpha = 0.98$ in [31], in this study we use the look-up table provided by Mobley and Boss [33] which takes into account the sun angle and a wind ruffled surface.

(ii) $z_{0.415\text{-}L07}$, is calculated following [32], whose method for z_{eu}, can also be used for $z_{0.415}$.

2.2.5. Satellite-Float Matchup Criteria

For each float profile, the median value of the corresponding satellite data within a 3 × 3 pixel box centered on the profile's surface position was used if at least 5 values in the box were available and that the profile was within 3 hrs of satellite overpass (Figure 3), consistent with the criteria of Bailey and Werdell [45].

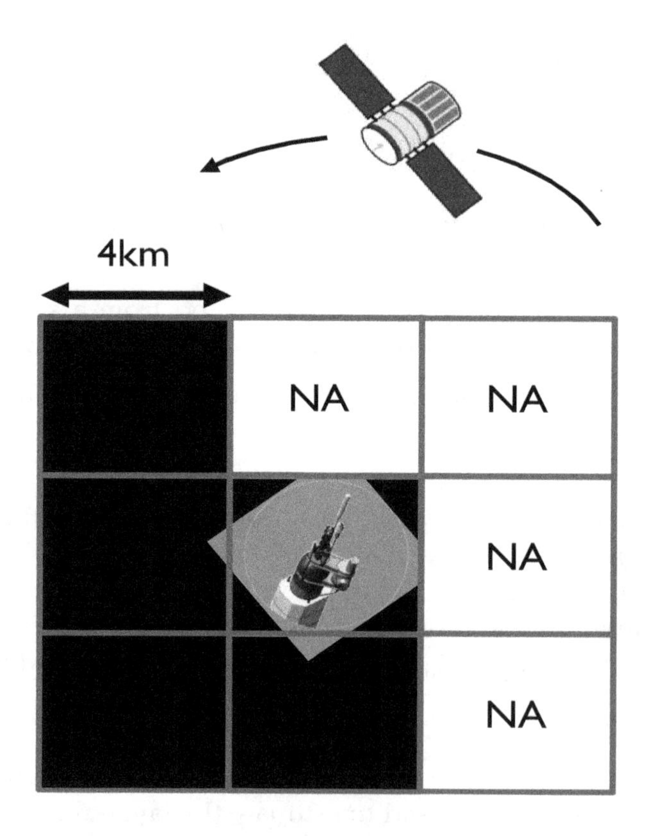

Figure 3. Diagram of a good match-up between satellite and float data. The 9 squares represent 9 pixels closest to the location of float surfacing, with 5 valid pixels (blue) and 4 invalid ones (white). The spatial resolution is 4 km. In such a condition or with more than 5 valid-value pixels out of 9, a pair of valid satellite-float matchup data is obtained.

2.3. Statistical Metrics

Three statistical metrics are used to evaluate the differences between products derived from the float database and satellite products: (1) Mean absolute difference (MAD), which represents the absolute differences between the measured (observation) values, and satellite-derived or model-estimated values; (2) Mean absolute percentage difference (MAPD), which represents the relative differences; (3) and Mean percentage difference (MPD), which represents the relative system bias. They are defined as:

$$MAD = \frac{1}{n}\sum_{i=1}^{n}|E_i - M_i| \quad (8)$$

$$\text{MAPD} = \frac{100}{n}\sum_{i=1}^{n}\frac{|E_i - M_i|}{M_i}\ (\%) \tag{9}$$

$$\text{MPD} = \frac{100}{n}\sum_{i=1}^{n}\frac{(E_i - M_i)}{M_i}\ (\%) \tag{10}$$

Here, M_i is the in-situ measured value, E_i represents the satellite-retrieved or model-estimated value, and n is the number of observations.

3. Results and Discussion

3.1. Distribution of $K_d(490)_{zpd}$, z_{eu}, and $z_{0.415}$ for the BGC-Argo Dataset

In the BGC-Argo dataset, $K_d(490)_{zpd}$ varies by over an order of magnitude, from 0.02 m^{-1} to 0.3 m^{-1} (Figure 4a). Regionally, the median $K_d(490)_{zpd}$ is observed to be lowest in subtropical gyres (0.024 m^{-1}) and highest in the Black Sea (0.112 m^{-1}), with the highest dynamic ranges in the Western Mediterranean (Med.) Sea, Southern Ocean, and subpolar regions (Figure 4b). z_{eu} spans about an order of magnitude in this dataset, from ~ 25 m in the (North Atlantic) subpolar gyre to ~ 250 m in subtropical gyres (Figure 4c), with a median value being shallowest (34.0 m) in the Black Sea (Figure 4d) where high CDOM concentration contributes significantly to attenuation [46]. $z_{0.415}$ has a similar dynamic range and regional distribution as z_{eu} (Figure 4e,f) with median values varying from 23.5 m in the Black Sea to 118.5 m in the subtropical gyres. It exhibits a few very shallow values (<10 m, see Figure 4e) corresponding to very cloudy days (daily PAR above sea surface are very low, e.g., <2 mol photons m^{-2} d^{-1}). In fact, $z_{0.415}$ is often shallower than z_{eu}, because the highest daily PAR at sea surface is about 70–80 mol photons m^{-2} d^{-1} for a clear summer day [33], and then the corresponding PAR(z_{eu}) is about 0.7–0.8 mol photons m^{-2} m^{-1}, slightly higher than 0.415 mol photons m^{-2} m^{-1} (as light intensity in water decreases with depth exponentially, about 5.2 m (6.6 m) are needed to decrease PAR from 0.7 (0.8) mol photons m^{-2} m^{-1} to 0.415 mol photons m^{-2} m^{-1} for a moderate K_d(PAR) as 0.1 m^{-1}). Thus, for a clear summer day, $z_{0.415}$ may be deeper than z_{eu} by about 5 m; but when clouds cause highly depressed daily PAR values, or in other seasons, z_{eu} is deeper than $z_{0.415}$. In the dataset analyzed here, the median z_{eu} is always deeper than the median $z_{0.415}$ in all regions (Figure 4d,f), by 3.64 m in subtropical gyres and by 14.7 m in the Southern Ocean.

3.2. Assessment of Satellite Algorithms for $K_d(490)$

The performance of classical B/G algorithm ($K_d(490)_{M\text{-}KD2M}$) and semi-analytical algorithm ($K_d(490)_{M\text{-}L13}$) outperform the others with this dataset (Figure 5a,b, Table 3). The former has the lowest relative differences (MAPD = 14.1%) and no significant mean bias (MPD = −0.3%), but it has a few points overestimated in the high-value range (>0.2 m^{-1}); the latter has the lowest absolute differences (MAD = 0.009 m^{-1}) and it does not exhibit the overestimation issue at high-values as the B/G algorithm. Note, however, that both algorithms underestimate $K_d(490)$ in the Black Sea. It is likely due to high colored dissolved organic matter (CDOM) concentration in this region [16,46], which affects the performance of both band-ratio and semi-analytical algorithms that have been devised using open-ocean data. $K_d(490)_{M\text{-}L05a}$ (Figure 4c) also performs well in these highly attenuating waters with similar but slightly worst results than $K_d(490)_{M\text{-}L13}$. Chla-based $K_d(490)_{M\text{-}M07}$ has the highest MAPD (19.1%) among all 4 MODIS-Aqua products, with an overestimation bias (MPD = 8.7%) mainly in oligotrophic waters (Figure 4d). As for the GlobColour products, $K_d(490)_{GC\text{-}L05a}$ exhibits an obvious overestimation in all waters (Figure 4e), having high relative and absolute differences (MAPD = 52.4%, MPD = 51.5%). Comparing MODIS-Aqua and GlobColour with the same Chla-based empirical algorithm (M07), we find little difference (Table 3); however, for the semi-analytical algorithm (L05a), $K_d(490)_{M\text{-}L05a}$, is much closer to float-observed $K_d(490)_{zpd}$, than $K_d(490)_{GC\text{-}L05a}$, suggesting that the overestimation of $K_d(490)_{GC\text{-}L05a}$ is associated with GlobColour, rather than the model of Lee et al. itself.

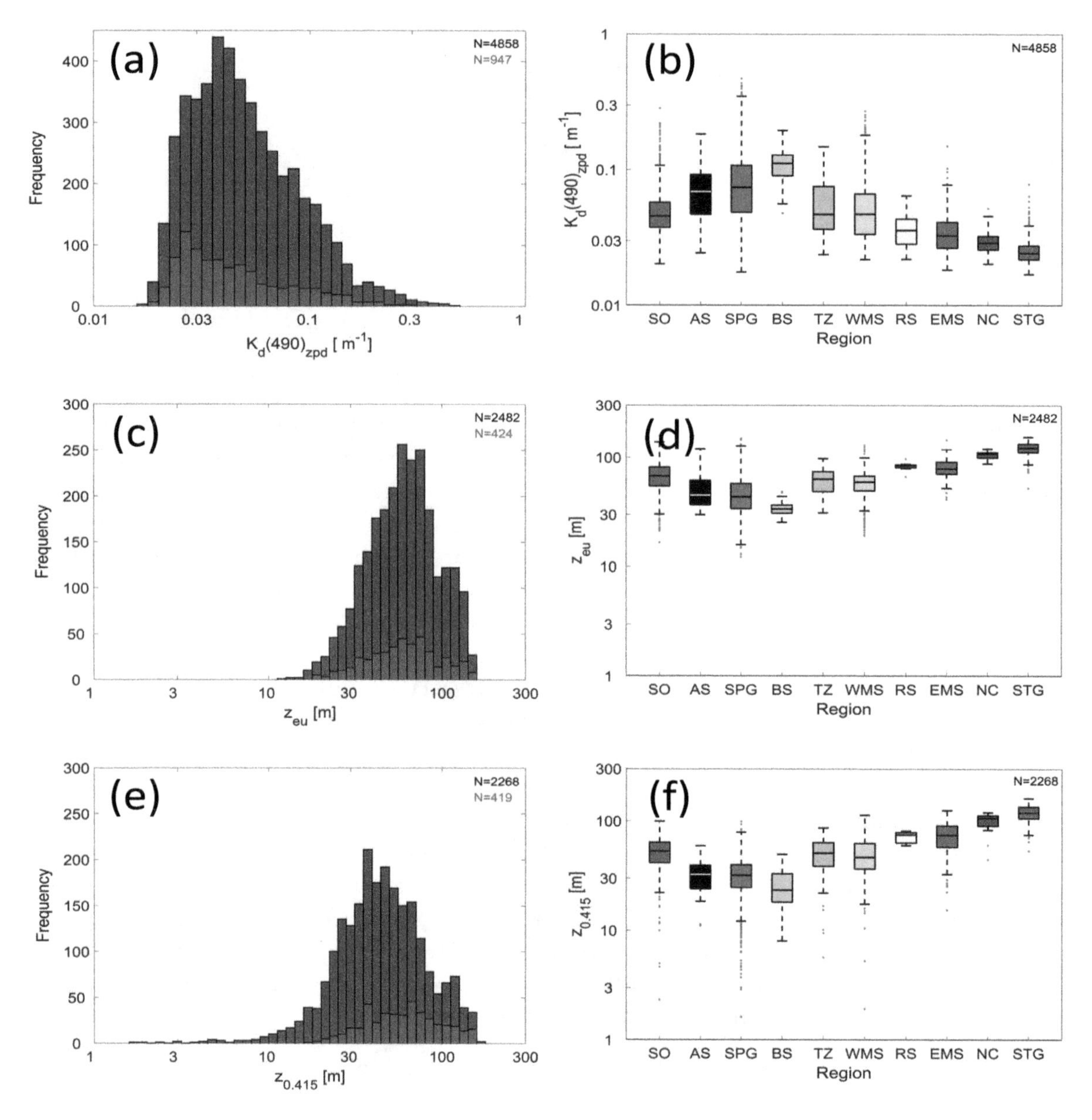

Figure 4. Histogram and boxplot of the distribution of the (**a**,**b**) near surface diffuse attenuation coefficient at 490 nm ($K_d(490)_{zpd}$), (**c**,**d**) euphotic layer depth (z_{eu}), (**e**,**f**) and isolume depth ($z_{0.415}$), for each of the 10 regions (SO: Southern Ocean; AS: Arctic Sea; SPG: Subpolar Gyre; BS: Black Sea; TZ: Transition Zone; WMS: West Med. Sea; RS: Red Sea; EMS: East Med. Sea; NC: New Caledonia; STG: Subtropical Gyre). In the histogram, black and red columns (as well as numbers) represent the full dataset and the satellite-matched ones, respectively. In the boxplot, red points beyond the end of the whiskers represent outliers beyond the 1.5 × IQR (IQR = interquartile range) threshold.

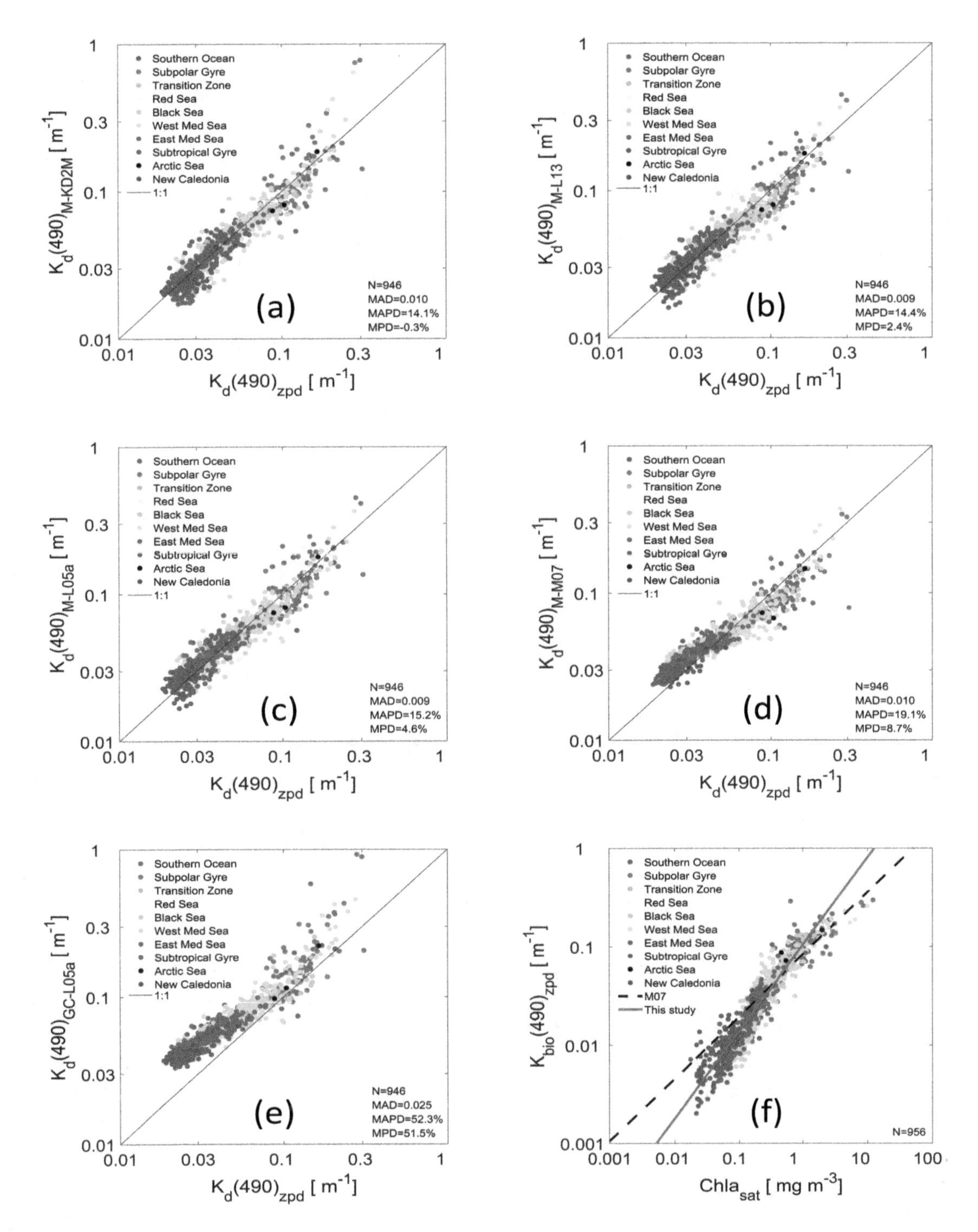

Figure 5. Assessment of satellite $K_d(490)$ products based on BGC-Argo dataset. Scatter plot of float-observed $K_d(490)_{zpd}$ v.s. MODIS-Aqua B/G-based empirical $K_d(490)_{M\text{-}KD2M}$ (**a**) semi-analytical $K_d(490)_{M\text{-}L13}$, (**b**) semi-analytical $K_d(490)_{M\text{-}L05a}$, (**c**) Chla-based empirical $K_d(490)_{M\text{-}M07}$, (**d**) GlobColour semi-analytical $K_d(490)_{GC\text{-}L05a}$, (**e**) Scatter plot of float-observed $K_d(490)_{zpd}$ vs. MODIS-Aqua-retrieved Chla, (**f**) black dashed line represents the empirical equation (Equation (5)) used in $K_d(490)_{GC\text{-}M07}$ and $K_d(490)_{M\text{-}M07}$, and the red solid line represents the relationship obtained in this study (Equation (11)). Black solid lines are the 1:1 lines. Note that in Panel (**a**)–(**e**) we use only data for which all algorithms could be evaluated.

Table 3. Statistical results of evaluation on the satellite-based $K_d(490)$ products *.

Product	Platform	Source	Algorithm	MAD	MAPD	MPD	Figure
$K_d(490)_{M\text{-}KD2M}$	MODIS-Aqua	Downloaded	B/G-based	$0.010\ m^{-1}$	14.1%	−0.3%	Figure 5a
$K_d(490)_{M\text{-}L13}$	MODIS-Aqua	Calculated	IOPs-based	$0.009\ m^{-1}$	14.4%	2.2%	Figure 5b
$K_d(490)_{M\text{-}L05a}$	MODIS-Aqua	Calculated	IOPs-based	$0.009\ m^{-1}$	15.2%	4.6%	Figure 5c
$K_d(490)_{GC\text{-}L05a}$	GlobColour	Downloaded	IOPs-based	$0.025\ m^{-1}$	52.4%	51.5%	Figure 5e
$K_d(490)_{M\text{-}M07}$	MODIS-Aqua	Calculated	Chla-based	$0.010\ m^{-1}$	19.1%	8.7%	Figure 5d
$K_d(490)_{GC\text{-}M07}$	GlobColour	Downloaded	Chla-based	$0.011\ m^{-1}$	19.5%	11.7%	/

* We compared 946 matchup values for which all products were available.

In addition, we find that the relationship of float-observed $K_d(490)_{zpd}$ and $Chla_{sat}$ (Figure 5f) is different than that in [29], having a larger scale factor and a higher exponent (Figure 5f and Equation (11)):

$$K_d(490)_{zpd} = 0.0166 + (0.1056 \pm 0.002) \times Chla_{sat}{}^{(0.886\pm0.010)}\ (R^2 = 0.94) \quad (11)$$

We applied this relationship to retrieve $K_d(490)_{zpd}$ from $Chla_{sat}$ and found it to have similar statistical differences to $K_d(490)_{M\text{-}KD2M}$ (MAD = $0.013\ m^{-1}$, MAPD = 15.1%, MPD = 2.1%). Although our results show that the updated Chla-$K_d(490)$ equation has limited improvement on the $K_d(490)$ retrieval from satellite, it suggests the need to re-consider the bio-optical relationship with more in-situ data.

In summary, the GlobColour-provided semi-analytical $K_d(490)$ product seems to have a retrieval problem while the classical B/G-based KD2M algorithm performs best except in highly turbid waters. The semi-analytical algorithm [30] also performs well in the open ocean, and even better in the high-value range as it is designed for solving both clear and turbid waters.

3.3. *Assessment of the Satellite Algorithm for $K_d(PAR)$*

The IOPs-$K_d(PAR)$ algorithm [10] was the only one to date to estimate the layer-averaged $K_d(PAR)$ at any depth. It was developed on the basis of Hydrolight simulation to provide a layer-averaged $K_d(PAR)$ at any depth from remote sensing. However, the Hydrolight simulation conducted by Lee et al. [10] assumed a homogeneous distribution of IOPs in the water column. Since the surface a(490) and $b_b(490)$ only represent the IOPs within the upper mixed layer, changes of IOPs below the mixed layer depth (MLD) are not accounted for in the IOPs-$K_d(PAR)$ algorithm. Satellite-retrieved $K_d(PAR)_{z\text{-}L05b}$ is evaluated with the BGC-Argo dataset from z_{pd} to $6z_{pd}$ (Figure 6 and Table 4). It performs well at all six optical depths in the high latitude (e.g., Southern Ocean, Subpolar Gyre, and Arctic Ocean), where most water-columns are well-mixed. However, in the stratified waters (including almost all profiles in subtropical gyres, most in the Eastern Mediterranean Sea, and many in the Western Mediterranean Sea), the algorithm increasingly and gradually underestimates $K_d(PAR)$ from $2z_{pd}$ to $4z_{pd}$ (from 5.7% to 0.8% of relative light intensity for PAR). In subtropical gyres, MPD decreases from 4.4% at z_{pd} to −14.6% at $4z_{pd}$ (Table 4). Since Chla increases with depth from MLD to the deep chlorophyll maximum (DCM) in oligotrophic waters [47], both a(490) and $b_b(490)$ are expected to increase with depth as well. As a consequence, the IOPs-$K_d(PAR)$ algorithm inevitably underestimates $K_d(PAR)$, even if the QAA algorithm could retrieve sea-surface a(490) and $b_b(490)$ correctly. z_{eu} corresponds roughly to 3–4 times z_{pd} (see Appendix B), and is near the DCM in oligotrophic waters [4,47]. It follows that the largest underestimation of $K_d(PAR)$ is at $4z_{pd}$. At $5z_{pd}$ and $6z_{pd}$ (correspondingly 0.43% and 0.28% of relative light intensity for PAR), there are fewer valid BGC-Argo observations and the underestimation issue is not clear. In addition, given that below the DCM Chla, a(490) and $b_b(490)$ are observed to decrease with depth, the underestimation of $K_d(PAR)$ at such depths will weaken.

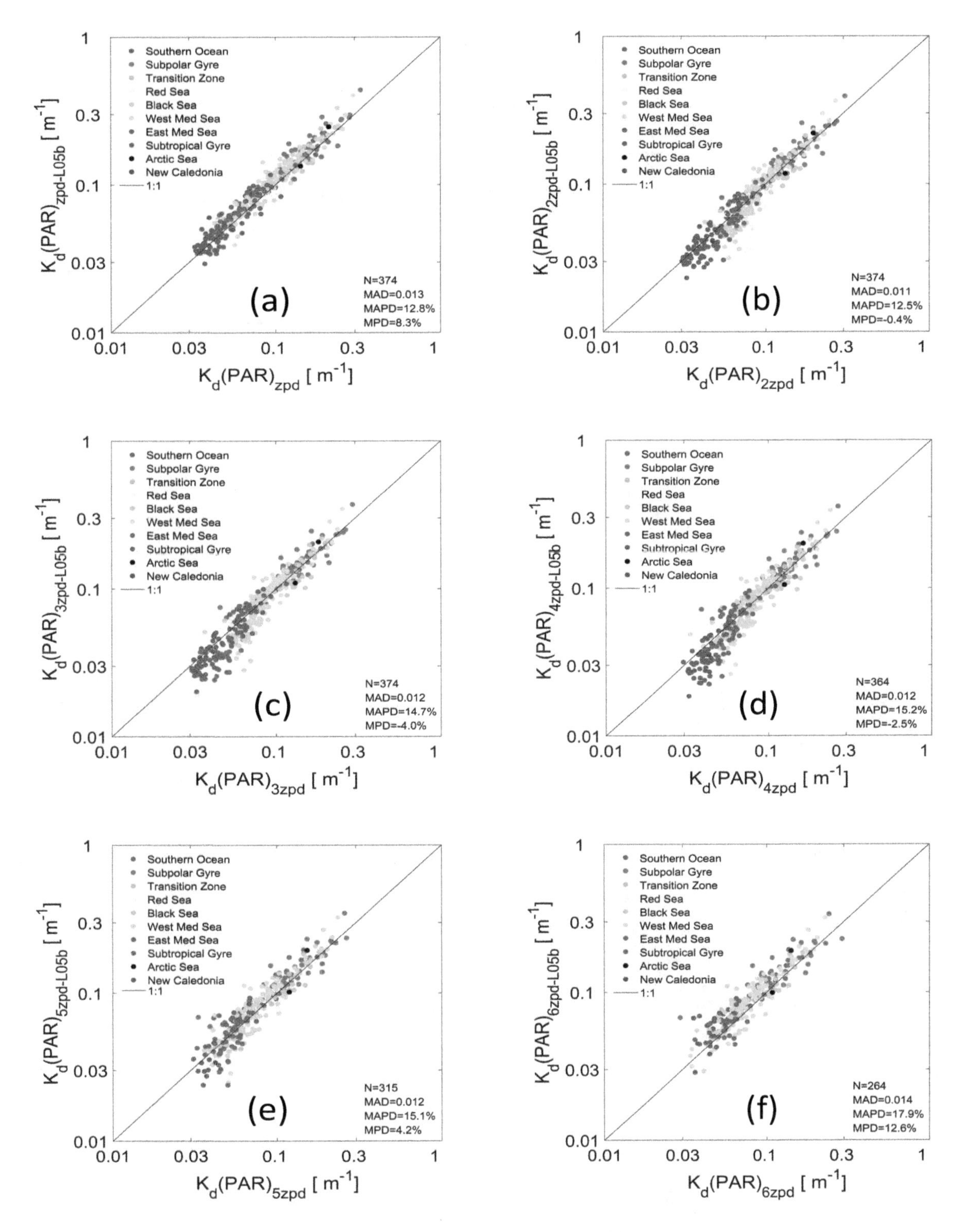

Figure 6. Assessment of satellite K_d(PAR) products based on BGC-Argo dataset. Scatter plots of float-observed K_d(PAR) and satellite-derived $K_d(PAR)_{z\text{-L05b}}$ [10] at (**a**) z_{pd}, (**b**) $2z_{pd}$, (**c**) $3z_{pd}$, (**d**) $4z_{pd}$, (**e**) $5z_{pd}$, and (**f**) $6z_{pd}$, respectively. All black solid lines represent 1:1 lines.

Table 4. Statistical results of evaluation on the satellite-based K_d(PAR) and related products (all products compared here are calculated for the MODIS-Aqua platform).

Product	Algorithm	Number #	MAD#	MAPD #	MPD #	Figure
$K_d(PAR)_{zpd\text{-}L05b}$	IOPs-based	374(49)	0.013 m^{-1} (0.005 m^{-1})	12.8%(11.1%)	8.3%(4.4%)	Figure 6a
$K_d(PAR)_{2zpd\text{-}L05b}$	IOPs-based	374(49)	0.011 m^{-1} (0.004 m^{-1})	12.5%(10.9%)	−0.4%(−1.3%)	Figure 6b
$K_d(PAR)_{3zpd\text{-}L05b}$	IOPs-based	374(49)	0.012 m^{-1} (0.006 m^{-1})	14.7%(14.7%)	−4%(−10.6%)	Figure 6c
$K_d(PAR)_{4zpd\text{-}L05b}$	IOPs-based	364(40)	0.012 m^{-1} (0.007 m^{-1})	15.2%(17.1%)	−2.5%(−14.3%)	Figure 6d
$K_d(PAR)_{5zpd\text{-}L05b}$	IOPs-based	315(14)	0.012 m^{-1} (0.005 m^{-1})	15.1%(12.5%)	4.2%(−2.9%)	Figure 6e
$K_d(PAR)_{6zpd\text{-}L05b}$	IOPs-based	264(5)	0.014 m^{-1} (0.006 m^{-1})	17.9%(15.5%)	12.6%(1.6%)	Figure 6f
$z_{eu\text{-}M07}$	Chla-based	374(49)	8.0 m (12.6 m)	12.5%(9.9%)	−1.7%(−8.4%)	Figure 7a
$z_{eu\text{-}L07}$	IOPs-based	374(49)	14.0 m (20.5 m)	19.7%(16.7%)	8.4%(12.0%)	Figure 7b
$z_{0.415\text{-}B10}$	Chla-based	374(49)	7.7 m (12.7 m)	12.2%(9.8%)	−1.1%(−8.2%)	Figure 7c
$z_{0.415\text{-}L07}$	IOPs-based	374(49)	16.4 m (32.7 m)	21.3%(24.7%)	10.2%(20.6%)	Figure 7d

The numbers in the parentheses represent the statistics in subtropical gyres.

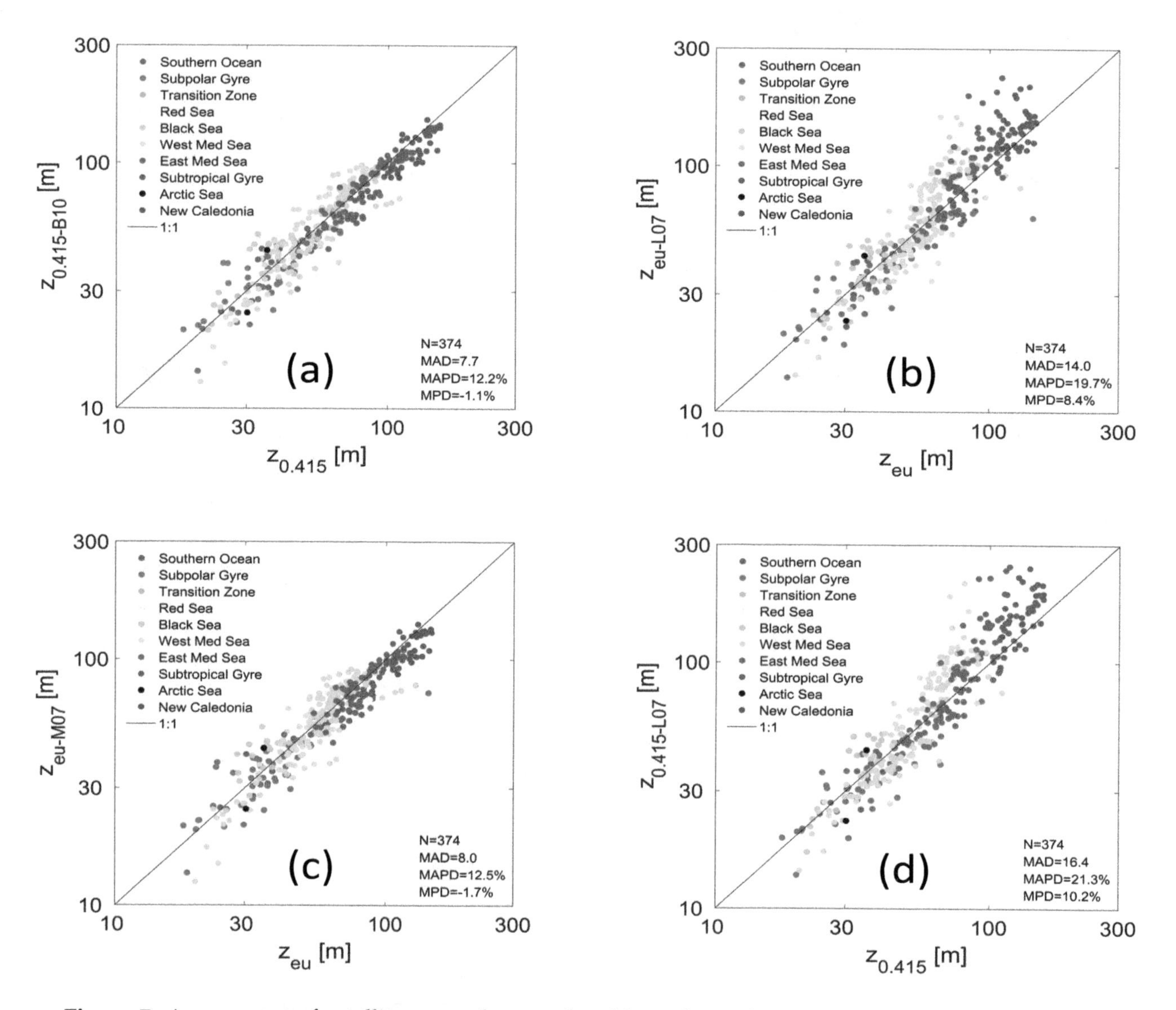

Figure 7. Assessment of satellite z_{eu} and $z_{0.415}$ algorithms through comparison with the BGC-Argo dataset. z_{eu} retrieved from a Chla-based algorithm ($z_{eu\text{-}M07}$; [29]) (**a**) and from the IOPs-based algorithm ($z_{eu\text{-}L07}$; [32]); (**b**) $z_{0.415}$ retrieved from a Chla-based algorithm ($z_{0.415\text{-}B10}$; [31]); (**c**) and from IOPs-based algorithm ($z_{0.415\text{-}L0}$ [32]); (**d**) Black solid lines are the 1:1 lines.

3.4. Assessment of Satellite Algorithms for z_{eu} and $z_{0.415}$

Next, we evaluate the satellite products of two important light-level horizons, the euphotic layer depth z_{eu}, and isolume depth $z_{0.415}$, which represent a relative and an absolute light horizon, respectively. Both z_{eu} and $z_{0.415}$ have optical and biological significance, and have been used to estimate primary production [48], retrieve the vertical distributions of Chla [47], and understand the mechanisms on formation and variability of the DCM [17,18]. Based on the BGC-Argo dataset, the Chla-based z_{eu} algorithm ($z_{eu\text{-}M07}$) performs better, with the lower absolute and percentage differences of 8.0 m and 12.5%, respectively (Figure 7a), than the IOPs-based one (Figure 7b). Regionally, $z_{eu\text{-}M07}$ exhibits a slight underestimation in the subtropical gyres (MPD = −8.4%) and the eastern Mediterranean Sea, and an overestimation in the Black Sea where CDOM deviates significantly from its global relationship to Chla [16,49]. The CDOM index (relative CDOM concentration to the global mean CDOM-Chla relationship) is extremely high in the Black Sea and low in oligotrophic gyres [50]. Therefore, the Black Sea (Subtropics) is expected to have a shallower (deeper) z_{eu} than the empirical Chla-z_{eu} relationship would provide (High CDOM shoals z_{eu} and low CDOM deepens z_{eu}). The overall performance of $z_{eu\text{-}L07}$ is still satisfactory (Figure 7b), with MAD of 14.0 m, but with a bias (MPD = 8.4%) which appears in the clearest waters (z_{eu} > 80 m). In subtropical gyres, the overestimation is most remarkable, with MPD reaching 12.0% (Table 4). This issue is likely due to the constant IOPs with depth assumed in this model design which does not include a DCM, consistent with the underestimation of $K_d(PAR)_{\text{-}L05b}$ at $3z_{pd}$ and $4z_{pd}$.

We assess two satellite $z_{0.415}$ algorithms (Figure 7c,d), which to our knowledge has never been done, although the equation (Equation 7) proposed by Boss and Behrenfeld [31] has been used in several subsequent studies (e.g., [51–54]). Overall, the patterns of two scatter plots of $z_{0.415}$ (Figure 7c,d) are similar to those for z_{eu} (Figure 7a,b), with similar statistics (Table 4). $z_{0.415\text{-}B10}$ displays a slight but obvious underestimation in Subtropical Gyres and Eastern Mediterranean Sea as it is based on $z_{eu\text{-}M07}$ (Figure 7c) and $z_{0.415\text{-}L07}$ exhibits the same overestimation in subtropical gyres (Figure 7d) as $z_{eu\text{-}L07}$ (Figure 7b).

In summary, the evaluation of satellite-retrieved z_{eu} and $z_{0.415}$ suggests that the Chla-based algorithms outperform the IOPs-based ones, with lower scatter and lower relative system bias. Both IOPs-based z_{eu} and $z_{0.415}$ exhibit the overestimation in subtropical gyres due to the underestimation of $K_d(PAR)_{z\text{-}L05b}$ at depth.

4. Final Remarks and Conclusions

The synergy and joint use of BGC-Argo and satellite remote sensing data contribute to studies using both observing assets [55]. BGC-Argo floats provide the largest dataset for validation and evaluation of satellite products in the global ocean, extend the satellite ocean color observations from surface to depth, and fill missing data in satellite coverage due to low sun angle, high latitude winters and clouds; remote sensing is helpful in guiding the deployment of BGC-Argo floats (e.g., in the subtropical gyres or seasonal bloom regions), identifying the spatial scale of float-observed phenomena (basin-, meso- or submeso-scale), extending the BGC-Argo observation from discrete locations to continuous temporal and spatial distributions, and even as a method for calibration of chlorophyll fluorometers deployed on floats (e.g., [23]).

In this study, we present the use of BGC-Argo data to assess existing satellite products. First, for the MODIS-Aqua $K_d(490)$ products, both the B/G algorithm and the semi-analytical algorithm [30] perform well, in the open ocean. Moreover, our dataset shows, the B/G algorithm, while having the lowest bias in the open ocean, exhibits an overestimation at large-values region ($K_d(490) > 0.2$ m^{-1}), not exhibited by the semi-analytical one. All $K_d(490)$ algorithms underestimate in the Black Sea due to extremely high CDOM. The GlobColour $K_d(490)_{GC\text{-}L05a}$ product has an abnormal overestimation for all the data, which is likely not related to Lee et al.'s model embedded in it. The agreement between float and satellite platforms can be regarded as a "consistency check": on one hand, it provides the validation of existing products, and on the other hand, it suggests that the quality-controlled float-observed $E_d(490)$ [26] is of high quality.

K_d(PAR) is a critical variable for the retrieval of z_{eu}, and $z_{0.415}$, and is also useful for marine ecosystem modeling (e.g., [56]) and the estimation of biological heat effect on the upper-layer oceans [5]. However, K_d(PAR) is affected by differences in diffuse attenuation coefficient at different wavelengths. Near the surface, the decrease of PAR is mainly dominated in the open ocean by the losses at the red band due to water absorption [57]. With the depth increasing, as red and near-UV light nearly disappear, the green and blue bands make increasing contributions to K_d(PAR). Finally, K_d(PAR) at deep waters will be close to K_d of the blue/green band (440–510 nm), which is most penetrative in the clear waters [5]. Owing to its spectral sensitivity, the layer-averaged $K_d(\text{PAR})_z$ at a certain depth (z) is not a linear average of $K_d(\lambda)_z$ spectra, but a weighted average, and the weight varies as function of both downwelling irradiance spectra ($E_d(z)$) and attenuation spectra $K_d(\lambda)_z$ [57]. This characteristic makes devising an algorithm for it challenging close to the sea surface [58]. While, in the deeper waters, the vertical change of IOPs becomes the main error sources of retrieval algorithm of $K_d(\text{PAR})_z$. Subsurface structure of IOPs can vary for the same surface conditions, therefore, it is difficult to predict the whole PAR profile accurately, when only relying on the sea-surface IOPs information. Even so, overall, the IOPs-based model [10] performs well near the sea surface until ~$2z_{pd}$, and also estimates the vertical change of K_d(PAR) well in well-mixed waters, but underestimate remarkably below $2z_{pd}$ in stratified waters (including almost all profiles in Subtropical Gyres, most in the Eastern Mediterranean Sea, and quite a few in the Western Mediterranean Sea), due to the presence of a DCM not accounted for in its design. In turn, affected by the underestimation of K_d(PAR) below $2z_{pd}$, both IOPs-derived z_{eu} and $z_{0.415}$ exhibit a bias in the oligotrophic waters. Chla-based algorithms to estimate z_{eu} and $z_{0.415}$ [29,31] perform better although both exhibit some underestimation in subtropical gyres and eastern Mediterranean Sea.

Besides the validation of remote sensing products performed here (and the identification of the significant bias associated with some regions), the statistics associated with our validation can be used to assign errors to the remote sensing products, for example, to propagate them when used in models. For example, many primary-production algorithms parametrize light attenuation using the euphotic depth or include the diffuse attenuation of light attenuation directly (e.g., [48,59]). Simple error propagation (e.g., [60]) can provide the uncertainty in primary production resulting from their uncertainties.

Author Contributions: Conceptualization, X.X.; Methodology, X.X., E.B.; Investigation, X.X., E.B.; Data Curation, X.X., J.Z.; Writing—Original Draft Preparation, X.X.; Writing—Review & Editing, E.B.; Visualization, J.Z.; Funding Acquisition, F.C., E.B. All authors have read and agreed to the published version of the manuscript.

Acknowledgments: The authors are grateful to Hervé Claustre (Laboratoire d'Océanographie de Villefranche, Sorbonne Université, Villefranche-sur-Mer, France) for providing the radiometry data of BIOSOPE cruise, all the BGC-Argo data providers, and the principal investigators of related BGC-Argo float missions and projects.

Appendix A. Independent Evaluation of Processing Methods of BGC-Argo Data

In Figure A1, the processing methods of BGC-Argo radiometry data are validated with traditional ship-borne measurements in the Biogeochemistry and Optics South Pacific Experiment (BIOSOPE) cruise [61], which was conducted in the Southeast Pacific during October-December, 2004. Its radiometry dataset had 39 spectral-irradiance profiles recorded by a Satlantic profiler and corresponding surface irradiance measurements recorded by the Satlantic TSRB (Tethered Spectral Radiometer Buoy), spanning various trophic environments, from eutrophic (west of Marquesas Island, and the upwelling conditions off Chile) to ultra-oligotrophic (center of South Pacific subtropical gyre). We first conducted the same quality-control procedure as Organelli et al. [26] on all BIOSOPE radiometry data, removing the noisy profiles and points, then the measured irradiance values above the sea surface (E_s) are converted to the ones just below sea surface $E_d(0^-)$, by multiplying the transmission coefficient α [33], and using a similar procedure to obtain iPAR(0^-). Following the same steps as Figure 2, z_{pd}, z_{eu}, $K_d(490)$, and $K_d(PAR)$ are calculated. All these values are denoted as "measured", as they are obtained based on measured E_s. We follow the same extrapolation methods mentioned above to estimate $E_d(490,0^-)$ and iPAR(0^-) from below water measurements. Then, extrapolation-based derived z_{pd}, z_{eu}, $K_d(490)$, and $K_d(PAR)$ are denoted as "estimated".

After quality control procedures similar to those used for radiometry measured with floats, there remain 23 $E_d(490)$ and 21 iPAR profiles from BIOSOPE for validation (due to the requirement of at least 5 samples within the top 10 m). We find no obvious differences between the linear and non-linear extrapolation methods for $E_d(490)$, although both have a slight overestimation (MPD = 6.1% and 4.9%) (Figure A1a). Since the attenuation coefficient at 490 nm is mainly determined by IOPs, $K_d(490)$ is expected to vary little within the top 10 m (it will vary some due to the adjustment in the mean-cosine at that wavelength and Raman scattering). Thus, $\ln(E_d(490))$ is expected to be a nearly linear function of depth. As for iPAR, a second-degree polynomial extrapolation is necessary due to the strong attenuation of red wavelengths especially in the top 10 m just below the sea surface due to strong attenuation of red wavelengths [57], causing the linear extrapolation to yield a significant underestimation (MPD = −15.6%, Figure A1b). Derived z_{pd}, z_{eu}, $K_d(490)_{zpd}$, and $K_d(PAR)_{zpd}$ are computed to evaluate the error due to propagation of estimated sea surface $E_d(490,0^-)$ and iPAR(0^-) (Figure A1c–f). The influence of extrapolation is very limited, although $K_d(490)_{zpd}$ and $K_d(PAR)_{zpd}$ exhibit a slight bias.

We conclude from this validation exercise that our extrapolation methods to obtain iPAR(0^-) and $E_d(490,0^-)$ and computing z_{pd}, z_{eu}, $K_d(490)_{zpd}$, and $K_d(PAR)_{zpd}$ from radiometers on profiling floats to have well constrained and small uncertainties.

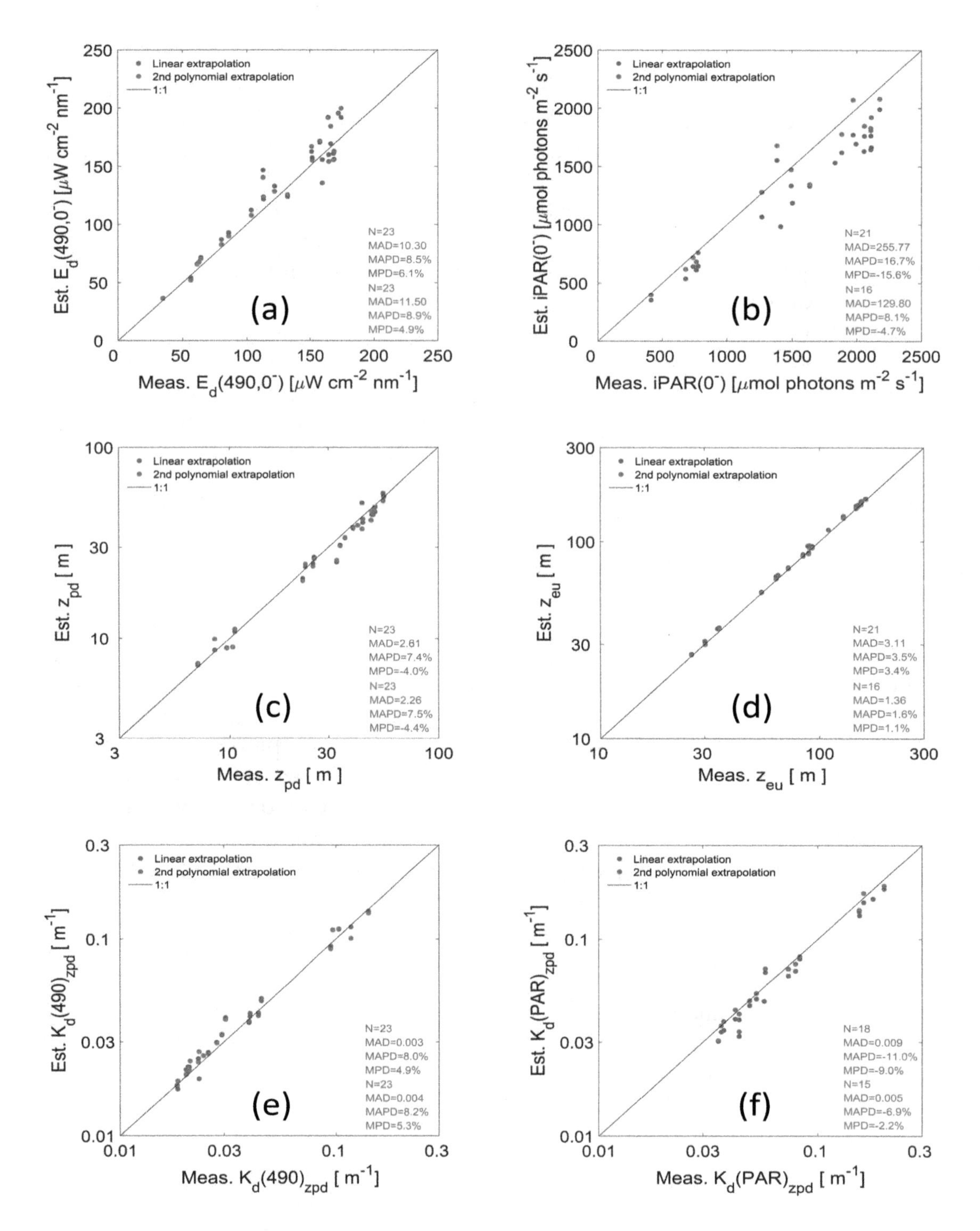

Figure A1. Evaluation of BGC-Argo data processing methods suggested here using the BIOSOPE data. Scatter plot of Measured and Estimated values, of $E_d(490,0^-)$ (**a**) iPAR(0^-), (**b**) z_{pd}, (**c**) z_{eu}, (**d**) $K_d(490)_{zpd}$, (**e**) and $K_d(PAR)_{zpd}$, (**f**) respectively. "Measured (meas.)" means determination of $E_d(490,0^-)$ and iPAR(0^-) based on measured radiometry above sea surface; "Estimated (Est.)" means determination of $E_d(490,0^-)$ and iPAR(0^-) is based on a linear (blue) or second-degree polynomial extrapolation (red) within the upper10 m of the ocean. The black solid lines represent the 1:1 lines.

Appendix B. Should We Use zpd = zeu/4.6?

Organelli et al. [15] proposed a method to derived $K_d(490)_{zpd}$ from BGC-Argo data by first determining z_{eu} from iPAR profile as we do here (Section 2.1), then determining z_{pd} based on the proposed equation: $z_{pd} = z_{eu}/4.6$ (derived based on the fact that $-\ln(0.01) \approx 4.6$). Finally, they applied a linear regression between $\ln(E_d(490))$ and depth to retrieve $K_d(490)_{zpd}$. In contrast, here we use Equation (3) to compute z_{pd}.

When we follow their method to derive $K_d(490)_{zpd\text{-}O17}$, it displays a slight underestimation in oligotrophic waters in comparison to ours (Figure A2a). The relationship $z_{pd} = z_{eu}/4.6$ is likely biased in most oligotrophic waters. On one hand, the diffuse attenuation of visible light is lowest in the blue band for such waters [62], and hence z_{eu} (the 1% light depth for PAR) must be shallower than $z_{1\%490}$ (the 1% light depth for $E_d(490)$, i.e., $\mathbf{z_{eu} < z_{1\%490}}$ in most waters [57], as shown in Figure A2b). On the other hand, from Equation (1):

$$K_d(490)_{z1\%490} \times z_{1\%490} = -\ln(0.01) \times K_d(490)_{zpd} \times z_{pd} \quad \text{(A1)}$$

which implies that the ratio of $z_{1\%490}$ to z_{pd} could reach $-\ln(0.01)$ (approximately 4.6) only when $K_d(490)$ is uniform from surface to $z_{1\%490}$ (i.e., $K_d(490)_{z1\%490} = K_d(490)_{zpd}$). However, due to the deep chlorophyll-a maximum (DCM) observed in highly clear and stratified waters [47], one would expect $K_d(490)$ to increase with depth to the DCM depth, which means $\mathbf{z_{1\%490}/4.6 < z_{pd}}$ (Figure A2c). It follows that $\mathbf{z_{eu}/4.6 < z_{pd}}$ in highly stratified (oligotrophic) waters, as shown in Figure A2d, consistent with the observed bias in [15] obtained in such waters (Figure A2a). Statistically, for all valid 2272 samples in our BGC-Argo dataset (Figure A2d), $z_{eu} = (3.53 \pm 0.83) \times z_{pd}$. Regionally, the ratio of (z_{eu}/z_{pd}) reaches its highest values in the North Atlantic subpolar gyre (4.05 ± 0.83), and its lowest values in the subtropical gyres (3.05 ± 0.46) and New Caledonia (2.98 ± 0.60). We note that Lee et al. [30] also showed that, $z_{1\%}$ of blue light (the arithmetic average of $z_{1\%412}$, $z_{1\%443}$, $z_{1\%488}$, and $z_{1\%531}$) was generally 30%–40% deeper than z_{eu} in open oceans.

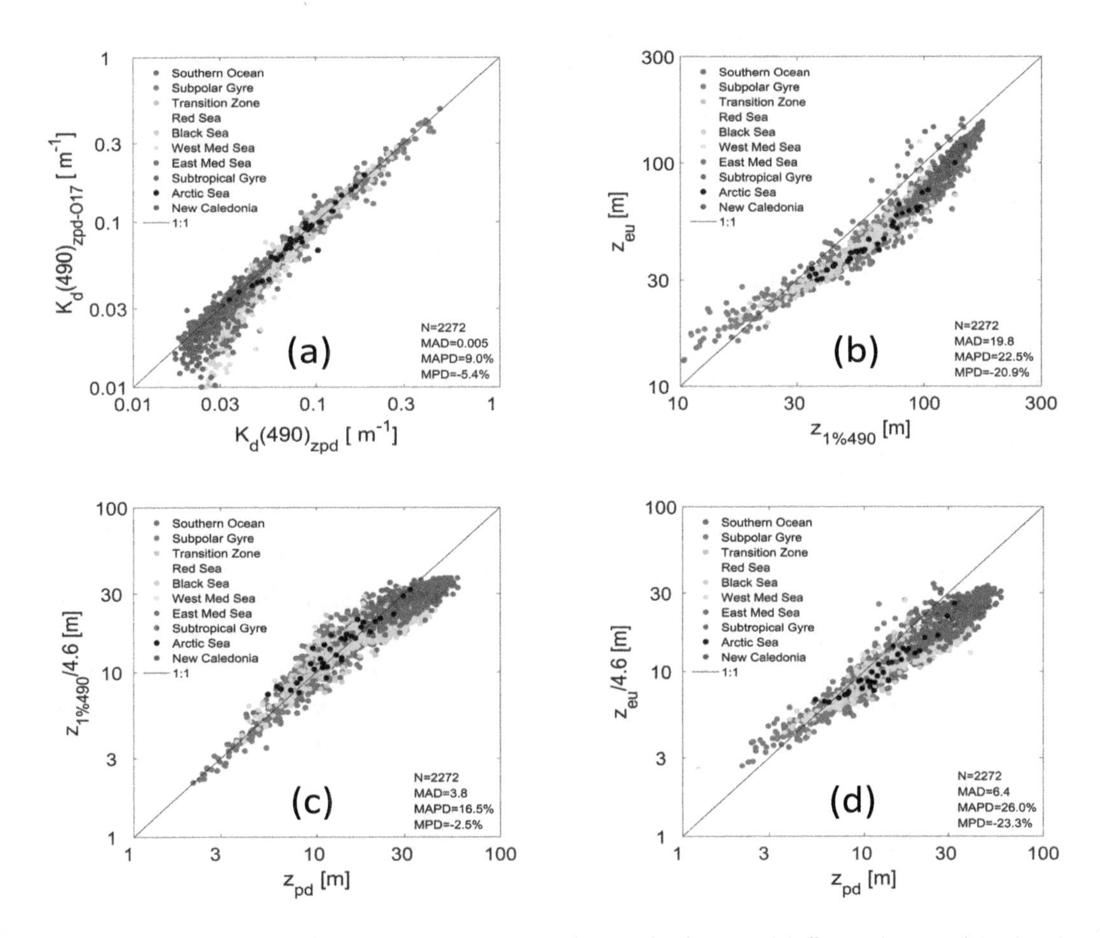

Figure A2. Comparison of different methodology of K_d calculation. (**a**) float-observed $K_d(490)_{zpd\text{-}O17}$ following the methodology of [15] vs. our determined $K_d(490)_{zpd}$; (**b**) z_{eu} vs. $z_{1\%490}$; (**c**) $z_{1\%490}/4.6$ vs. z_{pd}; (**d**) $z_{eu}/4.6$ vs. z_{pd}.

References

1. Kirk, J.T.O. A theoretical analysis of the contribution of algal cells to the attenuation of light within natural waters. III. Cylindrical and spheroidal cells. *New Phytol.* **1976**, *77*, 341–358. [CrossRef]
2. Mobley, C.D. *Light and Water: Radiative Transfer in Natural Waters*; Elsevier: New York, NY, USA, 1994.
3. Ryther, J.H. Photosynthesis in the ocean as a function of light intensity. *Limnol. Oceanogr.* **1956**, *1*, 61–70. [CrossRef]
4. Letelier, R.M.; Karl, D.M.; Abbott, M.R.; Bidigare, R.R. Light driven seasonal patterns of chlorophyll and nitrate in the lower euphotic zone of the North Pacific Subtropical Gyre. *Limnol. Oceanogr.* **2004**, *49*, 508–519. [CrossRef]
5. Chen, J.; Zhang, X.; Xing, X.; Ishizaka, J.; Yu, Z. A spectrally selective attenuation mechanism-based K_{par} algorithm for biomass heating effect simulation in the open ocean. *J. Geophys. Res. Oceans* **2017**, *122*, 9370–9386. [CrossRef]
6. Pimentel, S.; Tse, W.-H.; Xu, H.; Denaxa, D.; Jansen, E.; Korres, G.; Mirouze, I.; Storto, A. Modeling the near-surface diurnal cycle of sea surface temperature in the Mediterranean Sea. *J. Geophys. Res. Oceans* **2019**, *124*, 171–183. [CrossRef]
7. Skákala, J.; Bruggeman, J.; Brewin, R.J.W.; Ford, D.A.; Ciavatta, S. Improved representation of underwater light field and its impact on ecosystem dynamics: A study in the North Sea. *J. Geophys. Res. Oceans* **2020**, *125*, e2020jc016122. [CrossRef]
8. Frouin, R.; McPherson, J.; Ueyoshi, K.; Franz, B.A. A time series of photosynthetically available radiation at the ocean surface from SeaWiFS and MODIS data. In Proceedings of the SPIE Asia-Pacific Remote Sensing; SPIE-Intl Soc Optical Eng, San Jose, CA, USA, 12–16 February 2012; p. 85251. [CrossRef]
9. Lee, Z.; Du, K.; Arnone, R. A model for the diffuse attenuation coefficient of downwelling irradiance. *J. Geophys. Res. Oceans* **2005**, *110*, C02016. [CrossRef]
10. Lee, Z.; Du, K.; Arnone, R.; Liew, S.C.; Penta, B. Penetration of solar radiation in the upper ocean—A numerical model for oceanic and coastal waters. *J. Geophys. Res. Oceans* **2005**, *110*, C09019. [CrossRef]
11. Johnson, K.S.; Claustre, H. Bringing Biogeochemistry into the Argo Age. *Eos* **2016**, *97*, 11–15. [CrossRef]
12. Claustre, H.; Johnson, K.S.; Takeshita, Y. Observing the Global Ocean with Biogeochemical-Argo. *Annu. Rev. Mar. Sci.* **2020**, *12*, 23–48. [CrossRef]
13. Biogeochemical-Argo Planning Group. *The Scientific Rationale, Design, and Implementation Plan for a Biogeochemical-Argo Float Array*; Johnson, K., Claustre, H., Eds.; Ifremer: Issy-les-Moulineaux, France, 2016. [CrossRef]
14. Roemmich, D.; Alford, M.H.; Claustre, H.; Johnson, K.; King, B.; Moum, J.; Oke, P.; Owens, W.B.; Pouliquen, S.; Purkey, S.; et al. On the Future of Argo: A Global, Full-Depth, Multi-Disciplinary Array. *Front. Mar. Sci.* **2019**, *6*, 439. [CrossRef]
15. Organelli, E.; Barbieux, M.; Claustre, H.; Schmechtig, C.; Poteau, A.; Bricaud, A.; Boss, E.; Briggs, N.; Nencioli, F.; D'Ortenzio, F.; et al. Two databases derived from BGC-Argo float measurements for marine biogeochemical and bio-optical applications. *Earth Syst. Sci. Data* **2017**, *9*, 861–880. [CrossRef]
16. Organelli, E.; Claustre, H.; Bricaud, A.; Barbieux, M.; Uitz, J.; D'Ortenzio, F.; Dall'Olmo, G. Bio-optical anomalies in the world's oceans: An investigation on the diffuse attenuation coefficients for downward irradiance derived from Biogeochemical Argo float measurements. *J. Geophys. Res. Oceans* **2017**, *122*, 3543–3564. [CrossRef]
17. Mignot, A.; Claustre, H.; Uitz, J.; Poteau, A.; D'Ortenzio, F.; Xing, X. Understanding the seasonal dynamics of phytoplankton biomass and the deep chlorophyll maximum in oligotrophic environments: A Bio-Argo float investigation. *Glob. Biogeochem. Cycles* **2014**, *28*, 856–876. [CrossRef]
18. Barbieux, M.; Uitz, J.; Gentili, B.; De Fommervault, O.P.; Mignot, A.; Poteau, A.; Schmechtig, C.; Taillandier, V.; Leymarie, E.; Penkerch, C.; et al. Bio-optical characterization of subsurface chlorophyll maxima in the Mediterranean Sea from a Biogeochemical-Argo float database. *Biogeosciences* **2019**, *16*, 1321–1342. [CrossRef]
19. Terzić, E.; Lazzari, P.; Organelli, E.; Solidoro, C.; Salon, S.; D'Ortenzio, F.; Conan, P. Merging bio-optical data from Biogeochemical-Argo floats and models in marine biogeochemistry. *Biogeosciences* **2019**, *16*, 2527–2542. [CrossRef]
20. Xing, X.; Morel, A.; Claustre, H.; Antoine, D.; D'Ortenzio, F.; Poteau, A.; Mignot, A. Combined processing and mutual interpretation of radiometry and fluorimetry from autonomous profiling Bio-Argo floats: Chlorophyll a retrieval. *J. Geophys. Res. Oceans* **2011**, *116*, C06020. [CrossRef]

21. Xing, X.; Morel, A.; Claustre, H.; D'Ortenzio, F.; Poteau, A. Combined processing and mutual interpretation of radiometry and fluorometry from autonomous profiling Bio-Argo floats: 2. Colored dissolved organic matter absorption retrieval. *J. Geophys. Res. Oceans* **2012**, *117*, C04022. [CrossRef]
22. Xing, X.; Briggs, N.; Boss, E.; Claustre, H. Improved correction for non-photochemical quenching of in situ chlorophyll fluorescence based on the synchronous irradiance profile. *Opt. Express* **2018**, *26*, 24734–24751. [CrossRef]
23. Boss, E.; Swift, D.; Taylor, L.; Brickley, P.; Zaneveld, R.; Riser, S.; Perry, M.J.; Strutton, P.G. Observations of pigment and particle distributions in the western North Atlantic from an autonomous float and ocean color satellite. *Limnol. Oceanogr.* **2008**, *53*, 2112–2122. [CrossRef]
24. Johnson, K.S.; Plant, J.N.; Dunne, J.P.; Talley, L.D.; Sarmiento, J.L. Annual nitrate drawdown observed by SOCCOM profiling floats and the relationship to annual net community production. *J. Geophys. Res. Oceans* **2017**, *122*, 6668–6683. [CrossRef]
25. Haëntjens, N.; Boss, E.; Talley, L.D. Revisiting Ocean Color algorithms for chlorophyll a and particulate organic carbon in the Southern Ocean using biogeochemical floats. *J. Geophys. Res. Oceans* **2017**, *122*, 6583–6593. [CrossRef]
26. Organelli, E.; Claustre, H.; Bricaud, A.; Schmechtig, C.; Poteau, A.; Xing, X.G.; Prieur, L.; D'Ortenzio, F.; Dall'Olmo, G.; Vellucci, V. A Novel Near-Real-Time Quality-Control Procedure for Radiometric Profiles Measured by Bio-Argo Floats: Protocols and Performances. *J. Atmos. Oceans Technol.* **2016**, *33*, 937–951. [CrossRef]
27. Barbieux, M.; Organelli, E.; Claustre, H.; Schmechtig, C.; Poteau, A.; Boss, E.; Bricaud, A.; Briggs, N.; Dall'Olmo, G.; D'Ortenzio, F.; et al. A global database of vertical profiles derived from Biogeochemical Argo float measurements for biogeochemical and bio-optical applications. *SEANOE* **2017**. [CrossRef]
28. NASA (2020). Available online: https://oceancolor.gsfc.nasa.gov/atbd/kd_490/ (accessed on 22 July 2020).
29. Morel, A.; Huot, Y.; Gentili, B.; Werdell, P.J.; Hooker, S.B.; Franz, B.A. Examining the consistency of products derived from various ocean color sensors in open ocean (Case 1) waters in the perspective of a multi-sensor approach. *Remote. Sens. Environ.* **2007**, *111*, 69–88. [CrossRef]
30. Lee, Z.; Hu, C.; Shang, S.; Du, K.; Lewis, M.; Arnone, R.; Brewin, R. Penetration of UV-visible solar radiation in the global oceans: Insights from ocean color remote sensing. *J. Geophys. Res. Oceans* **2013**, *118*, 4241–4255. [CrossRef]
31. Boss, E.; Behrenfeld, M. In situ evaluation of the initiation of the North Atlantic phytoplankton bloom. *Geophys. Res. Lett.* **2010**, *37*, 18603. [CrossRef]
32. Lee, Z.; Weidemann, A.; Kindle, J.; Arnone, R.; Carder, K.L.; Davis, C. Euphotic zone depth: Its derivation and implication to ocean-color remote sensing. *J. Geophys. Res. Oceans* **2007**, *112*, C03009. [CrossRef]
33. Mobley, C.D.; Boss, E.S. Improved irradiances for use in ocean heating, primary production, and photo-oxidation calculations. *Appl. Opt.* **2012**, *51*, 6549–6560. [CrossRef]
34. Lee, Z.; Carder, K.L.; Arnone, R. Deriving inherent optical properties from water color: A multi-band quasi-analytical algorithm for optically deep waters. *Appl. Opt.* **2002**, *41*, 5755–5772. [CrossRef]
35. NASA (2020). Available online: https://oceancolor.gsfc.nasa.gov/atbd/chlor_a/ (accessed on 22 July 2020).
36. Gordon, H.R.; McCluney, W.R. Estimation of the depth of sunlight penetration in the sea for remote sensing. *Appl. Opt.* **1975**, *14*, 413–416. [CrossRef]
37. Cetinić, I.; Perry, M.J.; D'Asaro, E.; Briggs, N.; Poulton, N.; Sieracki, M.E.; Lee, C.M. A simple optical index shows spatial and temporal heterogeneity in phytoplankton community composition during the 2008 North Atlantic Bloom Experiment. *Biogeosciences* **2015**, *12*, 2179–2194. [CrossRef]
38. Schulien, J.A.; Behrenfeld, M.J.; Hair, J.W.; Hostetler, C.A.; Twardowski, M.S. Vertically- resolved phytoplankton carbon and net primary production from a high spectral resolution lidar. *Opt. Express* **2017**, *25*, 13577. [CrossRef]
39. Thushara, V.; Vinayachandran, P.N.M.; Matthews, A.J.; Webber, B.G.M.; Queste, B.Y. Vertical distribution of chlorophyll in dynamically distinct regions of the southern Bay of Bengal. *Biogeosciences* **2019**, *16*, 1447–1468. [CrossRef]
40. Banse, K. Should we continue to use the 1% light depth for estimating the compensation depth of phytoplankton for another 70 years? *Limnol. Oceanogr.* **2004**, *13*, 49–52. [CrossRef]
41. Behrenfeld, M.J.; Boss, E.S. Student's tutorial on bloom hypotheses in the context of phytoplankton annual cycles. *Glob. Chang. Biol.* **2018**, *24*, 1–23. [CrossRef]
42. Hu, C.; Lee, Z.; Franz, B. Chlorophyll a algorithms for oligotrophic oceans: A novel approach based on three-band reflectance difference. *J. Geophys. Res. Oceans* **2012**, *117*, C01011. [CrossRef]

43. Fanton d'Andon, O.; Mangin, A.; Lavender, S.; Antoine, D.; Maritorena, S.; Morel, A.; Barrot, G.; Demaria, J.; Pinnock, S. GlobColour—The European Service for Ocean Colour. In Proceedings of the 2009 IEEE International Geoscience & Remote Sensing Symposium, Cape Town, South Africa, 12–17 July 2009.
44. Zhang, X.; Hu, L.; He, M.-X. Scattering by pure seawater: Effect of salinity. *Opt. Express* **2009**, *17*, 5698–5710. [CrossRef]
45. Bailey, S.W.; Werdell, P.J. A multi-sensor approach for the on-orbit validation of ocean color satellite data products. *Remote Sens. Environ.* **2006**, *102*, 12–23. [CrossRef]
46. Churilova, T.; Suslin, V.; Krivenko, O.; Efimova, T.; Moiseeva, N.; Mukhanov, V.; Smirnova, L. Light absorption by phytoplankton in the upper mixed layer of the Black Sea: Seasonality and Parametrization. *Front. Mar. Sci.* **2017**, *4*, 90. [CrossRef]
47. Uitz, J.; Claustre, H.; Morel, A.; Hooker, S.B. Vertical distribution of phytoplankton communities in open ocean: An assessment based on surface chlorophyll. *J. Geophys. Res. Oceans* **2006**, *111*, C08005. [CrossRef]
48. Behrenfeld, M.J.; Falkowski, P.G. A consumer's guide to phytoplankton primary productivity models. *Limnol. Oceanogr.* **1997**, *42*, 1479–1491. [CrossRef]
49. Morel, A.; Claustre, H.; Gentili, B. The most oligotrophic subtropical zones of the global ocean: Similarities and differences in terms of chlorophyll and yellow substance. *Biogeosciences* **2010**, *7*, 3139–3151. [CrossRef]
50. Morel, A.; Gentili, B. A simple band ratio technique to quantify the colored dissolved and detrital organic material from ocean color remotely sensed data. *Remote Sens. Environ.* **2009**, *113*, 998–1011. [CrossRef]
51. Brody, S.R.; Lozier, M.S. Characterizing upper-ocean mixing and its effect on the spring phytoplankton bloom with in situ data. *Ices J. Mar. Sci.* **2015**, *72*, 1961–1970. [CrossRef]
52. Itoh, S.; Yasuda, I.; Saito, H.; Tsuda, A.; Komatsu, K. Mixed layer depth and chlorophyll a: Profiling float observations in the Kuroshio-Oyashio Extension region. *J. Marine Systems* **2015**, *151*, 1–14. [CrossRef]
53. Mayot, N.; D'Ortenzio, F.; Taillandier, V.; Prieur, L.; de Fommervault, O.P.; Claustre, H.; Bosse, A.; Testor, P.; Conan, P. Physical and Biogeochemical Controls of the Phytoplankton Blooms in North Western Mediterranean Sea: A Multiplatform Approach Over a Complete Annual Cycle (2012–2013 DEWEX Experiment). *J. Geophys. Res. Oceans* **2017**, *122*, 9999–10019. [CrossRef]
54. Balaguru, K.; Doney, S.C.; Bianucci, L.; Rasch, P.J.; Leung, L.R.; Yoon, J.-H.; Lima, I.D. Linking deep convection and phytoplankton blooms in the northern Labrador Sea in a changing climate. *PLoS ONE* **2018**, *13*, e0191509. [CrossRef]
55. Chai, F.; Johnson, K.S.; Claustre, H.; Xing, X.; Wang, Y.; Boss, E.; Riser, S.; Fennel, K.; Schofield, O.; Sutton, A. Monitoring ocean biogeochemistry with autonomous platforms. *Nat. Rev. Earth Environ.* **2020**, *1*, 315–326. [CrossRef]
56. Xiu, P.; Chai, F. Connections between physical, optical and biogeochemical processes in the Pacific Ocean. *Progr. Oceanogr.* **2014**, *122*, 30–53. [CrossRef]
57. Morel, A. Optical modeling of the upper ocean in relation to its biogenous matter content. *J. Geophys. Res.* **1988**, *93*, 10749–10768. [CrossRef]
58. Lee, Z. KPAR: An optical property associated with ambiguous values. *J. Lake Sci.* **2009**, *21*, 159–164. [CrossRef]
59. Westberry, T.; Behrenfeld, M.J.; Siegel, D.A.; Boss, E. Carbon-based primary productivity modeling with vertically resolved photoacclimation. *Glob. Biogeochem. Cycles* **2008**, *22*, GB2024. [CrossRef]
60. Joint Committee for Guides in Metrology. Evaluation of Measurement Data—Guide to the Expression of Uncertainty in Measurement. Available online: https://ncc.nesdis.noaa.gov/documents/documentation/JCGM_100_2008_E.pdf (accessed on 22 July 2020).
61. Claustre, H.; Sciandra, A.; Vaulot, D. Introduction to the special section: Bio-optical and biogeochemical conditions in the South East Pacific in late 2004—The BIOSOPE program. *Biogeosciences* **2008**, *5*, 679–691. [CrossRef]
62. Morel, A.; Maritorena, S. Bio-optical properties of oceanic waters: A reappraisal. *J. Geophys. Res. Oceans* **2001**, *106*, 7163–7180. [CrossRef]

2

A-Priori Calibration of a Structured Light Underwater 3D Sensor

Christian Bräuer-Burchardt [1,*], Christoph Munkelt [1], Ingo Gebhart [1], Matthias Heinze [1], Stefan Heist [1], Peter Kühmstedt [1] and Gunther Notni [1,2]

1 Fraunhofer Institute for Applied Optics and Precision Engineering IOF, Albert-Einstein-Str. 7, 07745 Jena, Germany; christoph.munkelt@iof.fraunhofer.de (C.M.); ingo.gebhart@iof.fraunhofer.de (I.G.); matthias.heinze@iof.fraunhofer.de (M.H.); stefan.heist@iof.fraunhofer.de (S.H.); peter.kuehmstedt@iof.fraunhofer.de (P.K.); gunther.notni@iof.fraunhofer.de (G.N.)

2 Department of Mechanical Engineering, Ilmenau University of Technology, Gustav-Kirchhoff-Platz 2, 98693 Ilmenau, Germany

* Correspondence: christian.braeuer-burchardt@iof.fraunhofer.de

Abstract: In this study, we introduce a new calibration method for underwater optical stereo scanners. It uses air calibration, additional underwater parameters, and extended camera modeling. The new methodology can be applied to both passive photogrammetric and structured light three-dimensional (3D) scanning systems. The novel camera model uses a variable principal distance depending on the radial distance to the principal point instead of two-dimensional distortion functions. This allows for an initial improvement of 3D reconstruction quality. In a second step, certain underwater-specific parameters—such as refraction indices, glass thickness, and view-port distances—are determined. Finally, a correction function for the entire measurement volume can be obtained from a few underwater measurements. Its application further improves the measurement accuracy. Measurement examples show the performance of the new calibration method in comparison to current underwater calibration strategies. A discussion of the possibilities and limits of the new calibration method and an outlook for future work complete this work.

Keywords: underwater 3D scanning; structured light projection; calibration; measurement accuracy

1. Introduction

Capturing objects under water and generating three-dimensional (3D) geometric models has become increasingly important in various fields of application, such as inspection tasks of industrial structures of energy production [1,2], biologic objects [3–6], shipwrecks [7,8], or archaeological sites and objects [8–12]. Depending on the specific application, several techniques have been established for underwater 3D object acquisition, i.e., underwater photogrammetry [3,5,7,12–15], laser scanning techniques [1,2,16–19], ultrasound sensors [20], and time-of-flight methodology [21]. Each of these techniques have advantages and disadvantages.

Photogrammetric measurements typically provide high spatial object point resolution and high measurement accuracy. The field of view of the cameras may be several square meters. However, preparation of the scene by placing landmarks onto or nearby the objects might require high effort. Additionally, photogrammetric underwater measurements typically require non-moving scenes and scanner fixation, e.g., on a tripod.

Laser scanning systems for underwater 3D acquisition are commercially available [19]. These systems can measure distances up to 40 m. Furthermore, they are suitable for application in polluted water. However, due to the line-scanning measurement principle, spatial and temporal resolution of the captured object points is limited.

Scanners using structured light projection techniques have been introduced for underwater 3D measurements in the past few years [21–26]. This technique provides some advantages over passive photogrammetry and laser scanning. More 3D points with higher spatial density can be captured, enabling a more detailed 3D reconstruction of the observed object. As no scene preparation is required, this measurement principle is versatile and allows for relative movements up to a certain velocity. Measurement data are obtained immediately and can be subsequently processed. The main disadvantage that have prevented wide commercial applicability is the necessity of a high-power lighting system. Consequently, recent systems have a small field of view and a short measurement distance.

Calibration of the 3D scanning system is essential for the accuracy and responsible for the systematic errors of the measurements. An excellent and extensive overview of camera calibration techniques for accurate underwater measurements is given by Shortis [27].

One strategy to obtain underwater calibration parameters is to analogously perform the calibration on classical air calibration, for instance using a chessboard [28] or other calibration targets [27] in a water basin under comparable conditions (e.g., temperature, salinity, pressure) to the place of measurement. The pinhole camera model can be used and the loss of a single viewpoint (projection center) is neglected. Refraction effects lead to distortions that can be described in the same manner as in air calibration. However, remaining deviations and errors will be larger than in air case due to ray refraction, especially when the measurement depth is extensive.

Considering the refraction at the media boundary, the camera model must be modified. Here, the works of Telem and Filin [15], as well as Sedlacek and Koch [14,29] provide solutions under consideration of all underwater environment conditions.

In previous studies [26,30], we developed a calibration method using air calibration according to the classical pinhole model with distortion correction and a subsequent determination of underwater parameters using at least four underwater measurements. In this work, we further developed this model by replacing the radial distortion function by a variable principal distance and a-priori estimation of underwater parameters. Finally, a refinement of the calibration was achieved by determining a 3D correction function over the measurement volume.

The goal of this work was to develop a methodology for an easy-to-handle pre-calibration of underwater optical stereo 3D scanners based on photogrammetric principles using conventional air calibration and an adapted camera modeling. In our experiments, we showed that the new methodology can be successfully applied to such scanners. However, in order to achieve highest measurement accuracy, additional calibration effort is necessary.

We used data of a handheld 3D underwater stereo scanner with structured light projection for a measurement volume of 250 mm × 200 mm × 100 mm.

2. Materials and Methods

Recent photogrammetric underwater applications of 3D scanning required high effort with regard to both preparation of the measurements and calibration of the scanner. In order to achieve high measurement accuracy, refraction of the vision rays at the boundary between the different media (air to glass and glass to water) must be taken into account. Recently, a number of different models were developed [14,15,31].

The goal of this work was to generate a set of calibration parameters for achieving the best measurement accuracy without any underwater measurements or, optionally, with minimal effort of underwater measurements. The strategy includes a novel geometric modeling of radial lens distortion and estimation of several parameters.

The new methodology was developed for a-priori calibration of underwater 3D stereo scanners. Our scanner is a hand-held underwater 3D scanner that covers a measurement volume of approximately 250 mm × 200 mm × 100 mm [26]. It has plane viewports adjusted in perpendicular direction to the principal rays of the cameras. This property is used as a constraint to the extended model described in the following section.

2.1. *Extended Pinhole Camera Model*

Typically, for stereo-photogrammetric 3D reconstruction, the common pinhole model (PM) is extended by so-called distortion functions. These distortion functions are mainly lens-dependent but also take inhomogeneities of the image sensor and adjustment deviations into account. Generally, distortion functions are described by several functional parameters or matrices of correction vectors. Considering the pinhole model, distortion correction leads to deviation of the vision rays in the projection center. Physically, however, vision rays do not pass through the projection center but close nearby. Pinhole-style modeling implies that correct 3D reproduction is achieved only for one certain plane in the measurement volume. At other distances, systematic errors appear in the reconstruction of the 3D points. These errors are typically neglected because their amount is below the noise level. However, at short measurement distances, these errors may become significant. A typical limit for significance is a magnification factor larger than 1/30, i.e., a measurement distance below 30 times the focal length of the lens for high-accuracy measurements.

For extended modeling, we assumed that vision rays passing close to the projection center O_0 intersected the optical axis in the point $O(r)$. The higher the radial distance r of the point p to the principal point p_0, the farther the intersection point $O(r)$ to O_0, i.e., in direction to the sensor image plane.

Common modeling uses the PM and a 2D correction function for distortion effects and assumes optimal parameter determination via calibration, thus leading to ideal results for just one object plane in the measurement volume. Reconstructed points outside this plane have certain systematic errors. The size of these errors correlates with the distortion function and can be approximated from the parameters of the distortion function. Assuming a typical measurement volume, the systematic errors are quite small and significantly smaller than the typically occurring random measurement errors. Consequently, these systematic errors can be neglected. In case of underwater measurements, however, they may increase due to refraction effects. Hence, the errors must be considered and compensated, or a new model avoiding these errors must be used. This new model will be introduced in the following section. It should be called "extended pinhole model with variable principal distance".

2.2. *Alternative Modeling of Radial Distortion*

The common radial distortion function is typically described by a polynomial in the sensor image plane is replaced by a variable principal distance. All other non-radial parts of distortion are neglected, because the amount of the radial parts typically is about 95% of the total distortion. The modeling by variable principal distance is close to the physical truth at image generation and leads to the effect that no distance-dependent errors occur at distortion correction. This is illustrated in Figure 1. Only points at distance d_0 are properly corrected by Δr (e.g., point P_1 in Figure 1), whereas other points such as P_2 are erroneously corrected with reconstruction error ΔP_2. Using variable principal distance, all points are correctly reconstructed.

Instead of a polynomial over r in the image plane, a variable principal distance $c(r)$ and an appropriately shifted projection center $O(r)$ is used. Determination of $c(r)$ may be realized in different ways and calculated according to Equation (6) (see Section 2.3 (calibration)). In case of calculating 3D points by triangulation using a stereo camera setup, every image point has its own variable principal distance $c(r)$ and its own projection center $O(r)$.

A precondition for calculating 3D coordinates is the determination of proper point correspondences. They cannot be directly obtained from rectified images [32], as this would require the common pinhole camera model. However, if function $c(r)$ is known, appropriate correction values can be calculated. Thus, correspondence finding can be realized iteratively in rectified images and an estimated pinhole model can be used as first approximation.

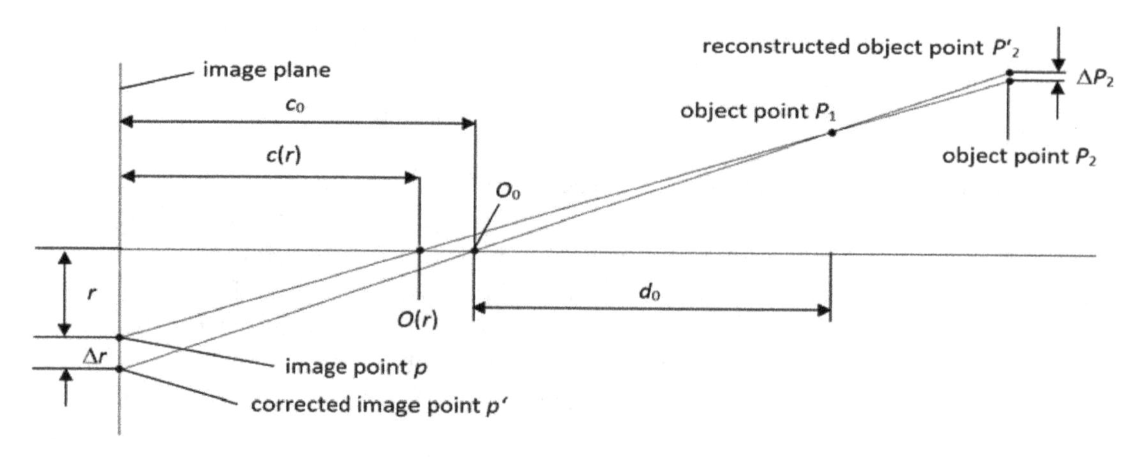

Figure 1. Compensation of deviating vision ray direction by 2D distortion vectors Δr in the image plane, accurate only for distance d_0 (sketch is not true to scale).

2.2.1. Transition to the Underwater Case

As previously shown by the authors [26], vision rays are refracted at the boundary between air and glass in the housing, and between glass and water outside the housing according to certain parameters, like refraction indices, glass thickness, distance between camera, and vision glass. In order to correctly reconstruct the scene, changed ray courses must be determined and inserted into the reconstruction calculation. Accordingly, considering a variable principal distance $c(r)$, radially dependent virtual principal distance $c'(r)$ for the underwater case leads to:

$$c'(r) = \frac{r}{\tan\left(\arcsin\left(\frac{\sin(\arctan(r/c(r)))}{n_w}\right)\right)} \tag{1}$$

For the translation of the projection center this means:

$$l(r) = l_w(r) + l_g(r) \tag{2}$$

with $l_w(r) = d\left(1 - \frac{r}{c\cdot\tan\left(\arcsin\left(\frac{\sin(\arctan(r/c(r)))}{n_w}\right)\right)}\right)$, $l_g(r) = th\left(1 - \frac{c}{r}\cdot\tan\left(\arcsin\left(\frac{\sin(\arctan(r/c(r)))}{n_g}\right)\right)\right)$, where th is the thickness of the glass, n_w refraction index of water, and n_g refraction index of glass. The refraction index of air is set to $n_a = 1.0$ and is omitted in the equations. For the new projection centers $O'(r)$ it holds:

$$O'(r) = O(r) - l(r)\cdot R\cdot e \tag{3}$$

where $O(r)$ is the initially determined projection center of the camera outside housing according to the pinhole model, e is the unit vector, and $\boldsymbol{R}$ is the rotation matrix describing the deviation of the principal ray of the camera according to the pinhole model calibration. Figure 2 shows the vision ray course for two image points with different distances to the principal point.

Application of Formulas (1)–(3) can be realized by using extended software modules according to the common methods (see Luhmann et al. [33]). Non-radial distortion effects can be compensated by application of additional functions. Finally, remaining errors may be compensated by a 3D correction function in the measurement volume [34]. It is expected that significant errors may remain, because some assumptions are very simplifying (e.g., the assumed orientations of the principal rays in relation to the glass surfaces of the housing ports). Determination of corresponding correction functions can be realized in a final calibration step (see Section 2.3 calibration).

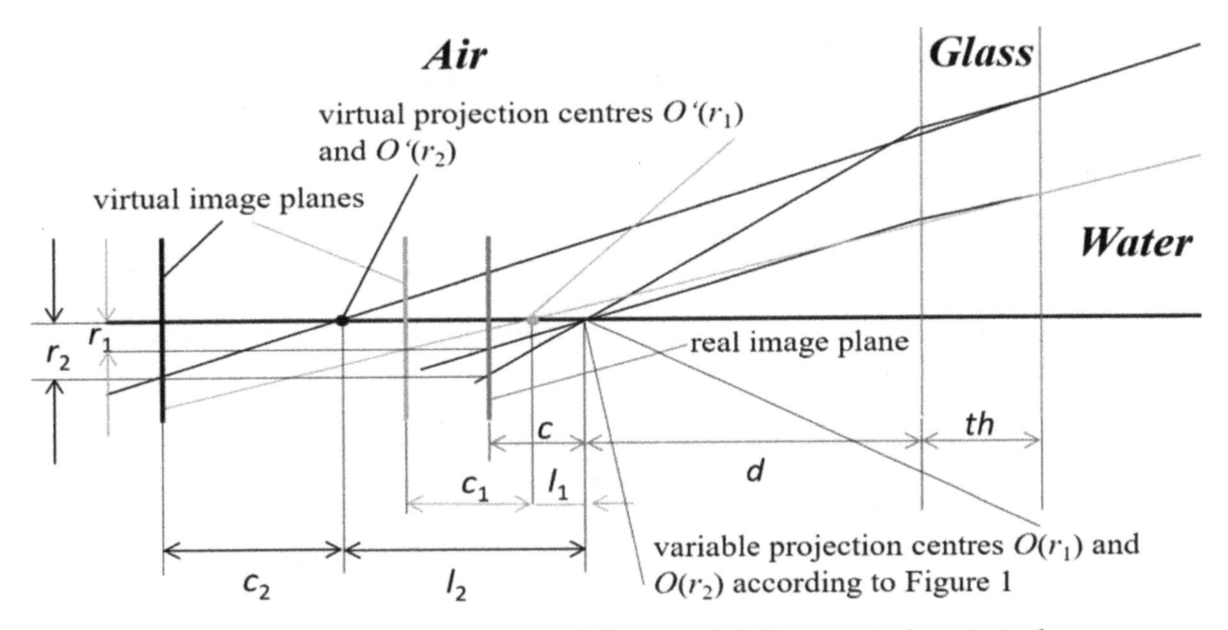

Figure 2. Course of vision rays with different radial distances to the principal ray.

2.2.2. Simplified Pinhole Model for Underwater Application

When using existing photogrammetric software for the underwater 3D reconstruction, the method described in the following should be applied.

As described before, every image point $p_i = (x_i, y_i)$ has its own individual principal distance and own projection center, depending on the radial distance to the principal point $p_0 = (x_0, y_0)$. Parameters are refraction indices, glass thickness, distance between camera and glass, and variable principal distances.

Both experiments and simulations using parameters close to real conditions have shown that the variance of $c'(r)$ and $O'(r)$ is relatively small. This is due to the fact that the variable principal distance in air is smaller for big radial distances, but larger in water. Hence, both effects partly compensate, which leads to the assumption that if using a fixed principal distance and a fixed projection center (if as using the classical pinhole model) only causes a small error in the 3D point calculation. For known parameters, the error can be estimated a-priori. Furthermore, it can be described and compensated for a plane at a certain distance in water by a classical radial distortion function or distortion matrix. Additionally, remaining errors due to deviation of the points from the reference plane can be appropriately predicted by calculation.

Fixed principal distance c' (and fixed projection center O' analogously) can be chosen as average values over all image points:

$$c' = \frac{1}{n}\sum_{i=1}^{n} c'(r_i); r_i = \sqrt{(x_i - x_0) + (y_i - y_0)} \tag{4}$$

Distortion function is valid for reference plane E_0, being in distance d_0 to the projection center O'. Accordingly, E_0 should be adjusted parallel to the image plane. If the camera sensor planes are not parallel, two different reference planes E_1 and E_2 may be chosen, and the common reference plane E_0 is obtained by averaging E_1 and E_2. Distortion will be calculated by:

$$\Delta r = r \cdot \left(\frac{c_0 \cdot (l(r) + d_0)}{c'(r) \cdot (len + d_0)} - 1 \right) \tag{5}$$

Here, *len* is a fix value for shifting the projection center which can be chosen as average value over all $l(r)$.

This model can be checked or replaced by a classical model by a pinhole calibration directly performed under water. It should be noted, however, that distance-dependent distortion effects are considerably larger than at air application of the scanner. However, the expected errors can be derived using Formulas (1)–(3).

2.3. Calibration

Calibration is performed in three steps and completely in air. Optionally, a fourth calibration step, which requires underwater measurements, can be added. The first step is identical to classical air calibration where intrinsic (principal distance, principal point, distortion) and extrinsic (relative orientation between the stereo cameras) parameters are determined in the world co-ordinate system (WCS).

Depending on the technical realization of the sensor hardware, air calibration is performed without or inside underwater housing. Accordingly, glass thickness and glass material (refraction index) must be considered. From the two-dimensional radial distortion function, the variable principal distance $c(r)$ can be estimated with c_0 as principal distance of the air calibration and assuming a main reference distance d_0 at air calibration:

$$c(r) = \frac{r \cdot (d_0 + c_0)}{r + d_0/c_0(r + \Delta r)} \tag{6}$$

Naturally, this estimation is erroneous, because the determination of the radial distortion within intrinsic calibration is erroneous, too. However, due to the new modeling concept, these errors are reduced. Alternatively, $c(r)$ can be determined separately by goniometric measurements. However, the risk of high random errors is high. The corresponding error cannot be determined because there does not exist any reference method. As we will see later in Section 3, extended modeling with variable principal distance actually leads to decreased systematic 3D measurement errors at air calibration.

The second step of calibration comprises determination of refraction indices, glass thickness, and distance between the camera and housing ports. The glass thickness can be measured and the refraction indices are known for the materials used. The refraction index of water depends both on temperature and salinity. The distances d_1 and d_2 between the two cameras to the housing ports may be estimated using the hardware design and possibly refined after performance of underwater measurements [26]. However, first experiments and simulations have shown a very weak dependence of the resulting 3D reconstruction error on the values of d_1 and d_2. Actually, an uncertainty of d_1 and d_2 of some millimeters lead to errors of few micrometers in the 3D measurement and can be typically ignored.

In the third step of underwater calibration the parameters for the 3D calculation according to Equations (1)–(3) are determined.

The quality of this calibration can be evaluated by four (or more) underwater measurements of planes and ball-bars in a certain arrangement. These measurements characterize the quality of the calibration by determination of certain characteristic quantities (e.g., length measurement deviation, flatness deviation [35]). Alternatively, they may serve for determination of a 3D correction function, which can be understood as the fourth step of the calibration process. This fourth step should be performed in accordance to the estimated remaining errors and the requirements to the measurement accuracy, if required.

The minimal number of evaluation measurements in order to estimate low-frequency systematic errors is four: two plane measurements M_1 and M_2 at the front and the back of the measurement volume (MV), and two (M_3 and M_4) ball-bed measurements in the same regions in the MV. These measurements can be used to estimate the distribution of the systematic error in the MV [34] and a correction function can be defined. This function may be a polynomial in R^3 or a field of 3D correction vectors at certain sampling points. It corrects the 3D measurement points depending on the distance and the radial displacement to the main axis of the sensor (which can be constructed as average of both optical axis of the cameras).

2.4. Material

In order to evaluate the proposed new methodology for a-priori calibration of optical underwater 3D scanners recordings of a hand-held underwater 3D scanner were used and compared to the results obtained from measurements using the calibration method proposed in [26]. Figure 3 shows the scanning device. Calibration was initially performed without housing using the commercially available software BINGO [36]. Afterwards, measurements were performed in order to evaluate the air calibration and to assess the effects of the use of the extended pinhole model with variable principal distance.

Figure 3. Handheld underwater 3D scanner, front view (**left**) and back view (**right**).

Subsequently, measurements inside housing and underwater measurements were performed. Measurement objects were a plane made of ceramics and a ball-bed with a calibrated distance between the sphere center points (see Figure 4).

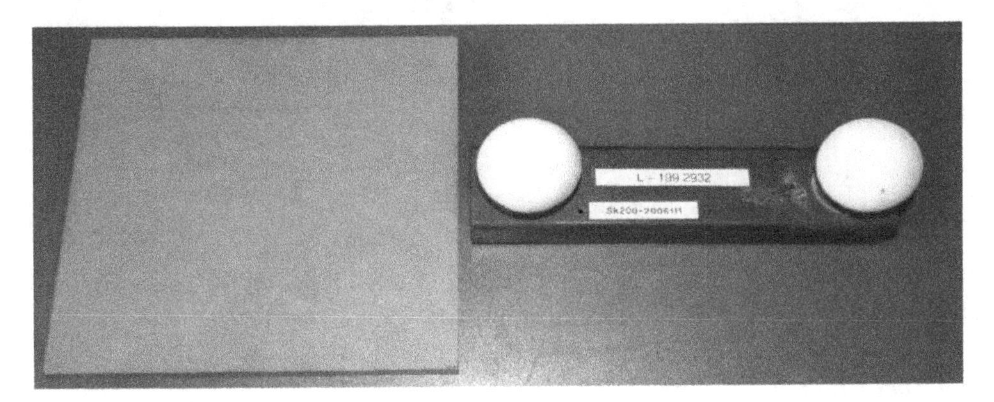

Figure 4. Measurement objects: ceramic plane (**left**) and ball-bed (**right**).

3. Experiments and Results

3.1. Calibration Evaluation

Calibration of the scanner was performed using a scene without landmarks including a frustum of a pyramid and using the software BINGO. This results in a set of intrinsic and extrinsic parameters for both cameras. Distortion functions were determined by application of a number (8 of 60) of one-parametric distortion functions.

The amount of the radial symmetric distortion was 98% (left camera) and 95% (right camera), respectively. The average distance of the measured points was about 500 mm. The function $c(r)$ was calculated at 1000 discrete values for r according to Equation (6). A third-degree polynomial was fitted to these sampling points for both cameras in order to obtain the functional description of $c(r)$.

Subsequently, the underwater parameters were determined (Table 1). According to Equations (1) to (3), the functions $c'(r)$ and $l(r)$ were obtained in the third step of the underwater calibration.

Table 1. Underwater parameters of the 3D scanner obtained from literature (refraction indices, water index is 1.0), design (viewport distance), and mechanical measurement (glass thickness).

Glass Thickness	Refraction Index Water	Refraction Index Glass	Distance to Viewport
3.0 mm	1.334	1.76	60 mm

Additionally, using the four measurements M_1 to M_4 (see Section 2.3) a 3D correction function $\Delta v = f(x,y,z)$ was determined and applied according to the description in Section 2.3 as follows. Plane measurements (approximated by two planes E_1 and E_2 in the WCS) were used to correct Z coordinates described by a squared polynomial in r according to the estimated symmetry points s_1 and s_2 of the plane deformation. Correction values for points outside the planes E_1 and E_2 are obtained by linear interpolation or extrapolation. Ball-bed measurements were used to estimate the correction function in X and Y. Here, linear scaling of the points according to the radius regarding s_1 and s_2 was applied for points in E_1 and E_2 and interpolation or extrapolation was processed analogously to Z correction.

Finally, a conventional pinhole model was generated according to the description of Section 2.2.2. See Figure 5 for the function plots of $c(r)$ in air and $c'(r)$ in water according to Equation (1) and calculated radial distortion function Δr according to Equation (5) for three different arbitrarily chosen values for the standard distance d_0 (d_{01} = 400 mm, d_{02} = 600 mm, and d_{03} = 800 mm).

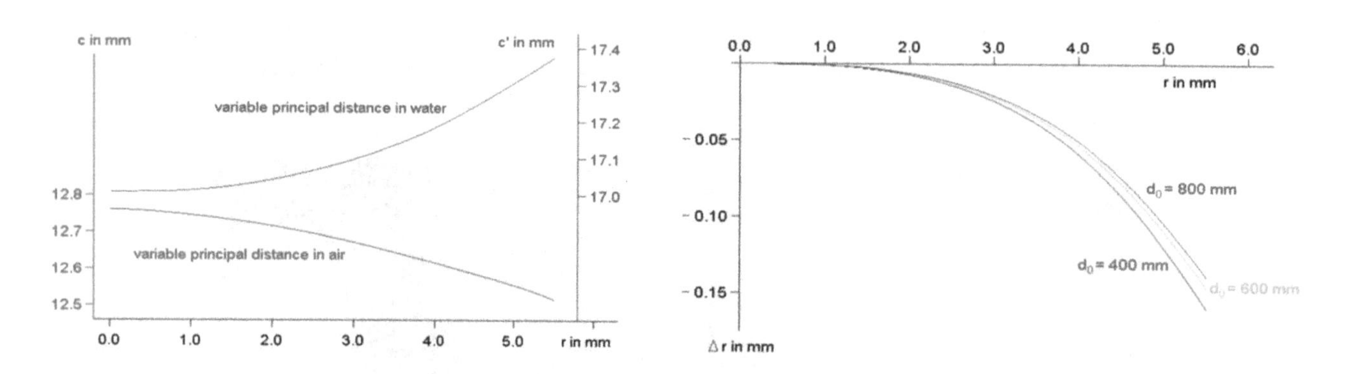

Figure 5. Plot of variable principal distance functions for both air and water case (**left**) and calculated distortion functions depending on reference distance (**right**).

3.2. Experimental Evaluation

In order to evaluate the different calibrations, several measurements of a plane and a ball-bed at different distances between 380 mm and 500 mm were performed. For evaluation, flatness deviation, and length measurement deviation were determined according to a prior description [26] and in accordance to the VDI/VDE guidelines of the "Verein Deutscher Ingenieure" [35]. First, measurements were realized without underwater housing in air. The results obtained by the common pinhole model with distortion correction was compared to the results obtained by the new modeling with variable principal length and projection center using identical measurement data. Table 2 shows the results of length and flatness deviation at air measurements.

In a second experiment, underwater measurements were performed. Characteristic quantities were determined for identical underwater measurement data using different calibrations, i.e., previous underwater calibration [26] (cal_0), new methodology with variable principal distance without (cal_1), 3D correction function (cal_2), and fix principal distance model (cal_3), according to Section 2.2.2. Table 3 shows the results of length and flatness deviation in underwater measurements.

Table 2. Comparison of air calibration cal_1 using variable principal distance (PD) and projection center to classical pinhole calibration cal_2 including distortion correction (DC) using identical measurement data.

Calibration	PD	DC	Length Error	Flatness Error
cal_1	variable	no	100 µm (0.05% [1])	212 µm (0.05% [2])
cal_2	fix	yes (radial and other)	145 µm (0.07% [1])	266 µm (0.07% [2])

[1] reference length was 200 mm, [2] plane diagonal was between 270 mm and 400 mm.

Table 3. New underwater calibration method without underwater measurements (cal_1) with 3D correction polynomial (cal_2) and as approximated classical pinhole calibration (cal_3) compared to previous calibration [26] (cal_0).

Calibration	Air/Water	PD	DC	Length Error	Flatness Error
cal_0	air + water	variable	yes	0.21 mm (0.11% [1])	2.5 mm (0.8%[2])
cal_1	air	variable	no	0.40 mm (0.20% [1])	2.0 mm (0.6%[2])
cal_2	air + water	variable	no	0.15 mm (0.08% [1])	1.0 mm (0.3%[2])
cal_3	air	fix	yes	0.87 mm (0.44% [1])	2.4 mm (0.8%[2])

[1] reference length was 200 mm, [2] plane diagonal was between 250 mm and 300 mm.

Although it is difficult to compare accuracy results of different calibration and 3D calculation methods, we compared our results with some (partly extracted) values from the literature (see Table 4). As can be seen, most error results transformed to percentage have similar magnitude. A more extensive analysis of underwater calibration methods including error quantities is presented by Shortis [27].

An application example of the scanner showing projected fringe pattern and 3D reconstruction result is shown in Figure 6. Figure 7 shows the difference of two ball-bed measurements in air and water, respectively. Different colors represent the deviation in the 3D comparison.

Figure 6. Projected fringe pattern during the measurement of a pipe (**left**) and 3D reconstruction result (**right**).

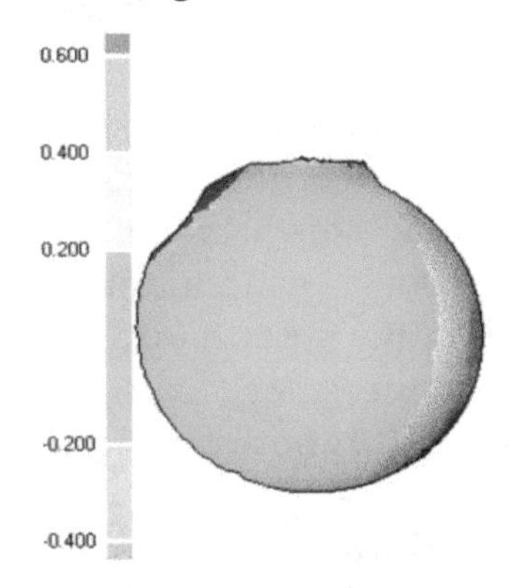

Figure 7. 3D difference image of two measurements of the ball-bed in air and water, respectively. Colors represent the deviation of the 3D coordinates (scale is in mm).

Table 4. Comparison of selected 3D accuracy results from the literature.

Author/Reference	Method	Validation	Percentage Error [1]
Li et al. 1996 [37]	Photogrammetry	Length error	0.8
Telem and Filin 2010 [15]	Photogrammetry	Flatness and length error	0.1–0.4
Zhang 2011 [24]	Fringe projection	Length error	0.5
Bruno et al. 2011 [23]	Fringe projection	Flatness error	0.5
Bianco et al. 2013 [25]	Fringe projection	Flatness and length error	0.1–0.3
Buschinelli et al. 2016 [38]	Fringe projection	Length error (ball-bar)	0.4

[1] percentage error was estimated by the authors from the published error values.

3.3. Error Consideration

The remaining errors of the new methodology may have multiple sources. As the typically used pinhole model has its deviations from reality, the new modeling with variable principal distance has deviations, too. One potential error source in this regard is the method to determine $c(r)$. The proposed method assumes a correct distortion determination for the distance of the reference plane. However, this is only an approximation. We assume that by using a better method for finding $c(r)$, the results can be further improved. Additionally, chromatic aberration effects may influence the determination of the distortion function or variable principal distance, respectively. Whereas at air calibration white light is typically present, blue light has greater impact to the image generation under water.

One next potential error source is the deviation from the perpendicularity between principal rays of the two cameras and glass ports of the housing. By using this as precondition, deviations may be compensated only by a 3D correction function. Deviations of the actual refraction indices from the used values seem to be negligible as well as small deviations of the parameters d_1 and d_2 (camera distances to the viewports). This was confirmed by simulations.

All deviations of the real situation from the assumed model properties lead to systematic errors of the reconstruction of the vision ray and subsequently to systematic errors of the determined 3D measurement points.

Other error sources may be inhomogeneities of media. For instance, the glass of the vision ports might contain material discontinuities.

4. Discussion

The results have shown the capability of the new methodology for calibration of a structured light optical underwater 3D stereo scanner. Additional effort with respect to classical air calibration is low. Measurement accuracy is comparable to former results and values from the literature (see [27]). Improvements regarding the 3D measurement accuracy are insignificant. Scanning object points in considerably different distances let expect to show the advantage of the new modeling concept.

By performing an additional calibration step using underwater measurements, high accuracy can be achieved using an additional correction function. The new method is advantageous for close range photogrammetric measurements, where the measurement distance is short and varying.

The presented experimental results obtained by complete air calibration (cal_1) almost achieve the results from the literature. However, for more comprehensive statements concerning measurement accuracy of the new methodology more experiments should be performed. This is one of the most important future tasks. It should be mentioned that the data used are from measurements using a very short distance to the measurement objects. In this context, local relative systematic errors are expected to be particularly high. Hence, for measurements at larger distances (e.g., 1 m), the new methodology, including the subsequent under water refinement of the calibration, provides more accurate 3D measurement results.

However, the introduced new methodology can be further improved. Certain error sources were neglected and simplifications were made. The method has not been applied to other underwater scanning devices yet.

Another shortcoming of the method is the use of a plane for determination of the additional correction function. Such a plane (material should be granite or ceramics) is easy to handle for small measurement volumes but may become cumbersome, heavy, and expensive for larger fields of view (e.g., 1 m^2).

In order to overcome these remaining weaknesses of the method, future work should aim to replace the plane measurements in water by calibration bodies which can be handled more easily.

Another focus of future work should be further verification by using more measurements of underwater 3D scanners of similar type in order to confirm results.

5. Conclusions

A-priori underwater calibration of optical 3D stereo scanners is possible using the described new method. This methodology follows a new geometric modeling of the radial lens distortion, leading to a variable principal distance depending on the radial distance of the points with respect to the principal point.

Application of the new calibration method provides satisfying 3D accuracy results with low effort. It provides a good alternative to complex calibration if no high-end precision is necessary. Varying external conditions, such as water temperature and salinity, can be considered in the calibration. A post-processing correction step using a small number of underwater measurements provides an improvement of the accuracy by factor two. It is expected that further improvements are possible, which will achieve a further reduction of the remaining errors.

Author Contributions: Conceptualization: C.B.-B.; methodology: C.B.-B. and C.M.; software: C.B.-B.; validation, C.B.-B.; formal analysis: C.B.-B.; investigation: C.B.-B. and I.G.; resources: I.G. and M.H.; data curation: C.B.-B. and M.H.; writing—original draft preparation: C.B.-B.; writing—review and editing: C.B.-B., S.H., and C.M.; visualization: C.B.-B. and C.M.; supervision: G.N. and P.K.; project administration: P.K.; All authors have read and agreed to the published version of the manuscript.

References

1. Tetlow, S.; Allwood, R.L. The use of a laser stripe illuminator for enhanced underwater viewing. In Proceedings of the Ocean Optics XII 1994, Bergen, Norway, 26 October 1994; Volume 2258, pp. 547–555.
2. McLeod, D.; Jacobson, J.; Hardy, M.; Embry, C. Autonomous inspection using an underwater 3D LiDAR. In *An Ocean in Common, Proceedings of the 2013 OCEANS, San Diego, San Diego, CA, USA, 23–27 September 2013*; IEEE: Toulouse, France, 2014.
3. Harvey, E.; Cappo, M.; Shortis, M.; Robson, S.; Buchanan, J.; Speare, P. The accuracy and precision of underwater measurements of length and maximum body depth of southern bluefin tuna (Thunnus maccoyii) with a stereo–video camera system. *Fish. Res.* **2003**, *63*, 315–326. [CrossRef]
4. Dunbrack, R.L. In situ measurement of fish body length using perspective-based remote stereo-video. *Fish. Res.* **2006**, *82*, 327–331. [CrossRef]
5. Costa, C.; Loy, A.; Cataudella, S.; Davis, D.; Scardi, M. Extracting fish size using dual underwater cameras. *Aquac. Eng.* **2006**, *35*, 218–227. [CrossRef]
6. Bythell, J.C.; Pan, P.; Lee, J. Three-dimensional morphometric measurements of reef corals using underwater photogrammetry techniques. *Coral Reefs* **2001**, *20*, 193–199.
7. Korduan, P.; Förster, T.; Obst, R. Unterwasser-Photogrammetrie zur 3D-Rekonstruktion des Schiffswracks "Darßer Kogge". *Photogramm. Fernerkund. Geoinf.* **2003**, *5*, 373–381.
8. Canciani, M.; Gambogi, P.; Romano, F.G.; Cannata, G.; Drap, P. Low cost digital photogrammetry for underwater archaeological site survey and artifact insertion. The case study of the Dolia wreck in secche della Meloria-Livorno-Italia. *Int. Arch. Photogramm. Remote Sens. Spat. Inf. Sci.* **2003**, *34 Pt 5*, 95–100.
9. Roman, C.; Inglis, G.; Rutter, J. Application of structured light imaging for high resolution mapping of underwater archaeological sites. In Proceedings of the OCEANS'10 IEEE SYDNEY, Sydney, Australia, 24–27 May 2010; pp. 1–9.

10. Drap, P. Underwater photogrammetry for archaeology. In *Special Applications of Photogrammetry*; Da Silva, D.C., Ed.; InTech: London, UK, 2012; pp. 111–136, ISBN 978-953-51-0548-0.
11. Eric, M.; Kovacic, R.; Berginc, G.; Pugelj, M.; Stopinsek, Z.; Solina, F. The impact of the latest 3d technologies on the documentation of underwater heritage sites. In Proceedings of the IEEE Digital Heritage International Congress 2013, Marseille, France, 28 October–1 November 2013; Volume 2, pp. 281–288.
12. Menna, F.; Agrafiotis, P.; Georopoulos, A. State of the art and applications in archaeological underwater 3D recording and mapping. *J. Cult. Herit.* **2018**, *33*, 231–248. [CrossRef]
13. Kwon, Y.H.; Casebolt, J. Effects of light refraction on the accuracy of camera calibration and reconstruction in underwater motion analysis. *Sports Biomech.* **2006**, *5*, 315–340. [CrossRef]
14. Sedlazeck, A.; Koch, R. Perspective and non-perspective camera models in underwater imaging—overview and error analysis. In *Theoretical Foundations of Computer Vision*; LNCS vol. 7474; Springer: Berlin/Heidelberg, Germany, 2011; pp. 212–242.
15. Telem, G.; Filin, S. Photogrammetric modeling of underwater environments. *ISPRS J. Photogramm. Remote Sens.* **2010**, *65*, 433ff. [CrossRef]
16. Moore, K.D. Intercalibration method for underwater three-dimensional mapping laser line scan systems. *Appl. Opt. No.* **2001**, *40*, 5991–6004. [CrossRef] [PubMed]
17. Narasimhan, S.G.; Nayar, S.K. Structured light methods for underwater imaging: Light stripe scanning and photometric stereo. In Proceedings of the OCEANS 2005 MTS/IEEE, Washington, DC, USA, 17–23 September 2005; Volume 3, pp. 2610–2617.
18. Tan, C.S.; Seet, G.; Sluzek, A.; He, D.M. A novel application of range-gated underwater laser imaging system (ULIS) in near-target turbid medium. *Opt. Lasers Eng.* **2005**, *43*, 995–1009. [CrossRef]
19. 2Grobotics. Available online: http://www.2grobotics.com/ (accessed on 30 June 2020).
20. ARIS-Sonars. Available online: http://soundmetrics.com/Products/ARIS-Sonars (accessed on 30 June 2020).
21. 3DatDepth. Available online: http://www.3datdepth.com/ (accessed on 30 June 2020).
22. Massot-Campos, M.; Oliver-Codina, G. Underwater laser-based structured light system for one-shot 3D reconstruction. In Proceedings of the 5th Martech Int. Workshop on Marine Technology, Girona, Spain, 2–5 November 2014.
23. Bruno, F.; Bianco, G.; Muzzupappa, M.; Barone, S.; Razionale, A.V. Experimentation of structured light and stereo vision for underwater 3D reconstruction. *ISPRS J. Photogramm. Remote Sens.* **2011**, *66*, 508–518. [CrossRef]
24. Zhang, Q.; Wang, Q.; Hou, Z.; Liu, Y.; Su, X. Three-dimensional shape measurement for an underwater object based on two-dimensional grating pattern projection. *Opt. Laser Technol.* **2011**, *43*, 801–805. [CrossRef]
25. Bianco, G.; Gallo, A.; Bruno, F.; Muzzupappa, M. A comparative analysis between active and passive techniques for underwater 3D reconstruction of close-range objects. *Sensors* **2013**, *13*, 11007–11031. [CrossRef] [PubMed]
26. Bräuer-Burchardt, C.; Heinze, M.; Schmidt, I.; Kühmstedt, P.; Notni, G. Underwater 3D surface measurement using fringe projection based scanning devices. *Sensors* **2016**, *16*, 13. [CrossRef] [PubMed]
27. Shortis, M. Camera calibration techniques for accurate measurement underwater. In *3D Recording and Interpretation for Maritime Archaeology*; Coastal Research Library, McCarthy, J., Benjamin, J., Winton, T., van Duivenvoorde, W., Eds.; Springer: Cham, Switzerland, 2019; Volume 31.
28. Zhang, S. Recent progresses on real-time 3D shape measurement using digital fringe projection techniques. *Opt. Lasers Eng.* **2010**, *48*, 149–158. [CrossRef]
29. Sedlazeck, A.; Koser, K.; Koch, R. 3D reconstruction based on underwater video from rov kiel 6000 considering underwater imaging conditions. In Proceedings of the OCEANS 2009-Europe, Bremen, Germany, 11–14 May 2009; pp. 1–10.
30. Bräuer-Burchardt, C.; Kühmstedt, P.; Notni, G. Combination of air- and water-calibration for a fringe projection based underwater 3D-Scanner. In *Computer Analysis of Images and Patterns, Proceedings of the CAIP 2015, Valletta, Malta, 2–4 September 2015*; Springer: Cham, Switzerland, 2015; Part II; Volume 9257, pp. 49–60.
31. Maas, H.G. New developments in multimedia photogrammetry. In *Optical 3-D Measurement Techniques III*; Grün, A., Kahmen, H., Eds.; Wichmann Verlag: Karlsruhe, Germany, 1995.
32. Oram, D. Rectification for any epipolar geometry. In Proceedings of the Proc British Machine Vision Conference 2001, Manchester, UK, 10–13 September 2001; pp. 653–662.

33. Luhmann, T.; Robson, S.; Kyle, S.; Harley, I. *Close Range Photogrammetry*; Wiley Whittles Publishing: Caithness, UK, 2006.
34. Bräuer-Burchardt, C.; Kühmstedt, P.; Notni, G. Improvement of measurement accuracy of optical 3D scanners by discrete systematic error estimation. In *Combinatorial Image Analysis, Proceedings of the IWCIA 2018, Porto, Portugal, 22–24 November 2018*; Barneva, R.P., Brimkov, V., Tavares, J., Eds.; Springer: Cham, Switzerland, 2018; Volume 11255, pp. 202–215.
35. VDI/VDE, VDI/VDE 2634. *Optical 3D-Measuring Systems*; VDI/VDE Guidelines; Verein Deutscher Ingenieure: Düsseldorf, Germany, 2008; Parts 1–3.
36. BINGO. Available online: http://www.bingo-atm.de (accessed on 29 June 2020).
37. Li, R.; Tao, C.; Curran, T.; Smith, R. Digital underwater photogrammetric system for large scale underwater spatial information acquisition. *Mar. Geod.* **1996**, *20*, 163–173. [CrossRef]
38. Buschinelli, P.D.V.; Matos, G.; Pinto, T.; Albertazzi, A. Underwater 3D shape measurement using inverse triangulation through two flat refractive surfaces. In Proceedings of the OCEANS 2016 MTS/IEEE, Monterey, CA, USA, 19–23 September 2016; pp. 1–7. [CrossRef]

3

A Brief Review of Mueller Matrix Calculations Associated with Oceanic Particles

Bingqiang Sun [1], George W. Kattawar [2], Ping Yang [1,*] and Xiaodong Zhang [3]

1 Department of Atmospheric Sciences, Texas A&M University, College Station, TX 77843, USA; sbq1418@gmail.com
2 Department of Physics and Astronomy and Institute for Quantum Science and Engineering, Texas A&M University, College Station, TX 77843, USA; kattawar@tamu.edu
3 Department of Earth System Science and Policy, University of North Dakota, Grand Forks, ND 58202, USA; xiaodong.zhang2@und.edu
* Correspondence: pyang@tamu.edu

Featured Application: This paper provides guidance for selecting an appropriate method for calculating the Mueller matrix associated with oceanic particles of arbitrary morphologies and refractive indices.

Abstract: The complete Stokes vector contains much more information than the radiance of light for the remote sensing of the ocean. Unlike the conventional radiance-only radiative transfer simulations, a full Mueller matrix-Stokes vector treatment provides a rigorous and correct approach for solving the transfer of radiation in a scattering medium, such as the atmosphere-ocean system. In fact, radiative transfer simulation without considering the polarization state always gives incorrect results and the extent of the errors induced depends on a particular application being considered. However, the rigorous approach that fully takes the polarization state into account requires the knowledge of the complete single-scattering properties of oceanic particles with various sizes, morphologies, and refractive indices. For most oceanic particles, the comparisons between simulations and observations have demonstrated that the "equivalent-spherical" approximation is inadequate. We will therefore briefly summarize the advantages and disadvantages of a number of light scattering methods for non-spherical particles. Furthermore, examples for canonical cases with specifically oriented particles and randomly oriented particles will be illustrated.

Keywords: ocean optics; light scattering; Mueller matrix; volume and surface integral methods

---

1. Introduction

It is well known that the scattering of light by a particle is determined by the detailed characteristics of the scattering particle, particularly its size, chemical composition (thus, the index of refraction), the overall shape, and detailed surface texture (e.g., surface roughness). Oceanic particles vary greatly in size and morphology. While the Lorenz-Mie theory has been used frequently to simulate the optical properties of oceanic particles (e.g., [1–5]), these particles are predominately nonspherical. Significant differences exist in the optical properties simulated by using "equivalent" spheres and non-spherical shapes, such as spheroids (e.g., [6]). In addition, even the simplest biological cell has a membrane and plasma contained within the membrane. Previous studies have shown that accounting for the cell structure can better simulate the optical properties of various phytoplankton species, particularly the scattering at large scattering angles [5,7–11]. Advanced light scattering methods have been developed to deal with complex shape and structure. Here, we briefly summarize light-scattering computational methods for oceanic particles. Light scattering in an absorbing medium has been extensively discussed [12–14]. For generality, however, only a nonabsorbing medium is discussed here. Beginning with Maxwell's equations, in Section 2, we will show exact volume-/surface-integral equations for mapping the near field to the far field. Furthermore, we introduce both the amplitude scattering matrix and the scattering phase matrix. In Section 3, several scattering methods will be introduced. In Section 4, discussions are given that are based on oriented particles and particles in random orientation.

2. Fundamental Concepts for Mueller Matrix Calculations

2.1. Maxwell's Equations and the Volume/Surface-Integral Equations

Since all of the rigorous light-scattering computational methods should obey Maxwell's equations, we will first give a brief introduction to the role that both the volume/surface methods for mapping the near field to far field play in the final solutions. We will only consider time-harmonic electromagnetic waves and dielectric particles. The dielectric particles are assumed to be isotropic and have a linear response to an applied field. In this case, Maxwell's equations in the medium while using SI units are as follows:

$$\nabla\cdot\dot{\mathbf{E}}\,(\vec{r}) = 0,\ \nabla\times\mathbf{E}(\vec{r}) = -\mu\frac{\partial\mathbf{H}(\vec{r})}{\partial t}, \tag{1}$$

$$\nabla\cdot\dot{\mathbf{H}}\,(\vec{r}) = 0,\ \nabla\times\mathbf{H}(\vec{r}) = \varepsilon\frac{\partial\mathbf{E}(\vec{r})}{\partial t}, \tag{2}$$

where $\mathbf{E}$ and $\mathbf{H}$ are the electric and the magnetic fields, respectively; ε and μ are the permittivity and permeability of the medium. Using the Fourier transformation, an arbitrary incident field in the time-domain can be transformed into the summation of the fields in the frequency-domain, or the time-harmonic fields. Assuming that the time-harmonic field follows $\exp(-i\omega t)$, where ω is the angular frequency of the electromagnetic wave, Maxwell's equations in a time-independent form become:

$$\nabla\cdot\dot{\mathbf{E}}\,(\vec{r}) = 0,\ \nabla\times\mathbf{E}(\vec{r}) = i\omega\mu\mathbf{H}(\vec{r}), \tag{3}$$

$$\nabla\cdot\dot{\mathbf{H}}\,(\vec{r}) = 0,\ \nabla\times\mathbf{H}(\vec{r}) = -i\omega\varepsilon\mathbf{E}(\vec{r}). \tag{4}$$

Using Equations (3) and (4), the vector Helmholtz equations for the electric and magnetic fields are

$$\left(\nabla^2 + k^2\right)\mathbf{E}(\vec{r}) = 0, \tag{5}$$

$$\left(\nabla^2 + k^2\right)\mathbf{H}(\vec{r}) = 0, \tag{6}$$

where k is the wave number and $k^2 = \omega^2\mu\varepsilon$. For oceanic particles, the surrounding medium and the scattering particles are assumed to be nonmagnetic, thus $\mu = \mu_0$, where μ_0 is the vacuum permeability. The light speed c in vacuum is equal to $1/\sqrt{\mu_0\varepsilon_0}$, where ε_0 is the permittivity in vacuum. Consequently, the refractive index m of the medium is $m = c/v = \sqrt{\varepsilon/\varepsilon_0}$, where v is the light speed in the medium. Since the electric and the magnetic fields are dependent on each other, we will use the electric field to describe the electromagnetic field.

The volume integral and surface integral equations of the electric field can be deduced from Maxwell's equations and the vector Green function [15]. In the far-field regime, they can be expressed in the form

$$\mathbf{E}^{\text{sca}}\,(\vec{r})\Big|_{r\to\infty} = \frac{\exp(ikr)}{-ikr}\frac{ik^3}{4\pi}\int_V d^3\vec{r}'\left\{(m^2-1)\left[\hat{r}\times\hat{r}\times\mathbf{E}(\vec{r}')\exp(-ik\hat{r}\cdot\vec{r}')\right]\right\}, \tag{7}$$

$$\mathbf{E}^{\rm sca}\left(\vec{r}\right)\Big|_{r\to\infty} = \frac{\exp(ikr)}{-ikr}\frac{k^2}{4\pi}\hat{r}\times\oint_S d^2\vec{r}'\left\{\left[\hat{n}_s\times\mathbf{E}(\vec{r}')\right] - \frac{\omega\mu_0}{k}\hat{r}\times\left[\hat{n}_s\times\mathbf{H}(\vec{r}')\right]\right\}, \tag{8}$$

where the parameters are given in Figure 1; $\hat{n}_s$ is the outward normal to the surface. It is evident that the scattered far field only depends on the scattered directions with an outgoing spherical wave factor $\exp(ikr)/kr$.

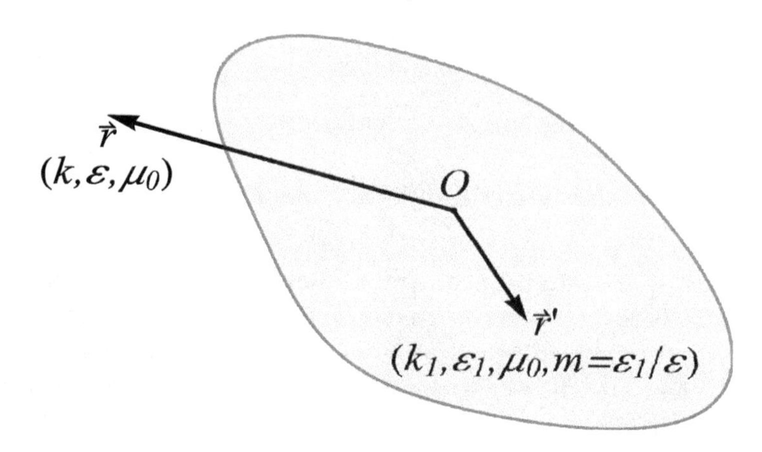

Figure 1. Parameters used for light scattering by a dielectric particle. The field point $\vec{r}$ is outside the scattering particle with wavenumber k, permittivity ε, and permeability μ_0 and the point $\vec{r}'$ is inside the particle with wave number k_1, permittivity ε_1, and permeability μ_0.

2.2. Amplitude Scattering Matrix and Mueller Matrix

Let the incident direction of the incoming wave be along the z-axis of the laboratory frame of reference. The incident direction and the scattered direction define a scattering plane and the incident and scattered fields can be expanded into parallel and perpendicular components with respect to the scattering plane. Consequently, the amplitude scattering matrix **S** can be given by [16]

$$\begin{pmatrix} \mathrm{E}_{\parallel}^{\rm sca} \\ \mathrm{E}_{\perp}^{\rm sca} \end{pmatrix} = \frac{\exp(ikr - ikz)}{-ikr}\mathbf{S}\begin{pmatrix} \mathrm{E}_{\parallel}^{\rm inc} \\ \mathrm{E}_{\perp}^{\rm inc} \end{pmatrix}, \tag{9}$$

where $\mathrm{E}_{\parallel}$ and $\mathrm{E}_{\perp}$ denote the parallel and perpendicular components of the electric field with respect to the scattering plane and **S** is a 2×2 complex matrix. The Stokes parameters in a non-absorbing medium are defined based on the measurable quantities, which are normally expressed in terms of a four-element column vector, the Stokes vector **I**, as follows:

$$\mathbf{I} = \begin{pmatrix} \mathrm{I} \\ \mathrm{Q} \\ \mathrm{U} \\ \mathrm{V} \end{pmatrix} = \begin{pmatrix} \mathrm{E}_{\parallel}\mathrm{E}_{\parallel}^* + \mathrm{E}_{\perp}\mathrm{E}_{\perp}^* \\ \mathrm{E}_{\parallel}\mathrm{E}_{\parallel}^* - \mathrm{E}_{\perp}\mathrm{E}_{\perp}^* \\ \mathrm{E}_{\parallel}\mathrm{E}_{\perp}^* + \mathrm{E}_{\perp}\mathrm{E}_{\parallel}^* \\ i\left(\mathrm{E}_{\parallel}\mathrm{E}_{\perp}^* - \mathrm{E}_{\perp}\mathrm{E}_{\parallel}^*\right) \end{pmatrix}. \tag{10}$$

In the above equation, i is the imaginary unit and a constant factor, $\sqrt{\varepsilon/\mu_0}/2$, is neglected since usually relative quantities are measured. The Mueller matrix (also called the scattering phase matrix in the literature) is the transformation matrix from the incident to the scattered Stokes parameters, as follows:

$$\mathbf{I}^{\rm sca} = \frac{1}{(kr)^2}\mathbf{P}\mathbf{I}^{\rm inc}, \tag{11}$$

where the 4×4 Mueller matrix **P** can be given as quadratic expressions of the amplitude scattering matrix **S,** as follows [17,18]:

$$\mathbf{P} = \mathbf{A}(\mathbf{S} \otimes \mathbf{S}^{*})\mathbf{A}^{-1}, \tag{12}$$

where asterisk denotes the complex conjugate and symbol $\otimes$ denotes the tensor product, and the constant matrix **A** is

$$\mathbf{A} = \begin{pmatrix} 1 & 0 & 0 & 1 \\ 1 & 0 & 0 & -1 \\ 0 & 1 & 1 & 0 \\ 0 & i & -i & 0 \end{pmatrix}, \quad \mathbf{A}^{-1} = \frac{1}{2}\mathbf{A}^{\dagger}, \tag{13}$$

in which the symbol † (sometimes called the dagger) is composed of two operations; namely, complex conjugating (the * symbol), and then transposing the original matrix and the order of these operations is unimportant. Note that the Stokes parameters have the units of irradiance [19], and on the other hand, the corresponding radiance is invariant over distance if no scattering or absorption occurs.

If the incident light is unpolarized, the scattering cross-section can be given in terms of the element $\mathbf{P}_{11}$ by

$$C_{\text{sca}} = \frac{1}{k^2}\int_{4\pi} d\Omega \mathrm{P}_{11}(\theta, \varphi). \tag{14}$$

The phase function is defined as:

$$p = \frac{4\pi}{k^2 C_{\text{sca}}}\mathrm{P}_{11}, \tag{15}$$

and the scattering phase matrix can be defined as:

$$\mathbf{F} = \frac{4\pi}{k^2 C_{\text{sca}}}\mathbf{P}. \tag{16}$$

The symmetry relations of the phase matrix have been extensively discussed in general and also for forward and backward scattering [20–22]. For an arbitrary particle without mirror symmetry in the scattering plane, the scattering phase matrix of a particle in random orientation is in the form

$$\mathbf{F} = \begin{pmatrix} a_1 & b_1 & b_3 & b_5 \\ b_1 & a_2 & b_4 & b_6 \\ -b_3 & -b_4 & a_3 & b_2 \\ b_5 & b_6 & -b_2 & a_4 \end{pmatrix}, \tag{17}$$

where there are only 10 independent parameters. For a particle with mirror symmetry in the scattering plane, the scattering phase matrix of a particle in random orientation is reduced to a block-diagonal matrix, as follows:

$$\mathbf{F} = \begin{pmatrix} a_1 & b_1 & 0 & 0 \\ b_1 & a_2 & 0 & 0 \\ 0 & 0 & a_3 & b_2 \\ 0 & 0 & -b_2 & a_4 \end{pmatrix}, \tag{18}$$

where there are only six independent parameters. Equations (17) and (18) represent the scattering phase matrix of a particle in random orientation. For a collection of particles with a size distribution, the collective scattering phase matrix can also be defined in terms of the distribution and the reader is referred to the book by Mobley [23].

All of the scattering quantities have been presented and we are now faced with the problem of obtaining the near fields or directly the far fields satisfying Maxwell's equations.

3. General Scattering Method for Suspended Particles

The governing principle for light scattering by particles is Maxwell's equations. The scattering solution is called Rayleigh scattering if $x \ll 1$ and $|mx| \ll 1$ [16,20], where the size parameter x is defined as $2\pi r_v/\lambda$ with r_v being the radius of a sphere or volume-equivalent sphere and λ the incident wavelength in the surrounding medium. The analytical solutions to Maxwell's equations are only effectively available for spheres [16,20]. For a prolate or oblate spheroid, the analytical solution is given in a series of the spheroidal wave functions by Asano and Yamamoto [24] and Asano and Sato [25]. However, the analytical solutions for a spheroid are only computationally effective for small particles due to numerical instability in computing the spheroidal wave functions for large particles. For an infinite circular cylinder, the analytical solution can be easily computed [16]. However, the infinite morphology does not exist in nature. For a particle with spherical symmetry, such as a homogeneous sphere or a multi-layered sphere, the analytical solution can be obtained by using the Lorenz-Mie theory for any size [16,20]. The advantage of using the spherical model is the computational efficiency, while the disadvantage is the appearance of spherical artifacts, such as the rainbow or glory, which have seldom been observed for ocean water (e.g., [26]). For a non-spherical particle, the solution of Maxwell's equations consists of two categories: rigorous and approximate solutions. The rigorous solutions can be further divided into numerically exact solutions and semi-analytical T-matrix solutions.

3.1. Numerically Exact Methods

As the name implies, numerically exact solutions use numerical methods to directly solve Maxwell's equations or the volume or surface integral equations derived from Maxwell's equations. The computational precision depends on the numerical resolution.

The finite difference time-domain method (FDTD) is based on the discretization of Maxwell's equation Equations (1) and (2) both in time and space [27]. The FDTD method uses the Yee grid to discretize the space, which was developed by Yee [28] and reviewed by Taflove [27], and Yang and Liou [29,30]. Since the computational space has to be confined to a finite region, a perfectly matched layer is used to absorb all of the electromagnetic waves in the computational boundary and avoid any artificially reflected electromagnetic waves back into the computational region [31]. The computational region usually with cuboid shape has to encompass the scattering particle so the computational region is always larger than the scattering particle in the FDTD application. The electromagnetic fields on the grids are updated with the advance of time so the FDTD is an initial value problem.

For a time-harmonic field or a field in the frequency domain, Maxwell's equations become the vector Helmholtz equations that are given in Equations (5) and (6). The vector Helmholtz Equations (5) and (6) can be discretized in space while using the finite-element method (FEM) [32]. The boundary condition on the particle surface and the continuity condition on the neighboring grid give a series of linear equations. The FEM is a boundary value problem. A major challenge for applying the FEM to light scattering is choosing the finite region covering the scattering particle so that the field in the computational region satisfies the radiation condition in the far field [33]. Like the FDTD method, the computational region for the FEM is also larger than the region that is occupied by the scattering particle, which can constrain the application regime of the FEM.

Using the vector Green function, the differential equations become the volume-integral or the surface-integral equations given by Equations (7) and (8). Even though the volume-integral and the surface-integral methods are equivalent, the volume-integral method is numerically more stable than the surface-integral method because the volume-integral equation in Equation (7) is a Fredholm integral equation of the second kind whose matrix equation is usually diagonally dominant [34].

The discrete-dipole approximation (DDA) method is a typical volume-integral method. The DDA was first proposed by Purcell and Pennypacker [35] and it was reviewed by Draine [36,37] and by Yurkin and Hoekstra [38]. In the DDA method, the particle volume is discretized into usually cubic cells, as shown in Figure 2. Each cubic cell is represented while using an electric dipole and the excited field at that cell is composed of the original incident field and the field from all other cells but excluding the cell itself. The dipoles generate a series of linear equations and the fields with respect to all dipoles can be obtained by solving the corresponding linear equations. Once the total field with respect to each cell is obtained, the amplitude scattering matrix and Mueller matrix can straightforwardly be computed. It is evident from Equation (7) that the computational region is equal to the volume of a scattering particle. That makes the DDA method computationally efficient when compared to other numerically exact methods.

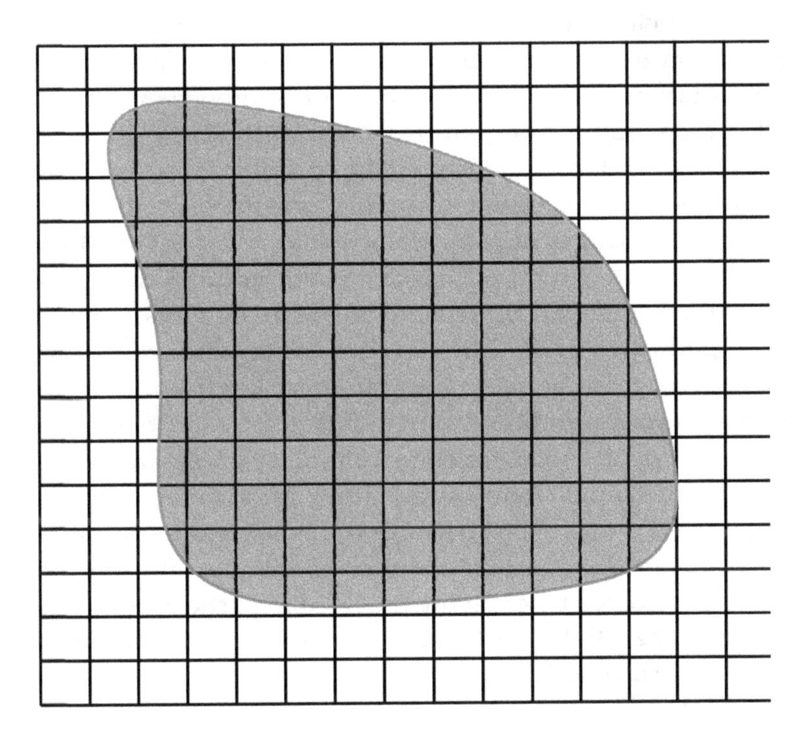

Figure 2. Discretization of a particle volume in the discrete-dipole approximation (DDA) method.

Two numerical implementations of the DDA are a FORTRAN implementation referred to as DDSCAT by Draine and Flatau [39] and a C implementation referred to as ADDA by Yurkin and Hoekstra [40]. The DDSCAT is parallelized for scattering only in different orientations so the memory requirement might restrict the computational capability for large particles. The ADDA is parallelized by distributing grids (dipoles) into different CPUs so the ADDA can handle particles of large sizes. The DDA method has been extensively used to simulate light scattering of oceanic particles. For instance, the light scattering of *Emiliania huxleyi* coccolithophore was simulated while using the DDSCAT by Gordon et al. [41] and using the ADDA by Zhai et al. [42]. Another example of the use of ADDA is the light scattering of dinoflagellates by Liu and Kattawar [43], where the chiral structure of the chromosomes is implemented by using discrete dipoles. This chiral structure leads to optical activity for certain dinoflagellates and another reason for measuring the complete single scattering Mueller matrix, which should be a fruitful area of research in remote sensing of the oceans.

The typical feature of a numerically exact method is that the error asymptotically approaches zero if the corresponding numerical grid that is associated with the method asymptotically reaches zero. Another feature for oceanic particles is that the convergence rate is much faster than the convergence rate of atmospheric particles because the relative refractive indices with respect to oceanic particles are close to unity. Moreover, the composition of a particle using the numerically exact methods can be arbitrary, homogeneous or inhomogeneous, or even different grid by grid.

For all of the numerically exact methods, the Mueller matrix is given in terms of the amplitude scattering matrix and they both depend on the incident direction. Consequently, the light scattering computation of a particle in random orientation while using these methods can usually be given by numerically summing the light scattering for different orientations. The convergence in the random orientation computation becomes significantly more difficult with increasing particle size so the computation will become time-consuming.

3.2. Semi-Analytical T-Matrix Method

The T-matrix method was originally proposed by Waterman [44,45]. The incident and scattered fields are expanded in a series of the vector spherical wave functions. The T-matrix connects the incident and scattered expansion coefficients because of the linearity of Maxwell's equations. The T-matrix of a particle only depends on the intrinsic properties of the particle, such as the refractive index, morphology, and the orientation of the particle frame of reference and its origin location, but not on the incident state. Corresponding to Equations (7) and (8), the T-matrix can be obtained using the surface-integral and volume-integral methods. The computational method of the T-matrix based on the surface integral is called the extended boundary condition method (EBCM) or the null field method, and it was reviewed by Tsang et al. [46], Mishchenko et al. [47,48], Mishchenko and Travis [49], and Doicu et al. [50]. The T-matrix method based on the volume integral equation is called the invariant-imbedding T-matrix method (IITM) and it was originally proposed by Johnson [51]. The IITM was reviewed and developed by Bi et al. [52]. For a particle with axial symmetry, the T-matrix is decoupled into a block-diagonal form, so the computation is significantly simplified. The applications of the EBCM on spheroids, cylinders, and Chebyshev shapes are exceptionally effective [49]. However, when a particle has a large size or an aspect ratio far from unity or one that is asymmetric, the matrices in the T-matrix computation are often ill-conditioned. The T-matrix method that is based on the volume integral is much more stable than the EBCM because the volume integral equation in Equation (7) is a Fredholm integral equation of the second kind, which is often less ill-conditioned [34]. The extreme stability of the IITM has been validated by applying the IITM to particles with large sizes, extreme aspect ratios, or asymmetric particles [52,53]. For instance, the IITM were used to compute the light scattering of oceanic particles, such as *Emiliania huxleyi* coccoliths and coccolithophores by Bi and Yang [54], and diatoms by Sun et al. [11].

When compared to the numerically exact methods, the significant advantage of the T-matrix solution is the analytical realization for a particle with a random orientation. The computational time of T-matrix methods is usually shorter than the numerically exact methods because T-matrix methods use matrix inversion instead of iterations. Moreover, in contrast to the relatively large refractive indices of atmospheric particles, such as ice crystals ($m \sim 1.33$) and aerosols (roughly $m > 1.5$), the relative refractive indices of oceanic particles are usually smaller than 1.2. For this reason, using the T-matrix method for oceanic particles normally yields faster convergence and higher computational efficiency than for atmospheric particles. However, the computational time and memory requirements of the T-matrix are strongly related to the radius of the circumscribed sphere of a particle and its morphology. For instance, for a needle particle with small volume but large circumscribed radius or a complex morphology, such as a porous particle, T-matrix methods are not as efficient as the numerically exact methods.

The numerically exact solutions can provide results for particle size parameters $x \sim 100$ or less; for the T-matrix solutions EBCM can reach $x \sim 180$, and the IITM can yield accurate results for $x \sim 300$. For even larger oceanic particles, approximate solutions must be used.

3.3. Physical-Geometric Optics Method

When the particle size is much larger than the incident wavelength, Maxwell's equations can be approximated by the eikonal equation [55]. The eikonal equation is the theoretical foundation of the geometric optics method. The key process for the geometric optics method is ray-tracing. The ray-tracing process of the geometric optics method consists of two parts: one is the diffracted rays and another is the transmitted rays, including external reflection, refraction without internal reflection, refraction with one internal reflection, and so on. The conventional geometric optics method (CGOM) considers the diffracted and the transmitted rays separately and it assumes equal contributions from the diffracted and transmitted rays under the assumption that the extinction efficiency is 2. Moreover, the CGOM does not consider the ray spreading effect from the near field to the far field, that is, there is no mapping process for the CGOM. The CGOM is applied to compute the light scattering of large particles with a large refractive index, such as ice particles (e.g., [56,57]). The CGOM can be improved by considering the ray spreading effect for a particle in random orientation. The improved geometric optics method (IGOM) can be used to compute light scattering of an intermediate particle or even a small particle [58,59].

For oceanic particles, the diffracted and transmitted rays have strong destructive interference so they cannot be separately handled. The physical-geometric optics method (PGOM) considers not only the interference between the diffracted and transmitted rays, but also the ray spreading effect in the far field [60]. Equations (7) and (8) are fundamental to the PGOM, which substantially extend the applicability of the principles of geometric optics in conjunction with physical optics to from large to moderate particles. For faceted particles, the ray-tracing process can be analytically accomplished since the phase change on a facet is linear [61–63]. The PGOM can be effectively used to compute the light scattering properties of oceanic particles.

4. Computational Results and Discussion

4.1. Dinoflagellate Simulation Using ADDA

Phytoplankton are one of two main categories of oceanic organisms and a significant component of the marine ecosystem that travel along the ocean currents. Many phytoplankton are positioned with preferred orientations due to the ocean flow [64]. Most phytoplankton are single-celled, such as dinoflagellates, diatoms, and coccolithophores. The bloomed phytoplankton can cause huge economic losses and influence environmental health, such as the red tide bloom of dinoflagellates in Florida [65]. Optical properties of an individual or bulked phytoplankton are essential to study phytoplankton populations (e.g., [66]). As mentioned in Section 3, dinoflagellates, diatoms, and coccolithophores have been simulated using the DDA and IITM [11,41–43,54]. Dinoflagellates have a large group of species so we take them as an example to describe the application of a scattering method.

Laboratory observation using transmission electron microscopy showed that the nucleus of dinoflagellates contains cylindrical chromosomes [67–69] and the chromosomes are arranged by ordered helical structures [69,70]. The helical structures are responsible for the strong circularly polarized effect that was observed in dinoflagellates [43,71,72]. The Mueller matrix element P_{14} reflects the circular polarization of a scattering particle and can be used as an index to indicate the strong circularly polarized effect [16,20]. Liu and Kattawar employed the ADDA code to fully simulate a single cell of a dinoflagellate and compute the 16 Mueller matrix elements [43], where the chromosomes are constructed using the plywood model [73]. For computational efficiency, only the nucleus with dozens of randomly positioned chromosomes is simulated. A chromosome is modeled as a cylindrical capsule with many layers, where every layer with fixed diameter contains parallel fibrils and the helical structure is described by making two adjacent layers with a constant rotation angle between them. The height with one period of rotation for the parallel fibrils is called the pitch. The chromosome simulation in the DDA method was performed by constructing fibrils in terms of dipoles and these fibrils were then arranged in layers and each layer was twisted a certain amount to make a helical shaped capsule to represent the chromosome. The diameter, the constant rotation angle, the number of helical periodicities, the pitch, the incident wavelength, and the incident directions can be changed to examine the circularly polarized effect of the helical structure. The important conclusions while using the ADDA code are given by Liu and Kattawar, as follows [43]:

- Strong back scattering signals from Mueller matrix element S_{14} are indeed from the helical structures of the chromosomes.

- Strong S_{14} back scattering signals are observed when the incident wavelength in the ocean is matched with the pitch of the helical structure, even if the chromosomes are under the random orientation condition.
- Strong S_{14} back scattering signals are observed when the incident direction is close to the main axis of the helical structure.
- The helical structure with constant rotation angle has stronger S_{14} back scattering signals than the helical structure with random rotation angle.

These conclusions suggest potential applications on the detection of the dinoflagellate and also the appropriate incident wavelength to match the pitch of helical structure.

4.2. Oceanic Particle Simulation Using ADDA, IITM, and PGOM

Section 4.1 describes an example of dinoflagellates while using the ADDA to compute the 16 Mueller matrix elements that were given by Liu and Kattawar [43], where the chromosome of the dinoflagellate has complex helical structure and is simulated mostly in fixed orientations. Generally, a simple nonspherical shape in random orientation is used to simulate the optical properties of oceanic particles. A hexahedron particle here is used as an example of an oceanic particle to show how Mueller matrix elements can be calculated by three typical methods: the ADDA, the IITM, and the PGOM. The relative refractive index of the particle is set to be 1.12 + i0.0005 and the incident wavelength is 0.658 μm. Only the Mueller matrix of the particle under the random orientation condition is given.

Figure 3 shows the comparisons of the non-zero Mueller matrix elements calculated by the IITM and the ADDA. The volume equivalent radius is 1 μm. The element P_{11} is normalized to give the normalized phase function while other elements are normalized by the element P_{11}. The simulation results calculated by the IITM and the ADDA are perfectly matched since they both are the exact solutions of Maxwell's equations. However, the computation using the IITM is much more efficient than the computation using the ADDA since the random orientation process is realized by ADDA through considering a large number of orientations. On the other hand, the ADDA for a fixed orientation in this case is more efficient than the IITM since the IITM has to compute the T-matrix of the particle, regardless of whether it is in a fixed orientation or under the random orientation condition.

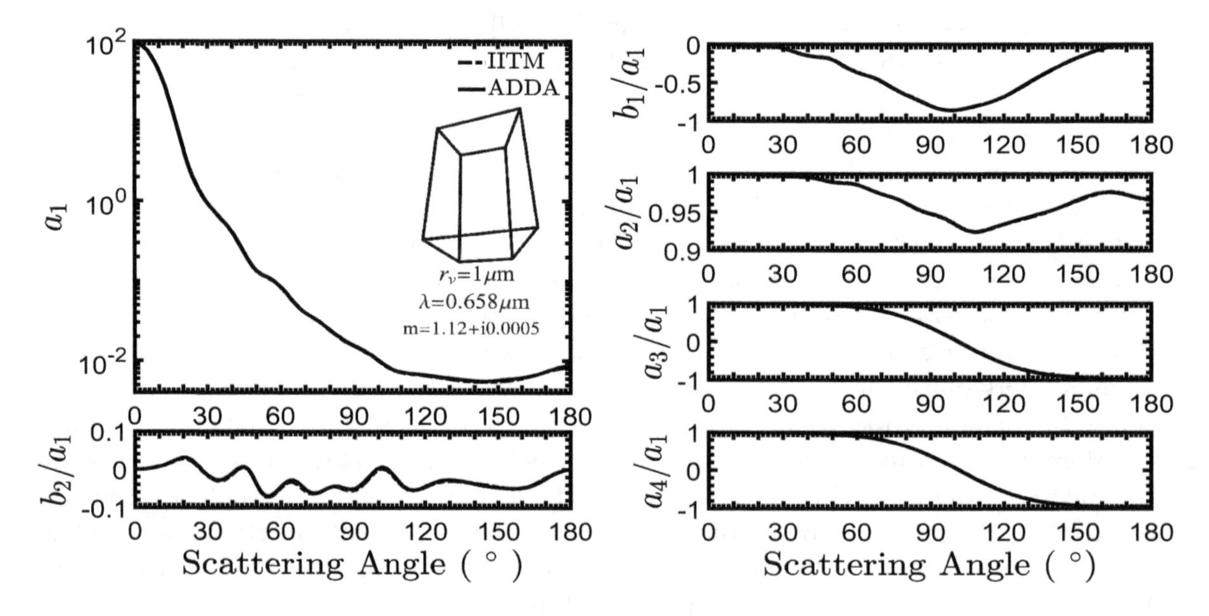

Figure 3. Comparisons of Mueller matrix elements of a hexahedron particle calculated by the invariant-imbedding T-matrix method (IITM) and the ADDA. The volume equivalent sphere radius is 1μm and the incident wavelength is 0.658 μm. The relative refractive index is 1.12 + i0.0005.

Figure 4 shows the comparisons of the Mueller matrix elements calculated by the IITM and the PGOM. The volume equivalent sphere radius is 8 μm. The PGOM results agree quite well with the IITM results, especially for the forward and backward scattering directions. Even though the PGOM is an approximate solution of Maxwell's equation, the process of including the interference between the diffracted and transmitted rays and mapping the near field to the far field significantly enhances its accuracy. The advantage of the PGOM is that it is computationally much more efficient than the IITM.

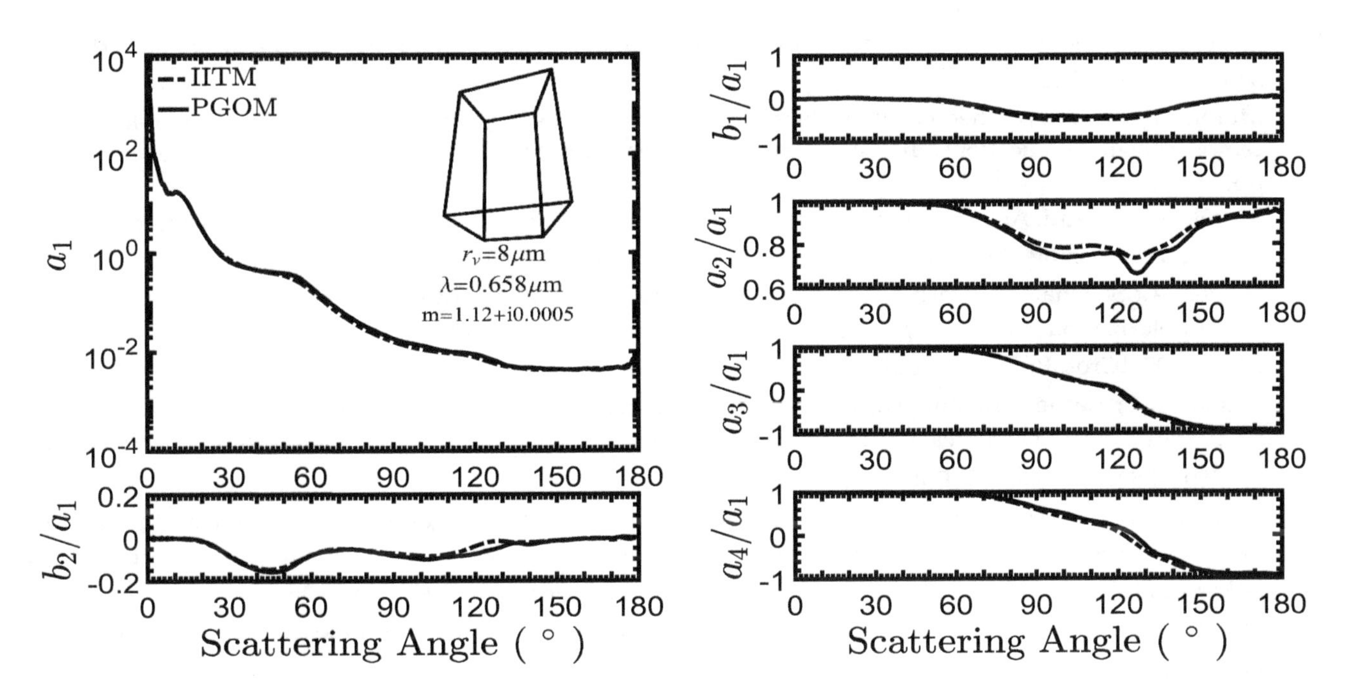

Figure 4. Comparisons of Mueller matrix elements of a hexahedron particle calculated by the IITM and the physical-geometric optics method (PGOM). The volume equivalent sphere radius is 8 μm and the incident wavelength is 0.658 μm. The relative refractive index is 1.12 + i0.0005.

5. Conclusions

A general introduction for calculating the Mueller matrix of suspended particles in the ocean is given. The surface and volume integral equations of the electromagnetic field can be given from Maxwell's equations. Also, the amplitude scattering matrix and the Mueller matrix with respect to light scattering can be defined to describe the polarization state of a suspended particle. To calculate the amplitude scattering matrix and Mueller matrix, the scattering methods are introduced based on the following categories: numerically exact methods, semi-analytical T-matrix methods, and geometric optics methods. For clarity, three typical methods: the DDA method, the IITM, and the PGOM, are briefly presented. Moreover, the Mueller matrix of an arbitrarily generated hexahedron particle under the random orientation condition is computed while using the ADDA and the IITM when the volume equivalent sphere radius is 1 μm and using the PGOM and the IITM for a volume equivalent sphere with a radius of 8 μm, while the incident wavelength is 0.658 μm. Perfect agreement between the ADDA and the IITM are given since both methods are considered to be the exact solutions. The IITM is more computationally efficient than the ADDA when the particle is under the random orientation condition. Excellent agreement between the PGOM and the IITM are obtained especially for the forward and the backward scattering directions. The PGOM is more computationally efficient than the IITM because of the ray-tracing process. Consequently, the Mueller matrix of suspended particles can be computed by using numerically exact methods, T-matrix methods, and the physical-geometric optics method to cover a complete size range.

Author Contributions: Conceptualization, G.W.K.; Methodology, B.S., P.Y., X.Z. and G.W.K.; Software, B.S.; Validation, B.S., P.Y. and X.Z.; Writing-Original Draft Preparation, B.S., P.Y. and X.Z.; Writing-Review & Editing, B.S., P.Y., X.Z. and G.W.K.; Visualization, B.S.; Supervision, P.Y. and G.W.K.; Project Administration, P.Y.; Funding Acquisition, P.Y. and X.Z.

Acknowledgments: This research were conducted with high performance research computing resources provided by Texas A&M University.

References

1. Stramski, D.; Kiefer, D.A. Light scattering by microorganisms in the open ocean. *Prog. Oceanogr.* **1991**, *28*, 343–383. [CrossRef]
2. Morel, A.; Bricaud, A. Theoretical results concerning the optics of phytoplankton, with special references to remote sensing applications. In *Oceanography from Space*; Gower, J.F.R., Ed.; Springer: Berlin, Germany, 1981; pp. 313–327.
3. Morel, A.; Bricaud, A. Inherent optical properties of algal cells, including picoplankton. Theorectical and experimental results. *Can. Bull. Fish. Aquat. Sci.* **1986**, *214*, 521–559.
4. Lerner, A.; Shashar, N.; Haspel, C. Sensitivity study on the effects of hydrosol size and composition on linear polarization in absorbing and nonabsorbing clear and semi-turbid waters. *J. Opt. Soc. Am. A* **2012**, *29*, 2394–2405. [CrossRef] [PubMed]
5. Tzabari, M.; Lerner, A.; Iluz, D.; Haspel, C. Sensitivity study on the effect of the optical and physical properties of coated spherical particles on linear polarization in clear to semi-turbid waters. *Appl. Opt.* **2018**, *57*, 5806–5822. [CrossRef] [PubMed]
6. Clavano, W.R.; Boss, E.; Karp-Boss, L. Inherent Optical Properties of Non-Spherical Marine-Like Particles—From Theory to Observation. In *Oceanography and Marine Biology: An Annual Review*; Gibson, R.N., Atkinson, R.J.A., Gordon, J.D.M., Eds.; Taylor & Francis: Didcot, UK, 2007; pp. 1–38.
7. Quinbyhunt, M.S.; Hunt, A.J.; Lofftus, K.; Shapiro, D.B. Polarized-light scattering studies of marine Chlorella. *Limnol. Oceanogr.* **1989**, *34*, 1587–1600. [CrossRef]
8. Meyer, R.A. Light scattering from biological cells: Dependence of backscattering radiation on membrane thickness and refractive index. *Appl. Opt.* **1979**, *18*, 585–588. [CrossRef] [PubMed]
9. Kitchen, J.C.; Zaneveld, J.R.V. A three-layered sphere model of the optical properties of phytoplankton. *Limnol. Oceanogr.* **1992**, *37*, 1680–1690. [CrossRef]
10. Quirantes, A.; Bernard, S. Light scattering by marine algae: Two-layer spherical and nonspherical models. *J. Quant. Spectrosc. Radiat. Transf.* **2004**, *89*, 311–321. [CrossRef]
11. Sun, B.; Kattawar, W.G.; Yang, P.; Twardowski, S.M.; Sullivan, M.J. Simulation of the scattering properties of a chain-forming triangular prism oceanic diatom. *J. Quant. Spectrosc. Radiat. Transf.* **2016**, *178*, 390–399. [CrossRef]
12. Mundy, W.C.; Roux, J.A.; Smith, A.M. Mie scattering by spheres in an absorbing medium. *J. Opt. Soc. Am.* **1974**, *64*, 1593–1597. [CrossRef]
13. Chylek, P. Light scattering by small particles in an absorbing medium. *J. Opt. Soc. Am.* **1977**, *67*, 561–563. [CrossRef]
14. Mishchenko, M.I.; Yang, P. Far-field Lorez-Mie scattering in an absorbing host medium: Theoretical formalism and FORTRAN program. *J. Quant. Spectrosc. Radiat. Transf.* **2018**, *205*, 241–252. [CrossRef]
15. Morse, P.M.; Feshbach, H. *Methods of Theoretical Physics*; McGraw-Hill: New York, NY, USA, 1953.
16. Bohren, C.F.; Huffman, D.R. *Absorption and Scattering of Light by Small Particles*; John Wiley & Sons: New York, NY, USA, 1983.
17. Parke, N.G., III. Optical Algebra. *J. Math. Phys.* **1949**, *28*, 131–139. [CrossRef]
18. Barakat, R. Bilinear constraints between elements of the 4 × 4 Mueller-Jones transfer matrix of polarization theory. *Opt. Commun.* **1981**, *38*, 159–161. [CrossRef]
19. Kattawar, G.W.; Yang, P.; You, Y.; Bi, L.; Xie, Y.; Huang, X.; Hioki, S. Polarization of light in the atmosphere and ocean. In *Light Scattering Reviews 10: Light Scattering and Radiative Transfer*; Kokhanovsky, A.A., Ed.; Springer: Berlin, Germany, 2016; pp. 3–39.

20. Van de Hulst, H.C. *Light Scattering by Small Particles*; John Wiley & Sons: New York, NY, USA, 1957.
21. Hu, C.; Kattawar, G.W.; Parkin, M.E.; Herb, P. Symmetry theorems on the forward and backward scattering Mueller matrices for light scattering from a nonspherical dielectric scatterer. *Appl. Opt.* **1987**, *26*, 4159–4173. [CrossRef] [PubMed]
22. Hovenier, J.W.; Mackowski, D.W. Symmetry relations for forward and backward scattering by randomly oriented particles. *J. Quant. Spectrosc. Radiat. Transf.* **1998**, *60*, 483–492. [CrossRef]
23. Mobley, C.D. *Light and Water: Radiative Transfer in Natural Waters*; Academic Press: San Diego, CA, USA, 1994.
24. Asano, S.; Yamamoto, G. Light scattering by a spheroidal particle. *Appl. Opt.* **1975**, *14*, 29–49. [CrossRef] [PubMed]
25. Asano, S.; Sato, M. Light scattering by randomly oriented spheroidal particles. *Appl. Opt.* **1980**, *19*, 962–974. [CrossRef] [PubMed]
26. Voss, K.J.; Fry, E.S. Measurement of the Mueller matrix for ocean water. *Appl. Opt.* **1984**, *23*, 4427–4439. [CrossRef] [PubMed]
27. Taflove, A.; Hagness, S.C. *Computational Electrodynamics: The Finite-Difference Time-Domain Method*; Artech House: Boston, MA, USA, 2000.
28. Yee, S.K. Numerical solution of initial boundary value problems involving Maxwell's equations in isotropic media. *IEEE Trans. Antennas Propag.* **1966**, *14*, 302–307.
29. Yang, P.; Liou, K.N. Finite-difference time domain method for light scattering by small ice crystals in three-dimensional space. *J. Opt. Soc. Am. A* **1996**, *13*, 2072–2085. [CrossRef]
30. Yang, P.; Liou, K.N. Finite difference time domain method for light scattering by nonspherical and inhomogeneous particles. In *Light Scattering by Nonspherical Particles*; Mishchenko, M.I., Hovenier, J.W., Travis, L.D., Eds.; Academic Press: San Diego, CA, USA, 2000; pp. 173–221.
31. Berenger, J.P. A perfectly matched layer for the absorption of electromagnetic waves. *J. Comput. Phys.* **1994**, *114*, 185–200. [CrossRef]
32. Silvester, P.P.; Ferrari, R.L. *Finite Elements for Electrical Engineers*; Cambridge University Press: Cambridge, UK, 1996.
33. Morgan, M.; Mei, K. Finite-element computation of scattering by inhomogeneous penetrable bodies of revolution. *IEEE Trans. Antennas Propag.* **1979**, *27*, 202–214. [CrossRef]
34. Press, W.H.; Flannery, B.P.; Teukolsky, S.A.; Vetterling, W.T. *Numerical Recipes*; Cambridge University Press: Cambridge, UK, 1989.
35. Purcell, E.M.; Pennypacker, C.R. Scattering and absorption of light by nonspherical dielectric grains. *Astrophys. J.* **1973**, *186*, 705–714. [CrossRef]
36. Draine, B.T. The discrete-dipole approximation and its application to interstellar graphite grains. *Astrophys. J.* **1988**, *333*, 848–872. [CrossRef]
37. Draine, B.T. *The Discrete Dipole Approximation for Light Scattering by Irregular Targets*; Academic Press: San Diego, CA, USA, 2000; pp. 131–144.
38. Yurkin, M.A.; Hoekstra, A.G. The discrete-dipole-approximation code ADDA: Capabilities and known limitations. *J. Quant. Spectrosc. Radiat. Transf.* **2011**, *112*, 2234–2247. [CrossRef]
39. Draine, B.T.; Flatau, P.J. User guide for the discrete dipole approximation code DDSCAT 7.3. *arXiv* **2013**, arXiv:1305.6497.
40. Yurkin, M.A.; Hoekstra, A.G. User Manual for the Discrete Dipole Approximation Code ADDA 1.3b4. 2014. Available online: http://a-dda.googlecode.com/svn/tags/rel_1.3b4/doc/manual.pdf (accessed on 6 May 2018).
41. Gordon, R.H.; Smyth, J.T.; Balch, M.W.; Boynton, C.G.; Tarran, A.G. Light scattering by coccoliths detached from *Emiliania huxleyi*. *Appl. Opt.* **2009**, *48*, 6059–6073. [CrossRef] [PubMed]
42. Zhai, P.W.; Hu, Y.; Trepte, C.R.; Winker, D.M.; Josset, D.B.; Lucker, P.L.; Kattawar, G.W. Inherent optical properties of the coccolithophore: *Emiliania huxleyi*. *Opt. Express* **2013**, *21*, 17625–17638. [CrossRef] [PubMed]
43. Liu, J.; Kattawar, G.W. Detection of dinoflagellates by the light scattering properties of the chiral structure of their chromosomes. *J. Quant. Spectrosc. Radiat. Transf.* **2013**, *131*, 24–33. [CrossRef]
44. Waterman, P.C. Matrix formulation of electromagnetic scattering. *Proc. IEEE* **1965**, *53*, 805–812. [CrossRef]
45. Waterman, P.C. Symmetry, unitarity, and geometry in electromagnetic scattering. *Phys. Rev. D* **1971**, *3*, 825. [CrossRef]
46. Tsang, L.; Kong, J.A.; Ding, K.H. Scattering of electromagnetic waves. In *Theories and Applications*; Wiley: Hoboken, NJ, USA, 2000; Volume 1.

47. Mishchenko, M.I.; Travis, L.D.; Mackowski, D.W. T-matrix computations of light scattering by nonspherical particles: A review. *J. Quant. Spectrosc. Radiat. Transf.* **1996**, *55*, 535–575. [CrossRef]
48. Mishchenko, M.I.; Travis, L.D.; Lacis, A.A. *Scattering, Absorption, and Emission of Light by Small Particles*; Cambridge University Press: Cambridge, UK, 2002.
49. Mishchenko, M.I.; Travis, L.D. Capabilities and limitations of a current FORTRAN implementation of the T-matrix method for randomly oriented, rotationally symmetric scatterers. *J. Quant. Spectrosc. Radiat. Transf.* **1998**, *60*, 309–324. [CrossRef]
50. Doicu, A.; Wriedt, T.; Eremin, Y.A. *Light Scattering by Systems of Particles: Null-Field Method with Discrete Sources: Theory and Programs*; Springer: Berlin, Germany, 2006; Volume 124.
51. Johnson, B.R. Invariant imbedding T matrix approach to electromagnetic scattering. *Appl. Opt.* **1988**, *27*, 4861–4873. [CrossRef] [PubMed]
52. Bi, L.; Yang, P.; Kattawar, G.W.; Mishchenko, M.I. Efficient implementation of the invariant imbedding T-matrix method and the separation of variables method applied to large nonspherical inhomogeneous particles. *J. Quant. Spectrosc. Radiat. Transf.* **2013**, *116*, 169–183. [CrossRef]
53. Bi, L.; Yang, P. Accurate simulation of the optical properties of atmospheric ice crystals with the invariant imbedding T-matrix method. *J. Quant. Spectrosc. Radiat. Transf.* **2014**, *138*, 17–35. [CrossRef]
54. Bi, L.; Yang, P. Impact of calcification state on the inherent optical properties of *Emiliania huxleyi* coccoliths and coccolithophores. *J. Quant. Spectrosc. Radiat. Transf.* **2015**, *155*, 10–21. [CrossRef]
55. Born, M.; Wolf, E. *Principles of Optics*; Cambridge University Press: Cambridge, UK, 1999.
56. Takano, Y.; Liou, K.N. Solar radiative transfer in cirrus clouds. Part I: Single-scattering and optical properties of hexagonal ice crystals. *J. Atmos. Sci.* **1989**, *46*, 3–19. [CrossRef]
57. Macke, A.; Mueller, J.; Raschke, E. Single scattering properties of atmospheric ice crystals. *J. Atmos. Sci.* **1996**, *53*, 2813–2825. [CrossRef]
58. Yang, P.; Liou, K.N. Geometric-optics—Integral-equation method for light scattering by nonspherical ice crystals. *Appl. Opt.* **1996**, *35*, 6568–6584. [CrossRef] [PubMed]
59. Muinonen, K. Scattering of light by crystals: A modified Kirchhoff approximation. *Appl. Opt.* **1989**, *28*, 3044–3050. [CrossRef] [PubMed]
60. Yang, P.; Liou, K.N. Light scattering by hexagonal ice crystals: Solutions by a ray-by-ray integration algorithm. *JOSA A* **1997**, *14*, 2278–2289. [CrossRef]
61. Bi, L.; Yang, P.; Kattawar, G.W.; Hu, Y.; Baum, B.A. Scattering and absorption of light by ice particles: Solution by a new physical-geometric optics hybrid method. *J. Quant. Spectrosc. Radiat. Transf.* **2011**, *112*, 1492–1508. [CrossRef]
62. Borovoi, A.G.; Grishin, I.A. Scattering matrices for large ice crystal particles. *JOSA A* **2003**, *20*, 2071–2080. [CrossRef] [PubMed]
63. Sun, B.; Yang, P.; Kattawar, G.W.; Zhang, X. Physical-geometric optics method for large size faceted particles. *Opt. Express* **2017**, *25*, 24044–24060. [CrossRef] [PubMed]
64. Nayak, A.R.; Mcfarland, M.N.; Sullivan, J.M.; Twardowski, M.S. Evidence for ubiquitous preferential particle orientation in representative oceanic shear flows. *Limnol. Oceanogr.* **2018**, *63*, 122–143. [CrossRef] [PubMed]
65. Heil, C.A.; Steidinger, K.A. Monitoring, management, and mitigation of Karenia blooms in the eastern Gulf of Mexico. *Harmful Algae* **2009**, *8*, 611–617. [CrossRef]
66. Kiefer, D.A.; Olson, R.J.; Wilson, W.H. Reflectance spectroscopy of marine phytoplankton. part I. optical properties as related to age and growth rate. *Limnol. Oceanogr.* **1979**, *24*, 664–672. [CrossRef]
67. Steidinger, K.A.; Truby, E.W.; Dawes, C.J. Ultrastructure of the red tide dinoflagellate *Gymnodinium breve*. I. General description 2.3. *J. Phycol.* **1978**, *14*, 72–79. [CrossRef]
68. Rizzo, P.J.; Jones, M.; Ray, S.M. Isolation and properties of isolated nuclei from the Florida red tide dinoflagellate *Gymnodinium breve* (Davis). *J. Protozool.* **1982**, *29*, 217–222. [CrossRef] [PubMed]
69. Gautier, A.; Michel-Salamin, L.; Tosi-Couture, E.; McDowall, A.W.; Dubochet, J. Electron microscopy of the chromosomes of dinoflagellates in situ: Confirmation of Bouligand's liquid crystal hypothesis. *J. Ultrastruct. Mol. Struct. Res.* **1986**, *97*, 10–30. [CrossRef]
70. Rill, R.L.; Livolant, F.; Aldrich, H.C.; Davidson, M.W. Electron microscopy of liquid crystalline DNA: Direct evidence for cholesteric-like organization of DNA in dinoflagellate chromosomes. *Chromosoma* **1989**, *98*, 280–286. [CrossRef] [PubMed]

71. Shapiro, D.B.; Quinbyhunt, M.S.; Hunt, A.J. Origin of the Induced circular-polarization in the light-scattering from a dinoflagellate. *Ocean Opt. X* **1990**, *1302*, 281–289.
72. Shapiro, D.B.; Hunt, A.J.; Quinby-Hunt, M.S.; Hull, P.G. Circular-polarization effects in the light-scattering from single and suspensions of dinoflagellates. *Underw. Imaging Photogr. Visibility* **1991**, *1537*, 30–41.
73. Bouligand, Y.; Soyer, M.O.; Puiseux-Dao, S. La structure fibrillaire et l'orientation des chromosomes chez les Dinoflagellés. *Chromosoma* **1968**, *24*, 251–287. [CrossRef] [PubMed]

4

Measurements of the Volume Scattering Function and the Degree of Linear Polarization of Light Scattered by Contrasting Natural Assemblages of Marine Particles

Daniel Koestner *, Dariusz Stramski and Rick A. Reynolds

Marine Physical Laboratory, Scripps Institution of Oceanography, University of California San Diego, La Jolla, CA 92093-0238, USA; dstramski@ucsd.edu (D.S.); rreynolds@ucsd.edu (R.A.R.)

* Correspondence: dkoestne@ucsd.edu

Abstract: The light scattering properties of seawater play important roles in radiative transfer in the ocean and optically-based methods for characterizing marine suspended particles from in situ and remote sensing measurements. The recently commercialized LISST-VSF instrument is capable of providing in situ or laboratory measurements of the volume scattering function, $\beta_p(\psi)$, and the degree of linear polarization, $DoLP_p(\psi)$, associated with particle scattering. These optical quantities of natural particle assemblages have not been measured routinely in past studies. To fully realize the potential of LISST-VSF measurements, we evaluated instrument performance, and developed calibration correction functions from laboratory measurements and Mie scattering calculations for standard polystyrene beads suspended in water. The correction functions were validated with independent measurements. The improved LISST-VSF protocol was applied to measurements of $\beta_p(\psi)$ and $DoLP_p(\psi)$ taken on 17 natural seawater samples from coastal and offshore marine environments characterized by contrasting assemblages of suspended particles. Both $\beta_p(\psi)$ and $DoLP_p(\psi)$ exhibited significant variations related to a broad range of composition and size distribution of particulate assemblages. For example, negative relational trends were observed between the particulate backscattering ratio derived from $\beta_p(\psi)$ and increasing proportions of organic particles or phytoplankton in the particulate assemblage. Our results also suggest a potential trend between the maximum values of $DoLP_p(\psi)$ and particle size metrics, such that a decrease in the maximum $DoLP_p(\psi)$ tends to be associated with particulate assemblages exhibiting a higher proportion of large-sized particles. Such results have the potential to advance optically-based applications that rely on an understanding of relationships between light scattering and particle properties of natural particulate assemblages.

Keywords: marine optics; inherent optical properties; volume scattering function; degree of linear polarization; marine particles; light scattering measurements; LISST-VSF instrument

---

1. Introduction

It has long been recognized that inherent light-scattering properties of natural waters are of crucial importance and have strong potential for wide-ranging applications in aquatic sciences, including oceanography. These properties are essential inputs to the radiative transfer models used to compute the ambient light fields in natural water bodies [1–4]. The variability in the light scattering properties of seawater is driven primarily by the concentration of suspended particles, particle size distribution, and composition through particle refractive index, internal structure, and shape. Hence, scattering measurements carry potentially useful information about characteristics of natural particle assemblages. For example, the scattering and backscattering coefficients of suspended particles have been shown to provide useful proxies of mass concentration of total suspended particulate matter (SPM), particulate inorganic carbon (PIC), and particulate organic carbon (POC) in the ocean [5–7]. Multi-angle light scattering measurements provide a means to estimate the particle size distribution [8–11], including the submicrometer size range [12–14]. The angular pattern of light scattering can also contain useful information about the composition of particulate assemblages, including the bulk refractive index of particles [15–19]. In addition, measurements of the scattering matrix that provide information about polarization effects of light scattering [15,20–23] have the potential for identifying and discriminating different types of particles, such as phytoplankton species or minerals, which are present in complex natural assemblages [24–34]. Despite the potential usefulness of information provided by light scattering measurements, the complexity and variability in composition of natural particulate assemblages impose significant challenges in achieving an understanding of bulk light-scattering properties of seawater in terms of detailed compositional characteristics of particulate matter [35].

The volume scattering function, $\beta_p(\psi, \lambda)$, and the degree of linear polarization, $DoLP_p(\psi, \lambda)$ of light scattered by marine particles are of primary interest in this study. Here, ψ denotes the scattering angle, λ the light wavelength in vacuum, and the subscript p indicates that the quantity is associated with particles. When the subscript p is omitted, the quantity describes the scattering by the entire suspension with additive contributions from both water molecules and suspended particles. The volume scattering function, $\beta(\psi, \lambda)$ [in units of $\text{m}^{-1}\ \text{sr}^{-1}$], is one of the fundamental inherent optical properties (IOPs) of seawater, which describes the scattered intensity as a function of scattering angle per unit incident irradiance per unit volume of small sample of water [2]. Several light-scattering related IOPs can be derived from $\beta(\psi, \lambda)$. For example, integrating $\beta(\psi, \lambda)$ over all scattering directions gives the total spectral scattering coefficient, $b(\lambda)$ [m^{-1}]. In this integration, it is commonly assumed that light scattering by an assemblage of randomly-oriented scatterers (molecules and particles) in natural waters is azimuthally symmetric about the incident direction of light beam. When $\beta(\psi, \lambda)$ is normalized by $b(\lambda)$, the resulting scattering phase function $\widetilde{\beta}(\psi, \lambda)$ [sr^{-1}] provides a useful indicator of the angular shape of the volume scattering function. In optical remote sensing applications based on measurements with above-water sensors (e.g., from satellites or aircraft), the spectral backscattering coefficient, $b_b(\lambda)$ [m^{-1}], is particularly useful. This coefficient can be obtained by integrating $\beta(\psi, \lambda)$ over the range of backward scattering angles [2].

The volume scattering function provides incomplete information, in the sense that it does not contain information about polarization effects associated with light scattering. A complete characterization of elastic incoherent interactions of light at arbitrary wavelength λ with a sample volume of seawater is provided by a 4×4 scattering matrix, often referred to as the phase matrix or Mueller matrix [20–23]. This matrix describes a linear transformation of irradiance and polarization of an incident beam described by a 4-component Stokes vector into the intensity and polarization of the scattered beam that is also described by its corresponding Stokes vector. $\beta(\psi, \lambda)$ is related to the first element of the scattering matrix, $p_{11}(\psi, \lambda)$, and can be obtained from a measurement using unpolarized light for illumination of sample and measuring the total scattered intensity. The degree of linear polarization of scattered light, $DoLP(\psi, \lambda)$, describes the proportion of linearly polarized light relative to total intensity of the scattered light beam. As described in greater detail below, for various assemblages of particles including suspended marine particles and when the incident light beam is unpolarized, this quantity can be derived from the first two elements of the scattering matrix, which requires measurements involving linear polarization [29,36,37].

Despite the relative importance of $\beta(\psi, \lambda)$ and $DoLP(\psi, \lambda)$ of seawater and the associated particulate components $\beta_p(\psi, \lambda)$ and $DoLP_p(\psi, \lambda)$, the ocean optics community has historically relied mostly on simplified theoretical models (such as Mie scattering theory for homogenous spheres) and a limited dataset of measurements made with custom-built light scattering instruments. For example, over the past several decades, a limited dataset of $\beta(\psi, \lambda)$ measurements made by Petzold [38] was widely used as a standard input for the particulate scattering phase function for radiative transfer modeling in the ocean. Comprehensive determinations of the scattering matrix for natural seawater have been very scarce [39–42]. These determinations showed that the off-diagonal matrix elements for seawater are very small or negligible, indicating very small effects associated with optical activity or orientational anisotropy of seawater scatterers [37,42]. More recently, several light scattering sensors have been developed for in situ deployments or laboratory use [43–46], but to our knowledge, none of these sensors are commercially available. While measurements with these new sensors have already significantly contributed to the increase of available datasets of $\beta(\psi,\lambda)$ (or $\beta_p(\psi,\lambda)$ which can usually be satisfactorily estimated by subtracting the contribution associated with water molecules) in various oceanic environments [44,47,48], the determinations of $DoLP_p(\psi,\lambda)$ for natural assemblages of marine particles remain very scarce, as indicated by the rarity of scattering matrix measurements of seawater.

Recently, a new light scattering instrument, the LISST-VSF (Sequoia Scientific, Inc., Bellevue, WA, USA), has become commercially available, and provides the capability of determining both the volume scattering function and the degree of linear polarization of scattered light at a single light wavelength (532 nm) with high angular resolution over the range ~0.1° to 155° [49]. It is capable of both in situ and benchtop measurements on water samples. This commercial instrument is expected to enable routine measurements by different groups of investigators, so it has the potential to enhance our understanding of light scattering properties of seawater and marine particles and advance the related applications. In this study, we report on LISST-VSF measurements of $\beta_p(\psi)$ and $DoLP_p(\psi)$ and size and compositional characteristics for contrasting natural particulate assemblages from marine coastal and offshore environments. The particulate scattering (b_p) and backscattering (b_{bp}) coefficients have also been determined from measured $\beta_p(\psi)$.

To fully realize the potential of such quantitative determinations for seawater samples from this new instrument, we also conducted an evaluation of the LISST-VSF performance through a series of laboratory experiments using samples of National Institute of Standards and Technology (NIST) certified standard polystyrene beads ranging in diameter between 100 nm and 2 μm. These measurements were compared with theoretical simulations of light scattering by bead suspensions using Mie scattering computations. With this approach, we developed corrections to the determinations of $\beta_p(\psi)$ and $DoLP_p(\psi)$ from LISST-VSF measurements. A validation of the corrected measurements was performed using independent measurements of multi-angle light scattering with another instrument, the DAWN-EOS (Wyatt Technology Corporation, Santa Barbara, CA, USA).

2. Methods

The description of methods includes two main parts: first, a description of laboratory experiments and Mie scattering calculations for standard polystyrene beads which were carried out to evaluate the performance of the LISST-VSF instrument and develop a calibration correction; second, a description of measurements on natural assemblages of marine particles from coastal and offshore oceanic environments.

2.1. Laboratory Experiments and Mie Scattering Calculations to Evaluate LISST-VSF

In order to evaluate the LISST-VSF instrument, light scattering and beam attenuation measurements were made in the laboratory on samples of nearly monodisperse standard polystyrene spherical beads with mean nominal diameters of 100, 200, 400, 500, 700, and 2000 nm, which were suspended in water (Table 1). In addition to LISST-VSF, two other instruments were used in these experiments, a DAWN-EOS for measuring multi-angle light scattering and a dual beam UV/VIS spectrophotometer Lambda 18 (Perkin-Elmer, Inc., Waltham, MA, USA) equipped with a 15-cm integrating sphere (Labsphere, Inc., North Sutton, NH, USA) for measuring the beam attenuation coefficient of particles in suspension. The use of standard beads ensures that Mie scattering calculations for homogeneous spherical particles can be used to calculate the Mueller matrix elements for these particles to determine reference (expected) values of the volume scattering function and the degree of linear polarization. The comparison of measurements with such reference values allows for evaluation of performance of LISST-VSF instrument and formulation of calibration correction functions for improved determinations of the volume scattering function and the degree of linear polarization from this instrument. This type of approach, which combines measurements on standard well-characterized particles with accurate scattering calculations, has been previously used for the evaluation, calibration, and characterization of light scattering instruments [29,43,44,50]. Although the evaluation results presented in this study are relevant to the specific version of the LISST-VSF instrument used in our laboratory, most methodological aspects are generally applicable to evaluation of other light scattering instruments.

Table 1. Information on the polystyrene bead size standards used to create laboratory sample suspensions for experiments. The nominal bead diameter (D), catalog number, and actual mean diameter $\overline{D}$ ($\pm$ standard error of estimate) and standard deviation of the mean (SD) provided by the manufacturer (Thermo Fisher Scientific, Inc.) is listed. The particulate beam attenuation coefficient at light wavelength 532 nm of the master sample as determined with a spectrophotometer, c_p^{SPEC}, is listed in addition to specific dilution names and factors (e.g., DF1, DF2, etc.) of the master suspension used for LISST-VSF measurements at different PMT gain settings. The dilution factors in italic font denote the experimental data used for generation of the final correction functions CF_f and BF_f, and those in boldface font denote the six examples used for statistical evaluation in Table 2.

Nominal D [nm]	Catalog No.	$\overline{D}$ [nm]	SD [nm]	c_p^{SPEC} [m^{-1}]	Dilution Factor (PMT 500)	Dilution Factor (PMT 550)
100	3100A	100 ± 3	7.8	58.63	DF1: 96, DF2: 48.5, *DF3: 32.7*	DF1: 96, ***DF2: 48.5***, *DF3: 32.67*
200	3200A	203 ± 5	5.3	46.26	DF1: 96, *DF2: 48.5*, *DF3: 32.7*	DF1: 96, *DF2: 48.5*, ***DF3: 32.7***
400	3400A	400 ± 9	7.3	51.44	DF1: 87.4, ***DF2: 44.2***, *DF3: 29.8*	DF1: 87.4
500	3500A	508 ± 8	8.5	20.64	**DF2: 20**	
700	3700A	707 ± 9	8.3	50.93	DF1: 96, **DF2: 48.5**	
2000	4202A	2020 ± 15	21	18.21	**DF2: 20**	

2.1.1. Instrumentation

A LISST-VSF instrument (S/N 1475) was equipped with a custom designed 2 L sample chamber for benchtop laboratory use. This chamber effectively rejects ambient light and promotes good mixing conditions to maintain particles in suspension. For sample illumination the LISST-VSF uses a frequency-doubled YAG laser to produce a beam of light at a wavelength of 532 nm with a Gaussian beam profile of 3 mm in diameter. A single measurement takes approximately 4 s and consists of two scans of a 15-cm path within the sample, each with a different linear polarization state of the incident beam, i.e., parallel and perpendicular to the scattering plane. Scattered intensity is measured at multiple scattering angles ψ from 0.09° to 15.17° with 32 logarithmically-spaced ring detectors and from 14° to 155° with 1° interval using a fixed axis Roving Eyeball sensor equipped with photomultiplier tubes (PMTs). For the Roving Eyeball, scattered light is split between two PMTs with a polarizing prism allowing for only parallel or perpendicularly polarized light to be detected by each PMT. To enable measurements of large dynamic range of scattered intensity with a single PMT, the laser power is dimmed by a factor of 8 for the angular range 14–63° and returned to full power for 64–155°. The beam attenuation coefficient, c, is also measured at light wavelength of 532 nm for the 15-cm path length of the sample.

For incoherent elastic scattering of light at a given wavelength λ by a collection of particles suspended in water, the Stokes vector of incident light beam, $S_i = [I_i\ Q_i\ U_i\ V_i]^T$, where T represents the transpose operation, is transformed into the Stokes vector of scattered beam, $S_s(\psi)$, by a scattering matrix, $P(\psi)$. For an ensemble of randomly-oriented particles exhibiting certain symmetry properties and no optical activity, the scattering matrix simplifies to 6 independent non-zero elements [20,36,51]

$$S_s(\psi) = \begin{bmatrix} I_s(\psi) \\ Q_s(\psi) \\ U_s(\psi) \\ V_s(\psi) \end{bmatrix} = P(\psi)S_i = C \begin{bmatrix} p_{11}(\psi) & p_{12}(\psi) & 0 & 0 \\ p_{12}(\psi) & p_{22}(\psi) & 0 & 0 \\ 0 & 0 & p_{33}(\psi) & p_{34}(\psi) \\ 0 & 0 & -p_{34}(\psi) & p_{44}(\psi) \end{bmatrix} \begin{bmatrix} I_i \\ Q_i \\ U_i \\ V_i \end{bmatrix}, \quad (1)$$

where λ has been omitted for brevity, C is a constant factor (for a given sample, light wavelength, and measurement geometry), $p_{11}(\psi)$ represents the scattering phase function, and the reference plane is the scattering plane containing the incident and scattered directions [20,51,52]. This form provides a reasonable description of the measured scattering matrix by suspensions of randomly-oriented marine particles, including various specific types of particles present in seawater [24,29,33,36,37,42]. In the case of unpolarized incident light (i.e., Q_i, U_i, and V_i are all zero), the volume scattering function $\beta(\psi)$ equals (to within a constant factor) $p_{11}(\psi)$, and the degree of linear polarization $DoLP(\psi)$ can be determined from [29,36,37,53]

$$DoLP(\psi) = \frac{-p_{12}(\psi)}{p_{11}(\psi)} = \frac{-Q_s(\psi)}{I_s(\psi)}. \quad (2)$$

Positive values of $DoLP(\psi)$ are for dominantly perpendicular polarization and negative values for dominantly parallel polarization. We note that this definition of $DoLP(\psi)$ has been widely used for characterizing the inherent scattering properties of various types of particles beyond aquatic particles, such as aerosol particles and cosmic dust [30,54–58].

The LISST-VSF measurements of forward scattering within the angular range 0.09–15.17° are made with two linear polarization states of the incident beam, but with no polarization analyzers of the ring detectors. For the ring detectors, the calibrated $\beta(\psi)$ in absolute units is a standard output of the manufacturer's processing software. The absolute calibration is based on the manufacturer-provided conversion from ring detector counts to physical units using radiant sensitivity of ring detectors [59,60]. Detection of scattered light within the angular range 14–155° using the Roving Eyeball sensor employs measurements made with two linear polarization states of the incident beam and the corresponding two linear polarization states of the scattered light. The four measurement configurations allow for the determination of relative values of $p_{11}(\psi)$, $p_{12}(\psi)$, and $p_{22}(\psi)$. The calibrated $\beta(\psi)$ values within the Roving Eyeball angular range are obtained by scaling the $p_{11}(\psi)$ data from the Roving Eyeball sensor. Specifically, the scattering measurements from the first angles of the Roving Eyeball sensor are forced to match the calibrated $\beta(\psi)$ values from the overlapping last ring detectors. The $DoLP(\psi)$ values are obtained from Equation (2) using $p_{11}(\psi)$ and $p_{12}(\psi)$, and are also included in the standard output of the manufacturer's processing code.

We also used a DAWN-EOS multi-angle light scattering instrument which provided independent measurements of $\beta(\psi)$ and $DoLP(\psi)$ of polystyrene beads suspended in water. These measurements were made with a sample placed in a 20 mL cylindrical glass vial. The DAWN-EOS instrument used in this study has been previously characterized and calibrated for such measurement configuration [61]. This instrument uses a diode-pumped frequency-doubled Nd-YAG laser at light wavelength 532 nm with a Gaussian beam profile of 62 μm in diameter. The interrogated sample volume is on the order of 10 nL. The incident beam can be linearly polarized both parallel and perpendicular to the scattering plane. The intensity of scattered light is measured simultaneously with eighteen photodiode detectors and no polarization analyzers, enabling measurements within a range of scattering angles from 22.5°

to 147°. To encompass the large dynamic range of scattered intensity, three selectable gain settings are available for each detector (gain factors of 1, 21, or 101).

As the DAWN-EOS detectors have no polarization analyzers, they only measure the first parameter of Stokes vector of the scattered light, $I_s(\psi)$. Here we define $I_{s\parallel}(\psi)$ for the parallel polarization of the incident beam and $I_{s\perp}(\psi)$ for the perpendicular polarization of the incident beam. The matrix elements $p_{11}(\psi)$ and $p_{12}(\psi)$ can be obtained (to within a constant factor) from DAWN-EOS measurements as

$$p_{11}(\psi) = \frac{I_{s\parallel}(\psi) + I_{s\perp}(\psi)}{2} \tag{3}$$

$$p_{12}(\psi) = \frac{I_{s\parallel}(\psi) - I_{s\perp}(\psi)}{2}, \tag{4}$$

which allows for determination of $DoLP(\psi)$ from Equation (2). The calibration procedure described in Babin et al. [61] allows for determination of $\beta(\psi)$ in absolute units. Importantly, the calibration procedure of DAWN-EOS is fundamentally different from the calibration procedure of LISST-VSF. The manufacturer's calibration of LISST-VSF is based on a nominal radiant sensitivity of ring detectors (amperes of photoelectric current per watt of optical power) traceable to the National Institute of Standards and Technology [59,60]. In contrast, the calibration of DAWN-EOS is based on measurements of light scattered at 90° by pure toluene with the incident beam having a linear perpendicular polarization [61]. This calibration relies on the known magnitude of molecular scattering by toluene. The two different methods employed in calibration of LISST-VSF and DAWN-EOS allow for comparisons of independent estimates of $\beta(\psi)$ obtained by these instruments. We also recall that the $DoLP(\psi)$ estimates obtained with the two instruments within the common range of scattering angles are based on different polarization measurement configurations used by these instruments.

A Lambda 18 spectrophotometer was used to collect independent measurements of the spectral beam attenuation coefficient, $c(\lambda)$, of polystyrene beads suspended in water. These measurements were made for comparisons with the beam attenuation data obtained with LISST-VSF, and also to aid in the preparation of samples with appropriate concentrations of polystyrene beads to ensure that measurements with LISST-VSF and DAWN-EOS were made within the single scattering regime. The spectrophotometric measurements were made in the spectral range from 290 nm to 860 nm with 1 nm interval, but only data at 532 nm are used in this study. The general applicability of laboratory spectrophotometers with proper modifications to enable measurements of beam attenuation of particle suspensions, including colloidal samples, has long been recognized [62,63]. In our study, a sample of particle suspension was measured in a 1-cm quartz cuvette placed at a significant distance from the detector (~25 cm from the entrance of the integrating sphere), and field stops were aligned within the light path to reduce the size of the beam and acceptance angle of the detector to less than 1°. This measurement geometry has been used in our previous studies of spectral beam attenuation by various particle assemblages [64,65].

2.1.2. Experimental Procedure

Baseline measurements of 0.2 μm filtered water were collected with all three instruments used in the experiments; LISST-VSF, DAWN-EOS, and Lambda 18 spectrophotometer. These baseline measurements were subtracted from subsequent measurements taken on particle suspensions to determine the optical properties associated with suspended particles only, i.e., the particulate volume scattering function, $\beta_p(\psi)$, the particulate degree of linear polarization, $DoLP_p(\psi)$, and the particulate beam attenuation coefficient, c_p.

Original manufacturer's stock samples of standard polystyrene beads (100, 200, 400, 500, 700, and 2000 nm in diameter) were used to generate master samples using 0.2 μm filtered, deionized, and degassed water as a medium (with the exception of 2000 nm beads which used 0.2 μm filtered seawater). In the process of preparation of master samples, the particle concentration was optimized to ensure that spectrophotometric measurements of beam attenuation coefficient can be performed either directly or with small dilution factor (~3) on these samples over 1-cm path length with sufficiently high signal but negligible multiple scattering effects. The c_p values for master samples ranged from about 18 m^{-1} to 58 m^{-1} (Table 1).

The master sample was diluted for measurements with the LISST-VSF to avoid oversaturation of PMT detectors and multiple scattering over the longer path length (15 cm). For baseline measurements, the LISST-VSF sample chamber was filled with 1900 mL of 0.2 μm filtered water. The final samples of particle suspensions were created by addition of 20 to 100 mL of master sample to the LISST-VSF chamber. For most beads examined in our experiments, more than one particle suspension differing in terms of particle concentration was measured with LISST-VSF (Table 1). The different particle concentrations were achieved by different dilution of master sample within LISST-VSF chamber. Owing to different dilution factors ranging from 20 to 96 (labeled as DF1, DF2, and DF3 in Table 1), the c_p values of LISST-VSF samples ranged from about 0.5 m^{-1} to 1.8 m^{-1}. For a single bead size, concentration, and PMT gain, a series of LISST-VSF measurements was composed of 200 measurements taken in rapid succession (recall that a measurement refers to two scans, each with a different polarization of incident beam). This measurement series was divided into five sets of 20 measurements and one set of 100 measurements to enable manual gentle mixing of sample before each set of measurements. In addition, for the 2000 nm bead suspensions a magnetic stir bar which operated on low speed and changed direction of rotation every 30 s was used to prevent particle settling during the measurement.

Several LISST-VSF baseline measurements of 0.2 μm filtered water were collected for each experiment, i.e., for each examined bead size. However, for reasons of consistency and out of the desire to use an optimal baseline representative of the least contaminated 0.2 μm filtered water, a single baseline was used for processing of all experimental data collected for various bead sizes and concentrations except for 2000 nm sized beads which used 0.2 μm filtered seawater. This baseline was determined on the basis of finding a measurement which exhibited minimal scattering signal detected by Roving Eyeball and ring detectors and maximum directly transmitted light detected by the laser transmission sensor. We note, however, that for each PMT gain setting of the Roving Eyeball sensor a separate baseline was determined.

Measurements using the DAWN-EOS instrument were collected for 100, 200, 400, and 700 nm beads. Dilution factors of master samples for DAWN-EOS measurements were between 300 and 3000, depending on bead size. The gain settings for each detector were adjusted to the highest setting that would avoid saturation of signal with incident perpendicular polarization of light. For 400 and 700 nm bead suspensions, two different dilutions were measured. For each polarization state (i.e., perpendicular and parallel) of incident light, we acquired 1440 measurements with a sampling frequency of 8 Hz over 3 min. For a given sample, this data acquisition protocol was repeated three times. Each of these three replications was made with a different randomly-chosen orientation of sample cylindrical vial within the instrument. The sample was gently mixed between these replicate measurements. The baseline measurements of 0.2 μm filtered water were acquired using the same protocol.

As mentioned above, the optical measurements were made on sufficiently-diluted samples to ensure negligible effects of multiple scattering over a pathlength used by a given instrument. A criterion for a single scattering regime is generally defined in terms of small optical thickness of the sample, $\tau \ll 1$, where τ is a product of the beam attenuation coefficient, c, and pathlength, r [51,66]. Also, a simple practical test for ensuring that multiple scattering effects are negligible is to verify a direct proportionality between the measured optical signal and the concentration of particles in suspension by conducting a series of measurements on the same sample with different dilutions [51]. Our measurements on bead samples with different dilutions showed an excellent 1:1 relationship between the LISST-VSF measurement and the bead concentration over the range of beam attenuation coefficient up to at least 2 m^{-1}. The single scattering regime can also be determined by the condition $\tau(1\text{-}g) \ll 1$, where g is the average cosine of the scattering angle of the volume scattering function [23,66]. For the 100 nm polystyrene beads, the g value is 0.115, which yields the most restrictive condition in our study, $\tau << 1.13$. For all bead samples measured with LISST-VSF, including all bead sizes and sample dilutions, τ was always less than about 0.3. This condition was also satisfied for samples measured with a spectrophotometer. For the measurements with DAWN-EOS, the τ values were even smaller. For the natural seawater samples examined in our study (which is described below in Section 2.2), the g values (for the total volume scattering function including the contribution by pure seawater) were about 0.9 or somewhat higher, which yields less restrictive criterion $\tau << 10$. Our measurements of natural samples clearly satisfied this single scattering condition, as the highest value of c for the natural samples measured with LISST-VSF was about 2.6 m^{-1}, so τ was always less than about 0.45, given that the maximum pathlength for LISST-VSF is 17.5 cm for the scattering angle of 150°.

2.1.3. Data Processing

Processing of LISST-VSF data was done with a standard processing code provided by manufacturer (version of 2013) to determine $\beta_p(\psi)$, $DoLP_p(\psi)$, and c_p, denoted hereafter as $\beta_p^{LISST*}(\psi)$, $DoLP_p^{LISST*}(\psi)$, and c_p^{LISST*} respectively (the asterisk indicates that the variable is derived from the standard processing code without additional corrections developed in this study). Some details specific to the processing and quality control of our experimental data are provided below.

As a first step in data processing, the baseline values in raw counts were subtracted from each LISST-VSF measurement of raw counts acquired on samples of bead suspensions. To account for light attenuation along the path between the scattering volume and the detector, an attenuation correction factor was calculated using the average c_p^{LISST*} from the series of measurements and the length of the path for each scattering angle. Further, to account for the difference in sensitivity of the two Roving Eyeball PMT detectors, a factor α is used to adjust the measured counts of one PMT detector relative to the other [67]. The value of $\alpha = 0.9335$ was determined by averaging all median values of α derived from each series of measurements for each bead size, particle concentration, and PMT gain. The α parameter was observed to be nearly constant over the period of experiments (~18 months, the coefficient of variation < 5%). For each series of measurements a specific scaling factor was determined to convert $p_{11}(\psi)$ in PMT counts to $\beta_p^{LISST*}(\psi)$ in absolute units [m^{-1} sr^{-1}] for scattering angles 14–155° measured by the Roving Eyeball sensor. First, for each measurement from a given series of measurements, a scaling factor was determined by matching the PMT counts measured with Roving Eyeball sensor between 15° and 16° with $\beta_p^{LISST*}(\psi)$ in absolute units obtained from measurements with the last two ring detectors at 13.01° and 15.17°. Then, using these determinations, the average scaling factor for a given series of measurements was calculated and used for further data processing. Note that this scaling was not needed for the determination of $DoLP_p^{LISST*}(\psi)$ for the Roving Eyeball angular range, which is calculated from $p_{11}(\psi)$ and $p_{12}(\psi)$ determined in PMT counts following Equation (2).

Quality control of data was performed by removing the first set of 20 measurements (the remaining four sets with 20 measurements each were retained) and the first 20 measurements from the set of 100 measurements. We observed that this was necessary to ensure reasonable stability in the measured scattering signal. The mean and standard deviation values for each angle based on all of the 160 remaining measurements in the series were determined, and the outlying single measurements within the series were identified and rejected from subsequent analysis. Typically, 120 to 130 measurements from a given series of 200 measurements passed the quality criteria.

Example data of uncorrected $\beta_p^{LISST*}(\psi)$ for 200 nm and 2000 nm bead suspensions are shown in Figure 1. The series of measurements that remained after quality control and the median values of $\beta_p^{LISST*}(\psi)$ derived from the series of measurements are shown. We also note that the median values were very close to mean values for our data (<1% difference for most scattering angles). The results for 2000 nm beads show a distinct pattern with several scattering maxima and minima due to constructive and destructive interference of the scattered light from a nearly monodisperse population of beads that are large relative to the wavelength of light. The 200 nm beads are smaller than the wavelength of light leading to a more featureless shape of $\beta_p^{LISST*}(\psi)$. The variability between the individual measurements is largest at very small scattering angles, i.e., approximately <4°, where the scattering signal for submicron particles is low relative to our baseline measurements. Apart from small scattering angles, the coefficient of variation (CV) for each scattering angle calculated from a series of measurements on 200 nm beads is generally very small, ranging from ~3% to <1%, with the smallest values at angles greater than 64° where full laser power is used. The measurements of 2000 nm beads exhibit somewhat higher CV, i.e., between about 3% and 6%. The higher values of CV are observed mostly near the angles where minima of $\beta_p^{LISST*}(\psi)$ occur.

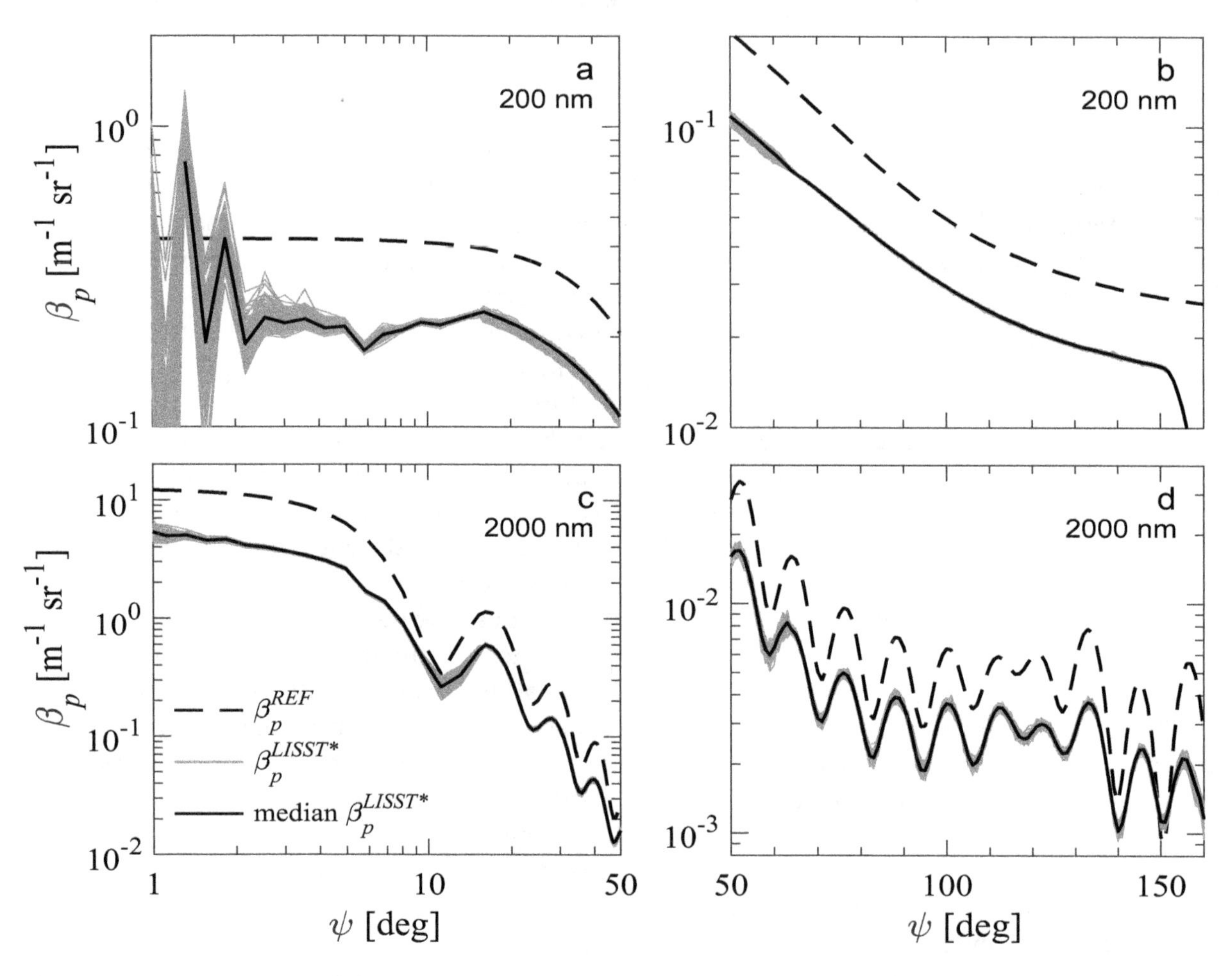

Figure 1. Measurements of the particulate volume scattering function, $\beta_p(\psi)$, at light wavelength of 532 nm for 200 nm (**a**,**b**) and 2000 nm (**c**,**d**) diameter polystyrene beads suspended in water. The left panels depict the angular range of 1–50° with logarithmic scaling, and the right panels depict the range 50–160° with linear scaling. The expected reference value, $\beta_p^{REF}(\psi)$, obtained from Mie scattering calculations is indicated as a dashed line. Quality-controlled but uncorrected measurements obtained with the LISST-VSF (gray lines, number of measurements N = 128) and the median value (solid black line) are shown.

The DAWN-EOS measurements for four bead sizes were also used to calculate $\beta_p(\psi)$ and $DoLP_p(\psi)$, denoted as $\beta_p^{\text{DAWN}}(\psi)$ and $DoLP_p^{\text{DAWN}}(\psi)$. First, for each time series of 1440 measurements with DAWN-EOS, the highest 2% of data was rejected, as these data are assumed to result from sample contamination with rare, larger particles. Each set of measurements then consists of 1411 measurements of both $I_{s\parallel}(\psi)$ and $I_{s\perp}(\psi)$ for a specific orientation of sample vial. These measurements were averaged to represent that orientation. Such results were then averaged for three vial orientations. This protocol was applied to both the sample and baseline measurements, with the exception that baseline values were calculated by averaging the lowest 5% of data. The final $I_{s||}(\psi)$ and $I_{s\perp}(\psi)$ for the beads were calculated by subtracting the average baseline from the average sample data. These particulate $I_{s||}(\psi)$ and $I_{s\perp}(\psi)$ were then used to determine $p_{11}(\psi)$ and $p_{12}(\psi)$ according to Equations (3) and (4), from which $\beta_p^{\mathrm{DAWN}}(\psi)$ [61] and $DoLP_p^{\mathrm{DAWN}}(\psi)$ (Equation (2)) were determined. Note that two dilutions of the master suspension for 400 and 700 nm beads were measured with DAWN-EOS and the average of the two was used to represent these bead sizes. As a final step, the determined $\beta_p^{\mathrm{DAWN}}(\psi)$ values were rescaled using relevant dilution factors to obtain final results representing the particle concentration in LISST-VSF samples and enable direct comparisons with LISST-VSF measurements. Note that such rescaling is not necessary for $DoLP_p^{\mathrm{DAWN}}(\psi)$.

With regard to processing of data acquired with a Lambda 18 spectrophotometer, the spectral data of measured optical density $\mathrm{OD}(\lambda)$ (i.e., measurements made in the absorbance mode of the spectrophotometer) were converted (after subtraction of baseline measurement) into the particulate beam attenuation coefficient [m^{-1}] using the relationship $c_p(\lambda) = \ln(10)\ \mathrm{OD}(\lambda)/0.01$, where ln is the natural logarithm and 0.01 is the path length in meters. The final particulate beam attenuation coefficient obtained from spectrophotometric measurements is denoted as c_p^{SPEC}. The estimates of c_p^{LISST*} from LISST-VSF measurements were calculated with the standard manufacturer's processing code. Because each LISST-VSF measurement consists of two linear polarization states of the incident beam, the average of these two is used as the final estimate of c_p^{LISST*}. As a final step, the determined c_p^{SPEC} values were multiplied by relevant dilution factors to obtain final results representing particle concentration in LISST-VSF samples and enable direct comparisons with LISST-VSF measurements.

2.1.4. Determination of Correction Functions

In addition to $\beta_p^{LISST*}(\psi)$, Figure 1 shows results for the 200 nm and 2000 nm polystyrene beads based on Mie scattering calculations (more details about these calculations are provided below). These results are significantly higher (nearly a factor of 2) than the measured values of $\beta_p^{LISST*}(\psi)$. We assume that the Mie scattering calculations for samples of spherical polystyrene beads are sufficiently accurate to provide reference values for such samples.

In order to correct for the mismatch between the measured and reference values, a calibration correction function $CF(\psi)$ is defined as

$$CF(\psi) = \frac{\beta_p^{REF}(\psi)}{\beta_p^{LISST*}(\psi)}, \tag{5}$$

where $\beta_p^{REF}(\psi)$ is a reference volume scattering function determined according to

$$\beta_p^{REF}(\psi) = \widetilde{\beta}_p^{Mie}(\psi) b_p^{REF}, \tag{6}$$

where $\widetilde{\beta}_p^{Mie}(\psi)$ is the scattering phase function [sr^{-1}] obtained from Mie scattering computations and b_p^{REF} is the reference particulate scattering coefficient [m^{-1}]. Note that all quantities in Equations (5) and (6) are for the LISST-VSF light wavelength of 532 nm.

For each examined suspension of standard polystyrene beads, $\widetilde{\beta}_p^{Mie}(\psi)$ was determined from Mie scattering computations for homogeneous spherical particles. We used the Mie scattering code for homogeneous spheres of Bohren and Huffman [20], which included our modifications to account for polydispersity of the sample, i.e., to use particle size distribution as input to the code rather than just a single particle diameter as in the original code. The computations were performed assuming a relative particle size distribution (PSD) of Gaussian shape, with 300 evenly spaced size bins about the nominal mean diameter $\pm$ 3 standard deviations, as provided by the manufacturer for each bead size (Table 1). The use of such PSDs allows us to account for the realistic, small degree of polydispersity of each sample. The Mie computations also require input of the refractive index of particles. Based on the study of Ma et al. [68] we assumed that the complex refractive index of polystyrene relative to water at

532 nm is $m = 1.193 + 0.0003i$, where the first component is the real part and the second component is the imaginary part of refractive index. Note that the imaginary part is very small because polystyrene is a weakly absorbing material in the examined spectral region.

Equation (6) also requires b_p^{REF}, which was determined from the combination of beam attenuation measurements and Mie scattering calculations as

$$b_p^{REF} = c_p^{LISST*} \frac{Q_b^{Mie}}{Q_c^{Mie}}, \quad (7)$$

where Q_b^{Mie} and Q_c^{Mie} are the single-particle scattering and attenuation efficiency factors, respectively, obtained from Mie computations. Because the populations of examined beads exhibit a slight degree of polydispersity, the calculated Q_b^{Mie} and Q_c^{Mie} represent the average values of efficiency factors for a given particle population [69]. Given very weak light absorption of polystyrene beads at 532 nm, the ratio $\frac{Q_b^{Mie}}{Q_c^{Mie}}$ was found to be >95%. We also note that in addition to c_p^{LISST*}, we have another potential measurement of beam attenuation coefficient from the spectrophotometer (c_p^{SPEC}). Figure 2 shows that the measurements of c_p^{LISST*} and c_p^{SPEC} are consistent, and generally agree very well.

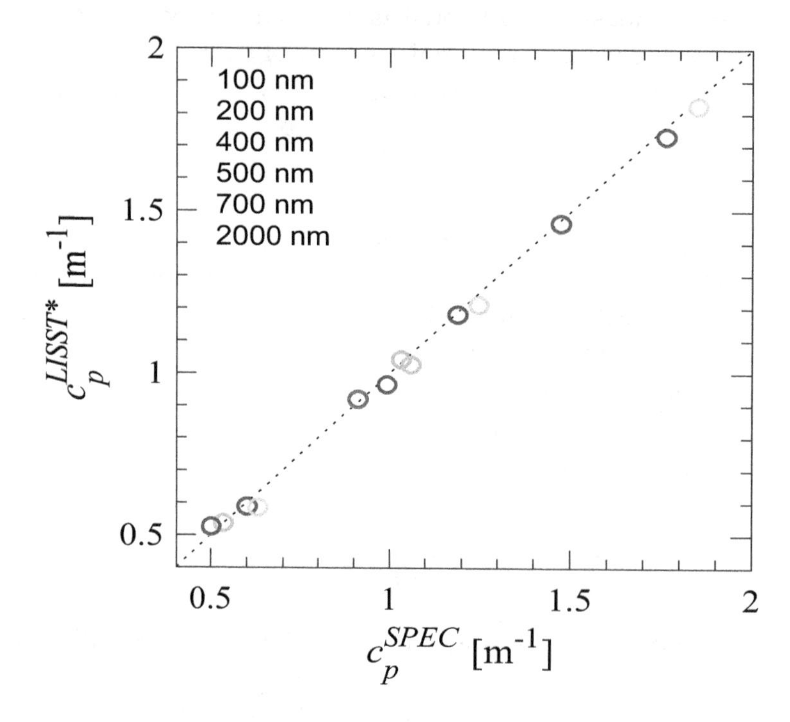

Figure 2. Comparison of measurements of the particulate beam attenuation coefficient, c_p, at 532 nm obtained with a spectrophotometer with measurements from the LISST-VSF. The comparison is depicted for suspensions of polystyrene beads of six different diameters as indicated in the legend, and the 1:1 line is plotted for reference (dotted black line). Appropriate dilution factors have been applied to account for the different particle concentrations used in measurements with each instrument. The presented values correspond to samples measured with the LISST-VSF.

By combining Equations (6) and (7), $\beta_p^{REF}(\psi)$ can be determined for each LISST-VSF measurement as

$$\beta_p^{REF}(\psi) = \widetilde{\beta}_p^{Mie}(\psi)\, c_p^{LISST*} \frac{Q_b^{Mie}}{Q_c^{Mie}}. \quad (8)$$

Note that the estimates of $\beta_p^{REF}(\psi)$ can vary between individual measurements because of variations in c_p^{LISST*}. Finally, by combining Equations (5) and (8), $CF(\psi)$ can be determined for each LISST-VSF measurement as

$$CF(\psi) = \frac{\widetilde{\beta}_p^{Mie}(\psi)\ c_p^{LISST*}\ \frac{Q_b^{Mie}}{Q_c^{Mie}}}{\beta_p^{LISST*}(\psi)}. \quad (9)$$

The application of this protocol to every individual measurement of $\beta_p^{LISST*}(\psi)$ helps to better capture the variability between individual measurements during a given series of LISST-VSF measurements on a given sample, for example due to imperfect mixing in the 2 L sample chamber or potential electronic fluctuations in the instrument.

The smaller-sized particle standards (<500 nm in diameter) appear as the best candidates for determination of $CF(\psi)$ because they produce a relatively featureless pattern of angular scattering (see the results for 200 nm beads in Figure 1). The measurements with larger beads (500 nm to 2000 nm) were not used in these determinations because the angular scattering pattern includes multiple maxima and minima (see the results for 2000 nm beads in Figure 1), which render the comparison of $\beta_p^{LISST*}(\psi)$ and $\beta_p^{REF}(\psi)$ particularly sensitive to even small uncertainties in measurements or theoretical calculations. The results obtained with beads of 100, 200, and 400 nm in diameter were considered in the determinations of final correction function $CF_f(\psi)$ within four angular ranges, as described below:

$$CF_f(\psi) = (CF_{100}(\psi) + CF_{200}(\psi) + CF_{400}(\psi))/3 \text{ for } \psi = 0.09\text{–}60^\circ \quad (10a)$$

$$CF_f(\psi) = CF_{200}(\psi) \text{ for } \psi = 61\text{–}128^\circ \quad (10b)$$

$$CF_f(\psi) = (CF_{200}(\psi) + CF_{400}(\psi))/2 \text{ for } \psi = 129\text{–}150^\circ \quad (10c)$$

$$CF_f(\psi) = (CF_{100}(\psi) + CF_{200}(\psi) + CF_{400}(\psi))/3 \text{ for } \psi = 151\text{–}155^\circ. \quad (10d)$$

The $CF_{200}(\psi)$ data obtained with 200 nm beads provide the main contribution to the determination of $CF_f(\psi)$. The $CF_{100}(\psi)$ data obtained with 100 nm beads are used partially because of increased uncertainty in the PSD of these beads (CV of nominal mean diameter is 7.8%, see Table 1). The $CF_{400}(\psi)$ data obtained with 400 nm beads are also used partially and cover the backscattering angles, where these particular data are useful for correction of an apparent artifact near 130–140°, which is rather minor but has been consistently observed with our LISST-VSF instrument for various natural particle assemblages. The calculations of $CF_{100}(\psi)$, $CF_{200}(\psi)$, and $CF_{400}(\psi)$ were made using data for particle concentrations and PMT gains which ensured sufficient signal for the ring detectors while avoiding PMT saturation of the Roving Eyeball sensor (see dilution factors in italic font in Table 1). For a given bead standard, the final values of correction function at different angles were determined as the median values of all the relevant determinations.

The final $CF_f(\psi)$ was smoothed in the angular range 2.56–155° with a 3-point and then a 5-point moving average. In addition, $CF_f(\psi)$ within the near-forward angular range 0.09–4.96° was set to a constant value of $CF_f(\psi_{32})$, where ψ_{32} = 15.17° corresponds to the last ring detector. The rationale for this assumption is that the scattering signal produced by the examined beads for the first 25 rings (ψ = 0.09–4.96°) is comparable to the baseline, while there is good signal relative to the baseline for the last ring detector.

The final correction simply involves the multiplication of uncorrected $\beta_p^{LISST*}(\psi)$ by the correction function $CF_f(\psi)$,

$$\beta_p^{LISST}(\psi) = \beta_p^{LISST*}(\psi)\ CF_f(\psi), \quad (11)$$

where $\beta_p^{LISST}(\psi)$ is the corrected LISST-VSF measurement of volume scattering function (note that the superscript * is removed from this symbol).

We also determined a correction function for $DoLP_p^{LISST*}(\psi)$,

$$BF(\psi) = DoLP_p^{LISST*}(\psi) - DoLP_p^{REF}(\psi), \quad (12)$$

where $BF(\psi)$ quantifies a correction for potential bias in $DoLP_p^{LISST*}(\psi)$ obtained from the standard processing code applied to LISST-VSF measurements and $DoLP_p^{REF}(\psi)$ is a reference degree of linear polarization determined from Mie scattering calculations of the two scattering matrix elements, $p_{11}^{Mie}(\psi)$ and $p_{12}^{Mie}(\psi)$, for a given sample of standard beads. The results for $BF_{100}(\psi)$, $BF_{200}(\psi)$, $BF_{400}(\psi)$, and the final correction function $BF_f(\psi)$ were obtained using a procedure similar to that for $CF_{100}(\psi)$, $CF_{200}(\psi)$, $CF_{400}(\psi)$, and $CF_f(\psi)$. The correction of $DoLP_p^{LISST*}(\psi)$ simply requires a subtraction of $BF_f(\psi)$,

$$DoLP_p^{LISST}(\psi) = DoLP_p^{LISST*}(\psi) - BF_f(\psi), \tag{13}$$

where $DoLP_p^{LISST}(\psi)$ is the corrected degree of linear polarization within the range of scattering angles from 16° to 150°. Because the $DoLP_p^{LISST*}(\psi)$ data output from standard processing of LISST-VSF measurements begins at $\psi = 16°$, no correction for the forward scattering angles of the ring detectors ($\psi < 16°$) was determined.

2.2. Measurements and Analysis of Natural Seawater Samples

Optical measurements with the LISST-VSF and ancillary analyses of natural particle assemblages were performed on seawater samples collected between summer 2016 and spring 2017 in contrasting marine environments, namely, in open ocean waters off the coast of Southern California, nearshore ocean waters at the pier of the Scripps Institution of Oceanography (SIO Pier) in La Jolla, and the tidal estuary of the San Diego River. Overall 17 samples representing a broad range of natural particle assemblages were analyzed. Most samples (number of samples $N = 11$) were collected at the SIO Pier. These samples were collected during typical dry weather conditions, phytoplankton bloom events, and after heavy rain. The tidal estuary samples ($N = 3$) include three tidal states between low and high tide. The offshore samples ($N = 3$) were collected in the Santa Barbara Channel, about 8 km off San Diego Bay, and about 2 km off SIO Pier. Seawater samples were collected just beneath the sea surface using either Niskin bottles or a bucket, except for one offshore sample (off San Diego Bay) that was collected at the subsurface chlorophyll-*a* maximum at a depth of 18 m. All samples were analyzed in the laboratory within 24 h of sampling.

To characterize the concentration and composition of particulate matter for each sample, we determined the dry mass concentration of total suspended particulate matter, SPM [g m^{-3}], mass concentration of particulate organic carbon, POC [mg m^{-3}], and mass concentration of the pigment chlorophyll-*a*, Chla [mg m^{-3}]. For these determinations, the particles were collected on glass-fiber filters (GF/F Whatman) by filtration of appropriate volumes of seawater (150–2100 mL depending on the sample). SPM was determined following a gravimetric method using pre-washed and pre-weighted filters [7,70]. The determinations of POC were made on precombusted filters with a standard CHN analysis involving high temperature combustion of sample filters [7,71,72]. Chla was determined spectrophotometrically using a Lambda 18 spectrophotometer and placing 1-cm cuvettes containing acetone extracts of the samples inside the integrating sphere. The measured absorbance values at 630, 647, 665, and 691 nm (after subtraction of acetone baseline values) were used in the calculation of Chla [73]. For each seawater sample, replicate determinations of SPM and POC were made on separate sample filters. The final SPM and POC are average values of replicate determinations. The replicates for SPM and POC agreed generally to within 15% and 10%, respectively. No replicates were taken for Chla. In addition to information about particle concentration, SPM, POC, and Chla provide useful proxies of bulk composition of particulate matter. The organic and inorganic fractions of SPM can be characterized using the ratio POC/SPM, and the contribution of phytoplankton to SPM using Chla/SPM [65]. These ratios are expressed on a [g/g] basis.

The measurements of particle size distribution (PSD) were made with a Coulter Multisizer 3 (Beckman Coulter, Brea, CA, USA) equipped with a 100 μm aperture, which allows particle counting and sizing in the range of volume-equivalent spherical diameter from 2 μm to 60 μm. Within this size range we used 300 log-spaced size bins to provide high resolution PSDs. For each experiment, 0.2 μm filtered seawater was used as a blank that was subtracted from sample measurements.

Approximately 10 to 15 replicate measurements of 2 mL subsamples of each seawater sample were collected. After removing outliers, the remaining measurements were summed and divided by the total analyzed volume to produce an average density function of PSD in particle number per unit volume per width of size bin. For each sample the power function fit with a slope parameter, ζ, was determined using these PSD data over the size range 2–50 μm. In these determinations, the linear regression analysis was applied to log-transformed data, and the last size bins with very low particle counts were ignored. Although the measured PSDs often showed significant deviations from the power function fits, we use the slope parameter ζ as a particle size metric, because this is the most common parameterization of size distribution of marine particles [23,74]. Additionally, assuming spherical particles, the particle volume distributions were determined from particle number distributions for each sample. From particle volume distributions, we calculated the percentile-based particle diameters such as the median diameter, D_V^{50}, and the 90th percentile diameter, D_V^{90}. These parameters have been shown to provide potentially useful metrics in the analysis of relationships between the optical and particle size properties in seawater [65].

Measurements and processing of data collected with LISST-VSF for natural seawater samples were made following a protocol similar to that described above for standard polystyrene bead samples. For each experiment, baseline measurements were taken on 0.2 μm filtered seawater obtained from a given seawater sample. However, a single baseline selected from the lowest measured baselines was used for data processing of all seawater samples to ensure a consistent baseline unaffected by possible variations associated with the imperfect purity of 0.2 μm filtered seawater prepared during different experiments. To ensure scattering measurements were acquired in a single-scattering regime, samples with an average c_p over 3.0 m^{-1} were diluted using 0.2 μm filtered seawater. Dilution was necessary only for the two most turbid samples collected in the San Diego River Estuary. Between four and eight sets of 50 measurements were collected for each seawater sample with gentle hand mixing between the measurement sets, while a magnetic stir bar was on very low speed changing direction of rotation every 30 s. All results from LISST-VSF measurements for natural seawater samples shown in this paper represent the $CF_f(\psi)$-corrected volume scattering function of particles, $\beta_p^{LISST}(\psi)$, and $BF_f(\psi)$-corrected degree of linear polarization of particles, $DoLP_p^{LISST}(\psi)$. For a given sample the final values of $\beta_p^{LISST}(\psi)$ and $DoLP_p^{LISST}(\psi)$ correspond to the median values of the series of measurements that passed the quality control criteria.

To determine the particulate scattering, b_p^{LISST}, and particulate backscattering, b_{bp}^{LISST}, coefficients, the corrected measured $\beta_p^{LISST}(\psi)$ was first extrapolated in the angular range 150–180°. The extrapolated portion of $\beta_p^{LISST}(\psi)$ was obtained by fitting a specific function to the data of $\beta_p^{LISST}(\psi)$ in the angular range 90–150°. We used two methods for fitting and extrapolating $\beta_p^{LISST}(\psi)$. The first method is based on a non-linear least squares best fit of the analytical function proposed by Beardsley and Zaneveld [75]. The second method is based on a linear mixing model that finds a non-negative least squares best fit for combined contributions of four end members representing shapes of volume scattering functions associated with scattering by small and large particles, as described in Zhang et al. [76].

A backscattering factor, κ, was determined for the fitted volume scattering function as

$$\kappa = \frac{b_{bp}^{fit}}{b_{bp,150}^{fit}}, \tag{14}$$

where b_{bp}^{fit} is the particulate backscattering coefficient determined by the integration of the fitted function in the angular range 90–180° and $b_{bp,150}^{fit}$ is the coefficient determined by the integration of the fitted function in the range 90–150°. The final estimate of backscattering coefficient, b_{bp}^{LISST}, was calculated as

$$b_{bp}^{LISST} = \kappa \, b_{bp,150}^{LISST}, \tag{15}$$

where $b_{bp,150}^{LISST}$ is obtained by the integration of $\beta_p^{LISST}(\psi)$ in the angular range 90–150°. The final estimate of scattering coefficient, b_p^{LISST}, was calculated as the sum of b_{bp}^{LISST} and the forward scattering coefficient obtained from the integration of $\beta_p^{LISST}(\psi)$ in the angular range 0.09–90°.

The calculations of b_p^{LISST} and b_{bp}^{LISST} were made for each seawater sample using the two methods for fitting and extrapolation. The particulate backscattering ratio, $\tilde{b}_{bp}^{LISST} = b_{bp}^{LISST}/b_p^{LISST}$ was also calculated. We note that the κ values for all examined seawater samples were found to range between 1.125 and 1.138 and 1.118–1.120 for the Beardsley and Zaneveld [75] and Zhang et al. [76] methods, respectively. An example illustration of fitting and extrapolation methods for one sample collected during high tide at the San Diego River estuary is depicted in Figure 3. As seen, both the Beardsley and Zaneveld [75] and Zhang et al. [76] fitted functions are in good agreement with the measured data of $\beta_p^{LISST}(\psi)$ in the angular range 90–150°. However, the extrapolated portion of the Beardsley and Zaneveld [75] function in the angular range 150–180° has somewhat higher values compared with the Zhang et al. [76] function. Nevertheless, the estimates of b_{bp}^{LISST} for this sample obtained from the two extrapolation methods differ only by 0.5%. For all other seawater samples the difference was also small, not exceeding 1.5%. The final results of b_p^{LISST} and b_{bp}^{LISST} for seawater samples presented in this study are based on the Zhang et al. [76] method.

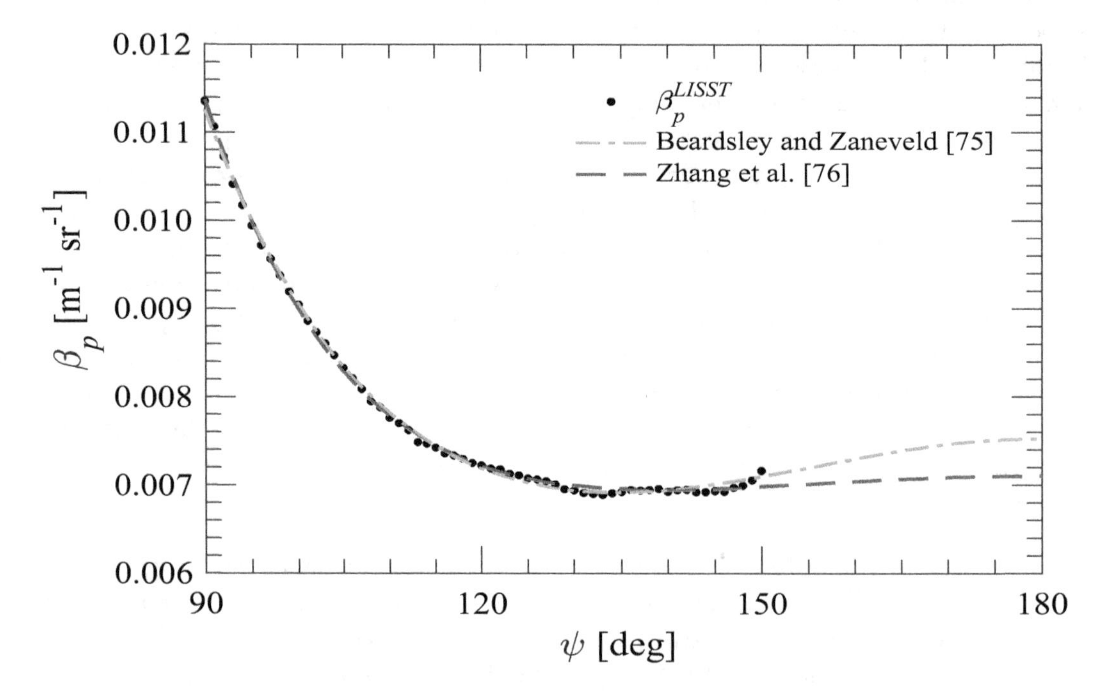

Figure 3. Measured values of the particulate volume scattering function $\beta_p(\psi)$ obtained with the LISST-VSF after correction (circles) for scattering angles 90–150° and illustration of the results of two model relationships (Beardsley and Zaneveld [75], Zhang et al. [76]) fitted to the data. The illustrated example measurement was made on a natural sample collected from the San Diego River estuary.

3. Results and Discussion

3.1. Correction Functions for LISST-VSF

The results for $CF_{100}(\psi)$, $CF_{200}(\psi)$, $CF_{400}(\psi)$, and $CF_f(\psi)$ are plotted in Figure 4. The final correction function $CF_f(\psi)$ indicates that $\beta_p^{LISST*}(\psi)$ is lower than $\beta_p^{REF}(\psi)$ by a factor of about 2, and also exhibits some angular variability. One consistent feature in the forward scattering region, which is independent of the bead size, is a sharp increase in $CF_f(\psi)$ with a peak at ring 26 (ψ = 5.84°). We observed a similar but inverse feature consistently in natural seawater samples, which suggests that the behavior of the correction function at these angles is credible. Within the angular range of data from the Roving Eyeball sensor (16–150°), the $CF_f(\psi)$ values remain generally in the range between

1.7 and 1.9. For angles larger than 150°, we did not obtain consistent results of the correction function for different bead sizes (not shown), so this angular range is omitted from our analysis of LISST-VSF measurements. Note also that $CF_{400}(\psi)$ differs greatly from $CF_{100}(\psi)$ and $CF_{200}(\psi)$ within the angular range between about 65° and 120°. This can be attributed to the uncertainty in the determinations of $CF_{400}(\psi)$ associated with a well-pronounced minimum in the volume scattering function for the 400 nm beads in this angular range. Therefore, the $CF_{400}(\psi)$ data in this angular range were not used in the determination of final $CF_f(\psi)$.

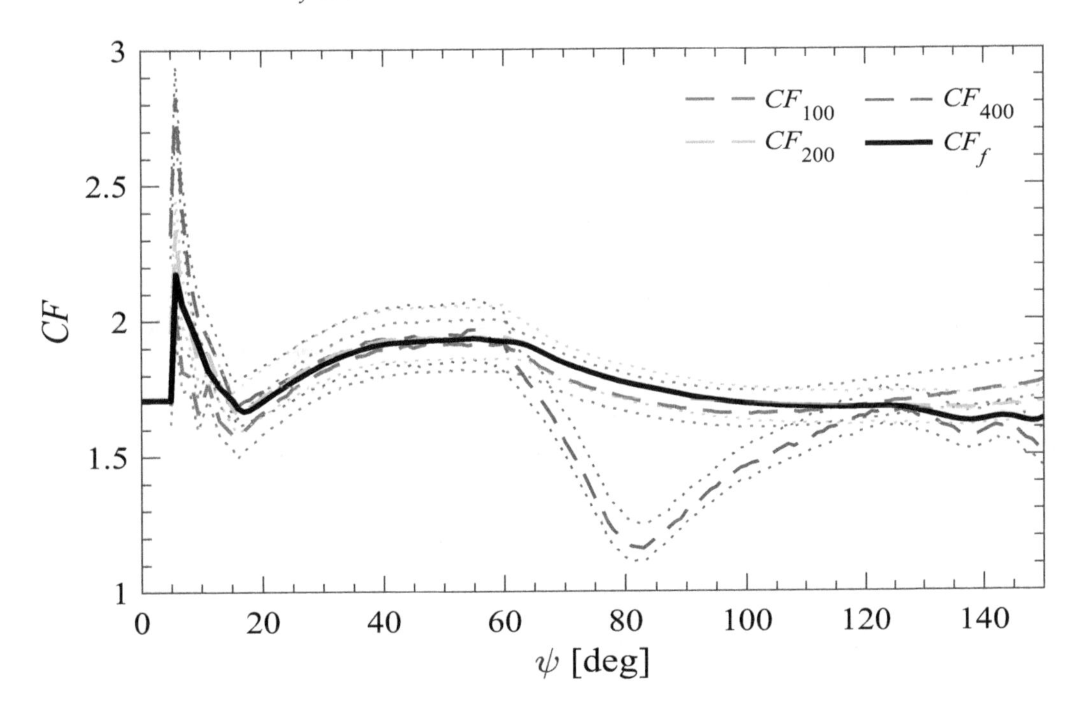

Figure 4. Correction functions, $CF(\psi)$, for the LISST-VSF measurements of particulate volume scattering function $\beta_p^{LISST*}(\psi)$ over the angular range 4.96–150° determined for 100, 200, and 400 nm polystyrene bead suspensions. For each individual bead size, dashed lines represent the median values and the dotted lines indicate the 25th and 75th percentiles determined from the series of measurements. The final computed correction function $CF_f(\psi)$ is shown in black, and includes the constant value used for the near-forward angular range from 0.09° to 4.96°.

The results for $BF_{100}(\psi)$, $BF_{200}(\psi)$, $BF_{400}(\psi)$, and $BF_f(\psi)$, are shown in Figure 5. As seen, $BF_f(\psi)$ is negative within the examined angular range and varies within a relatively narrow range of values between about −0.02 and −0.04. Similar to the results for $CF_{400}(\psi)$, the distinct feature of positive bias observed in the $BF_{400}(\psi)$ data around the scattering angle of 80° can be attributed to the uncertainty associated with a minimum in the volume scattering function for the 400 nm beads in this angular range. This portion of $BF_{400}(\psi)$ data was not used in the determination of final $BF_f(\psi)$.

The performance of the final correction function $CF_f(\psi)$ within the range of scattering angles from 0.09° to 150° was evaluated by comparing the corrected LISST-VSF measurements of volume scattering function, $\beta_p^{LISST}(\psi)$, with reference values of $\beta_p^{REF}(\psi)$ for six samples of polystyrene beads (100, 200, 400, 500, 700, and 2000 nm in diameter; see the dilution factors for these samples indicated in boldface in Table 1). The beads with diameters of 500, 700, and 2000 nm were not used in the generation of the final correction function, so they provide completely independent data for evaluating the performance of $CF_f(\psi)$. The evaluation with the data for 100, 200, and 400 nm beads is also useful because the final $CF_f(\psi)$ was determined by averaging the results obtained with multiple bead sizes and concentrations of these samples, and not from a single bead size and concentration. Results of independent measurements obtained with the DAWN-EOS on four bead suspensions (100, 200, 400, and 700 nm) are also included in the evaluation analysis for additional comparisons.

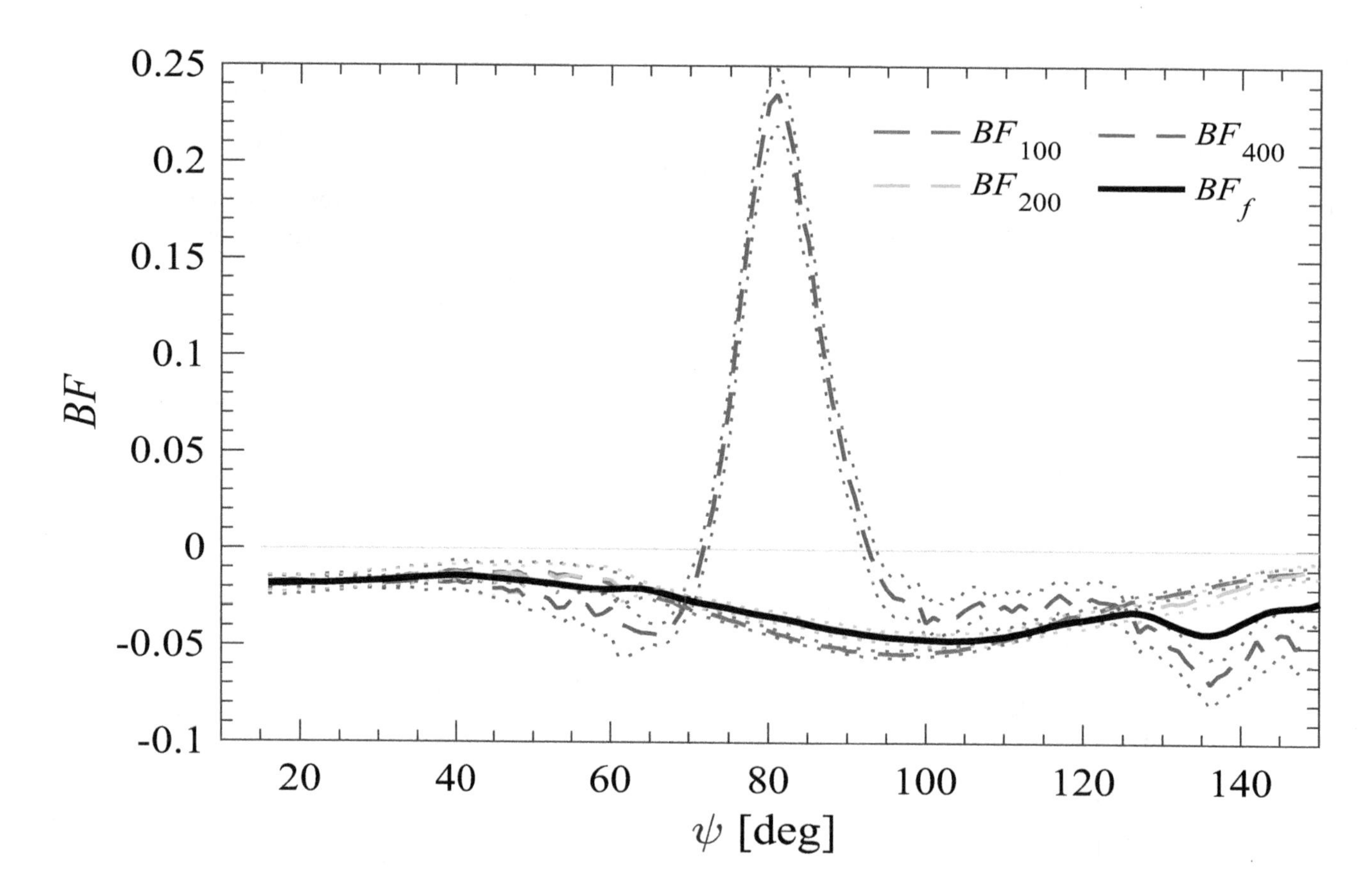

Figure 5. Correction functions, $BF(\psi)$, for LISST-VSF measurements of the degree of linear polarization of light scattered by particles, $DoLP_p^{LISST*}(\psi)$ over the angular range 16–150° determined for 100, 200, and 400 nm polystyrene bead suspensions. For each individual bead size, dashed lines represent the median values and the dotted lines indicate the 25th and 75th percentiles determined from the series of measurements. The final computed correction function $BF_f(\psi)$ is shown in black.

The comparisons of $\beta_p^{LISST}(\psi)$ and $\beta_p^{REF}(\psi)$ are shown in Figure 6 for the six polystyrene bead samples. The presented values of $\beta_p^{LISST}(\psi)$ are the median values for each angle from each measurement series. The measured values of $\beta_p^{\mathrm{DAWN}}(\psi)$ are additionally depicted for the 100, 200, 400, and 700 nm diameter beads. In general, the magnitude and angular dependence of $\beta_p^{LISST}(\psi)$ exhibits good agreement with reference values for all bead diameters. Notable differences occur within the minima of volume scattering function, for example near the angle of 80° for the 400 nm beads (Figure 6c). This issue has been mentioned above in the context of determinations of $CF_f(\psi)$ and $BF_f(\psi)$. The agreement observed between $\beta_p^{LISST}(\psi)$ and $\beta_p^{\mathrm{DAWN}}(\psi)$ lends additional credence to the determined correction function $CF_f(\psi)$ and its application to LISST-VSF measurements.

Figure 7a illustrates the relationship between $\beta_p^{LISST}(\psi)$ measured at all angles between 3.02° and 150° and $\beta_p^{REF}(\psi)$ for the corresponding angles for the six bead samples. The overall agreement is quite good over a range spanning nearly 4 orders of magnitude. The regions of largest disagreement correspond to angles measured with the ring detectors, as well as angles corresponding to sharp minima or maxima in volume scattering function which are observed for the larger beads. Although the measured minima and maxima occur essentially at the same angles as predicted by Mie scattering calculations, the measured magnitude of minima or maxima can differ by a few tens of percent from the reference values. This is illustrated by plots of percent differences between the measured and reference values (Figure 7b). The oscillations and peaks (both positive and negative) in percent differences correspond to the minima and maxima in the angular patterns of volume scattering function.

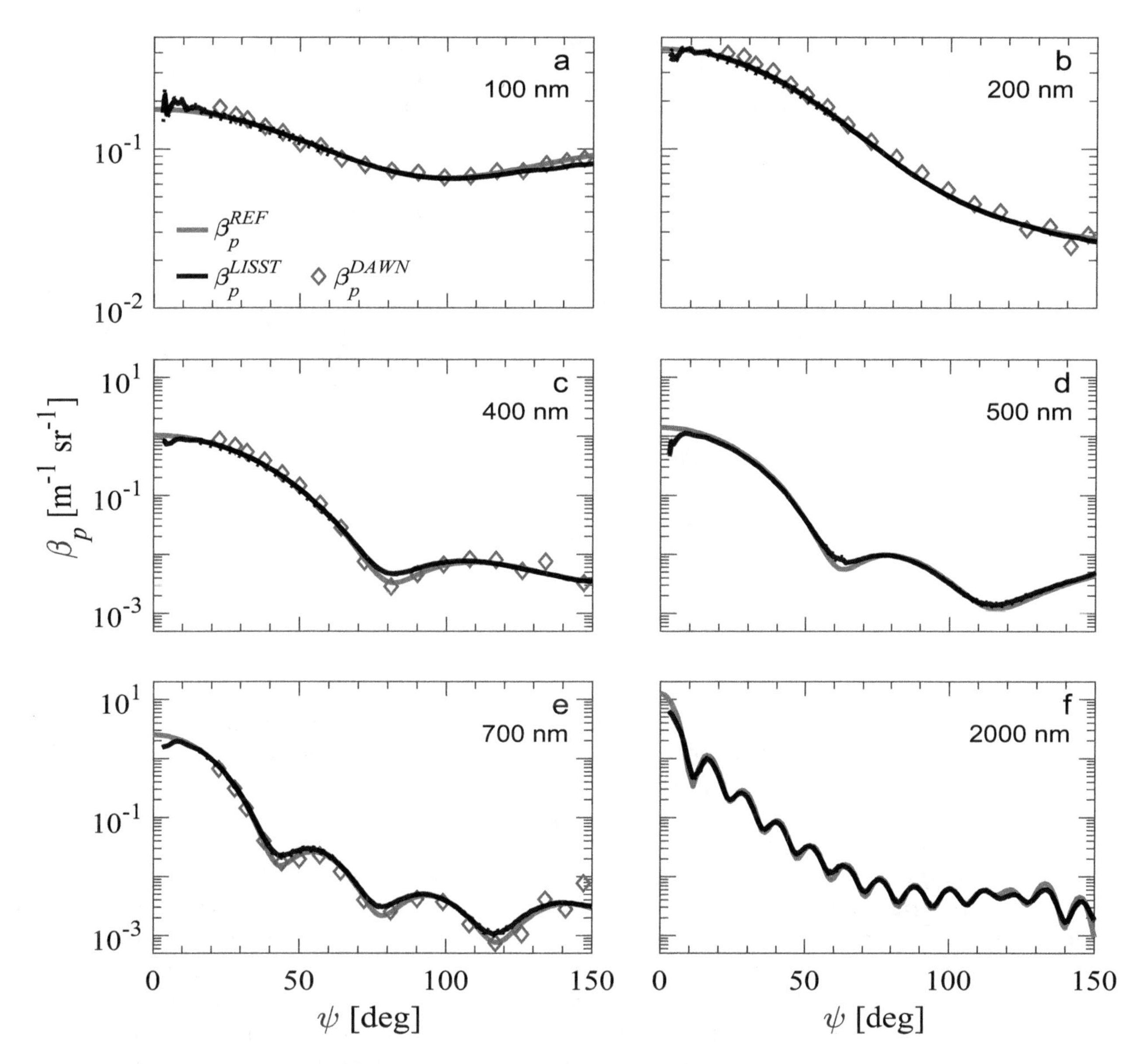

Figure 6. Comparison of $\beta_p(\psi)$ measurements on suspensions of polystyrene beads of varying diameter with reference values, $\beta_p^{REF}(\psi)$. The $\beta_p^{LISST}(\psi)$ data represent CF_f-corrected median values obtained from a series of measurements with the LISST-VSF. Independent measurements of $\beta_p(\psi)$ obtained with the DAWN-EOS instrument are also shown as diamonds in panels a, b, c and e. The bead diameters are indicated in the legend.

Table 2 includes several statistical parameters that quantify the agreement between the data of $\beta_p^{LISST}(\psi)$ and $\beta_p^{REF}(\psi)$ illustrated in Figure 7a. In this analysis we ignore $\psi < 3.02°$ due to generally low scattering signal relative to baseline for these ring detectors. The values of statistical parameters support the overall good agreement; for example, the median ratio (*MR*) of $\beta_p^{LISST}(\psi)$ to $\beta_p^{REF}(\psi)$ is very close to 1, and the median absolute percent difference (*MAPD*) between $\beta_p^{LISST}(\psi)$ and $\beta_p^{REF}(\psi)$ is only ~4%. These median values indicate no overall bias in the corrected measurements of $\beta_p^{LISST}(\psi)$ relative to the reference values of $\beta_p^{REF}(\psi)$ and small statistical differences between $\beta_p^{LISST}(\psi)$ and $\beta_p^{REF}(\psi)$. Table 2 also includes the statistical parameters for a subset of data presented in Figure 7a. In this subset, the forward scattering measurements with ring detectors were excluded, so the angular range is 16–150°. The statistical parameters for this subset are generally improved compared with the dataset covering the angular range 3.02–150°. For example, the root mean square difference (*RMSD*) is smaller (0.015 m^{-1} sr^{-1} vs. 0.21 m^{-1} sr^{-1}) and the slope of linear regression is closer to 1 (0.958 vs.

0.723). The improvements in the statistical parameters after removing the ring detector data are related primarily to much larger values of volume scattering function at forward scattering angles compared with larger angles, and a tendency to negative bias in $\beta_p^{LISST}(\psi)$ relative to $\beta_p^{REF}(\psi)$ at forward angles.

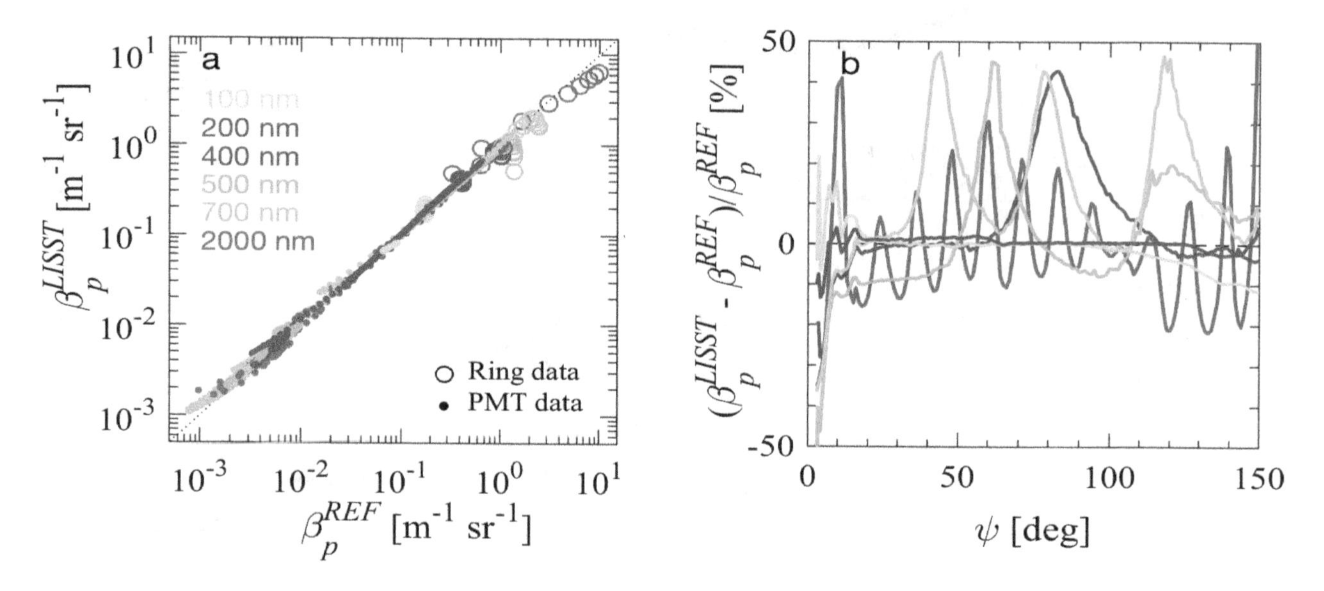

Figure 7. (**a**) Scatter plot of β_p^{LISST} vs. β_p^{REF} for polystyrene beads of varying diameters as indicated. Data obtained with the ring detectors and Roving Eyeball sensor are plotted separately, and the 1:1 line is plotted for reference (dotted black line). (**b**) Residuals expressed as percentages between β_p^{LISST} and β_p^{REF} for each bead size as a function of scattering angle.

To further validate the correction of LISST-VSF measurements with the $CF_f(\psi)$ function, we performed comparisons for approximate scattering and backscattering coefficients, $b_{p,150}$ and $b_{bp,150}$, respectively. The approximate scattering coefficient $b_{p,150}$ was obtained by integrating the volume scattering function within the angular range from 0.09° to 150°. The approximate backscattering coefficient $b_{bp,150}$ was obtained by the integration from 90° to 150°. These calculations were made for the uncorrected measured $\beta_p^{LISST*}(\psi)$, CF_f-corrected measured $\beta_p^{LISST}(\psi)$, and reference $\beta_p^{REF}(\psi)$. We also used Mie scattering calculations to estimate the underestimation of the scattering and backscattering coefficients for the examined polystyrene beads caused by the integration of $\beta_p^{REF}(\psi)$ up to 150° as opposed to 180°. We found that the approximate scattering coefficient, $b_{p,150}^{REF}$, can be lower by as much as 7% compared with the "true" scattering coefficient b_p^{REF}. This result was observed for 100 nm beads. For backscattering the approximate coefficient $b_{bp,150}^{REF}$ was found to be lower by as much as 24% for the 500 nm beads. Although the LISST-VSF measurements extend to 150° rather than 180°, the approximate coefficients are still useful for our validation exercise because most of the angular range and magnitude of total scattering and backscattering coefficients are included in the integration up to 150°. In addition, this validation analysis includes all 20 experiments conducted in this study, and not just the six example experiments presented in Figures 6 and 7.

Figure 8 compares the reference values of $b_{p,150}^{REF}$ and $b_{bp,150}^{REF}$ with LISST-VSF values determined from uncorrected $\beta_p^{LISST*}(\psi)$ and CF_f-corrected $\beta_p^{LISST}(\psi)$. In these calculations we used the median values of $\beta_p^{LISST*}(\psi)$ and $\beta_p^{LISST}(\psi)$ for each measurement series from all 20 experimental combinations of bead sizes, concentrations, and PMT gains listed in Table 1. For all experiments, the approximate coefficients, $b_{p,150}^{LISST*}$ and $b_{bp,150}^{LISST*}$, derived from uncorrected $\beta_p^{LISST*}(\psi)$ are nearly half of the reference values of $b_{p,150}^{REF}$ and $b_{bp,150}^{REF}$. After $CF_f(\psi)$ correction the approximate coefficients $b_{p,150}^{LISST}$ and $b_{bp,150}^{LISST}$ exhibit a nearly 1:1 relationship with $b_{p,150}^{REF}$ and $b_{bp,150}^{REF}$. The statistical parameters that quantify the overall good agreement between $b_{p,150}^{LISST}$ and $b_{p,150}^{REF}$ and between $b_{bp,150}^{LISST}$ and $b_{bp,150}^{REF}$ are listed in Table 2.

Table 2. Statistical results evaluating the comparison of corrected data from the LISST-VSF measurements with reference values obtained from Mie scattering calculations. For β_p^{LISST}, the results are shown for the angular range 3.02–150° which includes the ring data and for the range 16–150° without the ring data. R is the Pearson correlation coefficient and the coefficients A and B are the slope and y-intercept, respectively, determined from a type II linear regression between individual pairs of X_i and Y_i values where Y_i represents measured values and X_i reference values. The mean bias (MB) was calculated as $1/N \times \sum_{i=1}^{N} (Y_i - X_i)$ and MR represents the median ratio of Y_i / X_i. The root mean squared deviation, $RMSD$, was calculated as $\sqrt{\frac{1}{N}\sum_{i=1}^{N}(Y_i - X_i)^2}$, and the median absolute percent difference, $MAPD$, was calculated as the median value of $\left|\frac{Y_i - X_i}{X_i}\right| \times 100$. N is the number of data points used in the analysis.

Data	R	A	B	MB	MR	$RMSD$	$MAPD$	N
β_p^{LISST} (w/ rings)	0.987	0.72	0.031 m^{-1} sr^{-1}	−0.028 m^{-1} sr^{-1}	1.00	0.210 m^{-1} sr^{-1}	3.94%	876
β_p^{LISST} (w/o rings)	0.998	0.96	0.002 m^{-1} sr^{-1}	−0.002 m^{-1} sr^{-1}	1.00	0.015 m^{-1} sr^{-1}	3.39%	810
$b_{p,150}^{LISST}$	0.995	1.04	−0.048 m^{-1}	−0.007 m^{-1}	1.00	0.043 m^{-1}	2.30%	20
$b_{bp,150}^{LISST}$	0.999	0.99	0.0001 m^{-1}	−0.001 m^{-1}	1.00	0.006 m^{-1}	3.70%	20
$DoLP_p^{LISST}$	0.989	0.91	0.046	0.016	0.99	0.065	5.00%	810

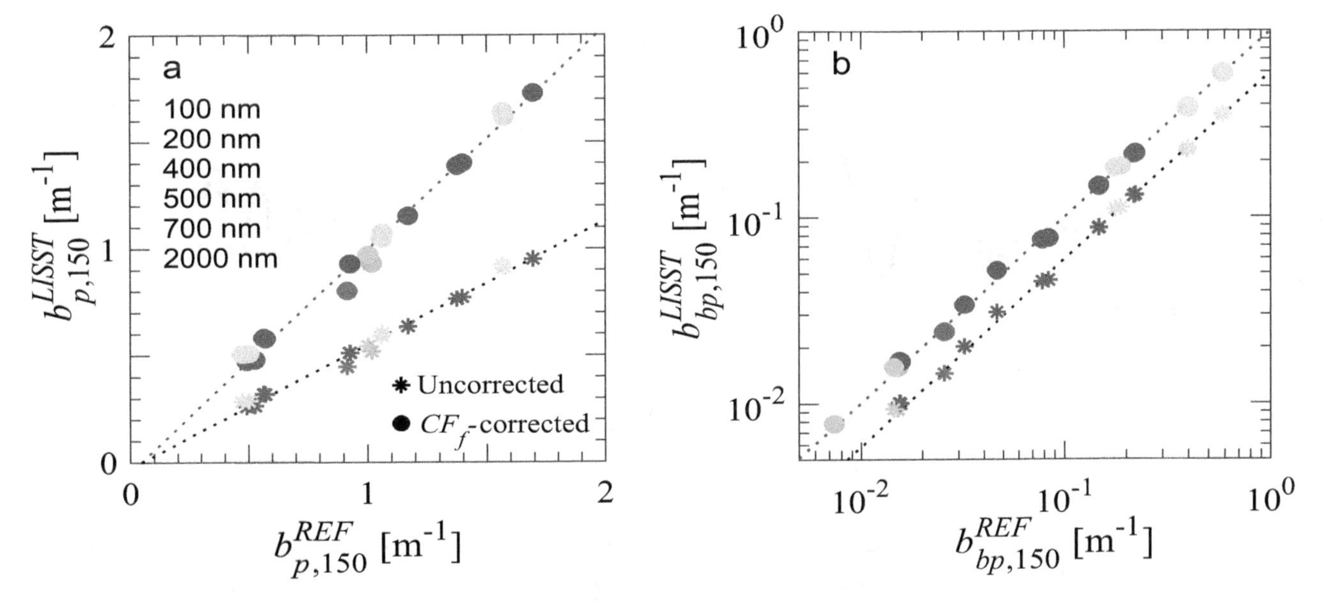

Figure 8. (**a**) Scatter plot comparing reference values of the particulate scattering coefficient computed over the angular range 0.09–150°, $b_{p,150}^{REF}$, with values determined from the LISST-VSF, $b_{p,150}^{LISST}$, before (asterisks) and after (circles) correction with CF_f. A type II linear regression model fit to the data is indicated by the dotted lines. (**b**) Similar to (**a**), but for the particulate backscattering coefficient computed over the range 90–150°.

Similarly to the validation analysis of $CF_f(\psi)$, the performance of the correction function $BF_f(\psi)$ was evaluated by comparing the corrected LISST-VSF measurements of the degree of linear polarization, $DoLP_p^{LISST}(\psi)$, with reference values of $DoLP_p^{REF}(\psi)$ for six samples of polystyrene beads (100, 200, 400, 500, 700, 2000 nm in diameter). Figure 9 depicts these comparisons. The values of $DoLP_p^{DAWN}(\psi)$ measured with DAWN-EOS are also depicted for the 100, 200, 400, and 700 nm beads. For all bead sizes, the magnitude and angular dependence of $DoLP_p^{LISST}(\psi)$ exhibits generally a very good agreement with both the reference values and DAWN-EOS measurements. For larger beads, notable differences occur within the minima of the degree of linear polarization, for example near the angle of 80° for the 400 nm beads (Figure 9c).

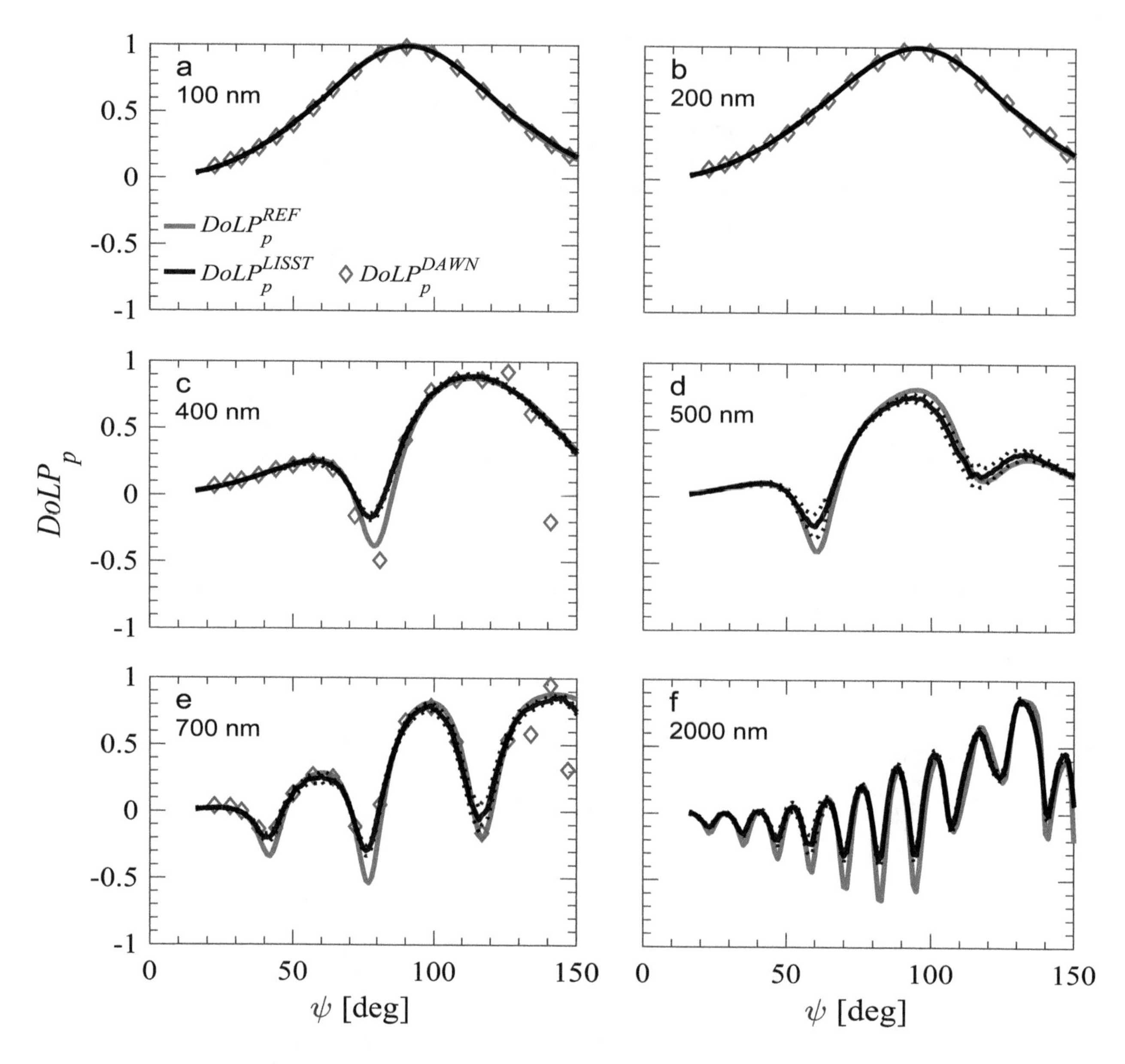

Figure 9. Similar to Figure 6, but for measured and reference values of particulate degree of linear polarization $DoLP_p$. Measurements obtained with the LISST-VSF were corrected with BF_f.

Figure 10a is a scatter plot of $DoLP_p^{LISST}(\psi)$ vs. $DoLP_p^{REF}(\psi)$ which includes all data for the six bead samples presented in Figure 9. In the region of negative values which correspond to the minima in the angular pattern of the degree of linear polarization, the $DoLP_p^{LISST}(\psi)$ exhibits a positive bias relative to $DoLP_p^{REF}(\psi)$. This bias is seen in the form of peaks in the angular pattern of the difference between $DoLP_p^{LISST}(\psi)$ and $DoLP_p^{REF}(\psi)$ for larger bead sizes (Figure 10b). The peak amplitudes generally range from 0.05 to 0.3. Importantly, however, aside from these features the data of $DoLP_p^{LISST}(\psi)$ vs. $DoLP_p^{REF}(\psi)$ are distributed close to the 1:1 line within the major part of the region of positive values (Figure 10a). This includes the region of maximum values of the degree of linear polarization of scattered light from natural seawater samples, which are observed at scattering angles near 90° or greater. The overall good agreement between BF_f-corrected measured $DoLP_p^{LISST}(\psi)$ and $DoLP_p^{REF}(\psi)$ is supported by the statistical parameters shown in Table 2 which are calculated on the basis of the entire dataset presented in Figure 10. For example, the *RMSD* and *MAPD* values are small, 0.065 and 5%, respectively. Also, despite some negative bias for data with negative values of the degree of linear polarization, the *MR* for the ratio of $DoLP_p^{LISST}(\psi)$ to $DoLP_p^{REF}(\psi)$ for the entire dataset is 0.993, indicating essentially no bias. These statistics would improve if the data within the minima in the angular pattern of the degree of linear polarization were removed from the analysis.

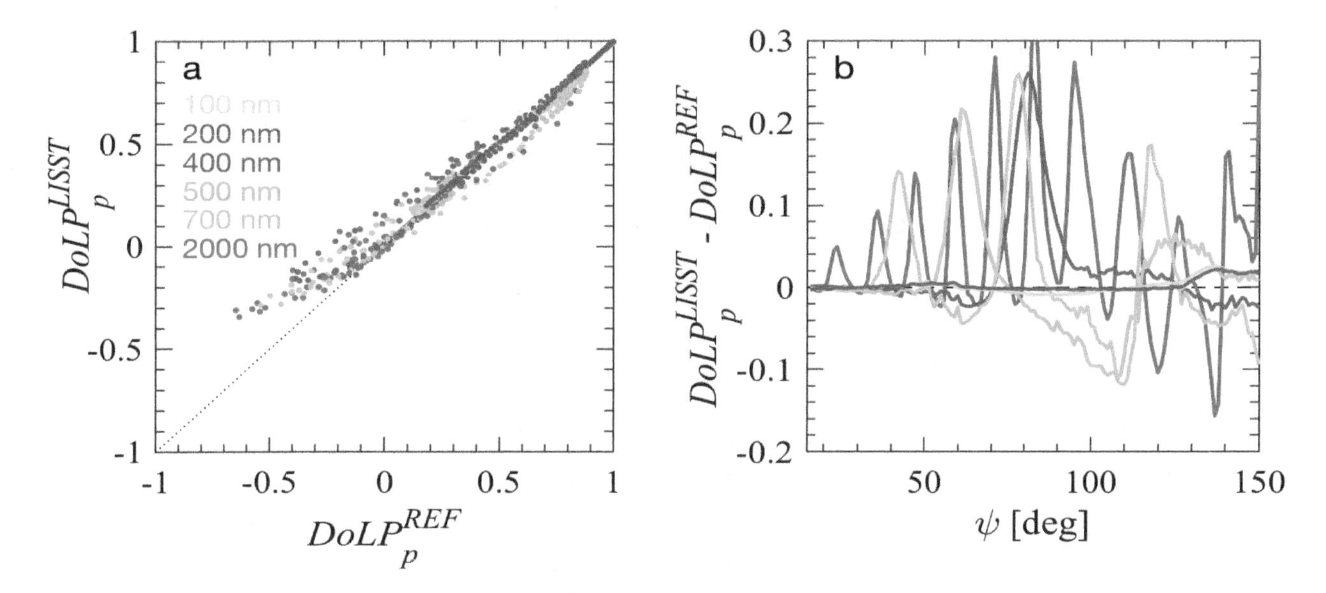

Figure 10. Similar to Figure 7, but for $DoLP_p$. All data are obtained with the Roving Eyeball sensor, and the residuals between $DoLP_p^{LISST}$ and $DoLP_p^{REF}$ in (**b**) are expressed as absolute differences.

3.2. Measured Light Scattering Properties of Natural Particulate Assemblages

Figure 11 depicts the corrected measured volume scattering function, $\beta_p^{LISST}(\psi)$, and the degree of linear polarization, $DoLP_p^{LISST}(\psi)$, for three contrasting natural assemblages of particles. The selected parameters describing the particulate and optical properties of these samples are provided in Table 3. Sample A was obtained ~8 km offshore from the subsurface chlorophyll-*a* maximum at a depth of 18 m, sample B was collected just beneath the sea surface at SIO Pier during a calm sunny summer day, and sample C was collected ~2 km inland at the San Diego River Estuary during low tide. The SPM range covers about one order of magnitude from 0.36 g m^{-3} for sample A to 3.18 g m^{-3} for sample C, which is reflected in significant range of the particulate scattering coefficient, b_p^{LISST}, from 0.36 m^{-1} to 2.23 m^{-1}. Chla was also lowest for sample A (0.75 mg m^{-3}) but highest for sample B (2.5 mg m^{-3}). Thus, whereas the offshore sample A represents a particle concentration that is within the range of observations in relatively clear open ocean waters, sample C is representative of more turbid coastal or nearshore waters [5,7]. Samples A and B have similarly high values of the ratio POC/SPM (0.43 and 0.47, respectively) and relatively high values of Chla/SPM (2.1 $\times$ 10^{-3} and 2.2 $\times$ 10^{-3}, respectively), indicating organic-dominated particulate assemblages with significant contribution of phytoplankton. In contrast, sample C has much lower values of POC/SPM (0.14) and Chla/SPM (3.8 $\times$ 10^{-4}), indicating inorganic-dominated particulate assemblage and relatively small role of phytoplankton, despite significant chlorophyll-*a* concentration (1.21 mg m^{-3}).

These differences in particle properties between the three samples are responsible for the differences in the magnitude and angular shape of $\beta_p^{LISST}(\psi)$ and $DoLP_p^{LISST}(\psi)$ presented in Figure 11 and the optical parameters listed in Table 3. These optical parameters include the particulate backscattering ratio, $\tilde{b}_{bp}^{LISST}$, the ratio of $\beta_p^{LISST}(45^\circ)$ to $\beta_p^{LISST}(135^\circ)$, and the maximum value of $DoLP_p^{LISST}(\psi)$ denoted as $DoLP_{p,max}^{LISST}$. This maximum value occurs at a scattering angle ψ_{max} which is also provided in Table 3. The offshore sample A has an intermediate value of $\tilde{b}_{bp}^{LISST}$ and the highest $DoLP_{p,max}^{LISST}$ of about 0.77 associated with the smallest ψ_{max} of 92°. Sample B from the SIO Pier has the lowest $\tilde{b}_{bp}^{LISST}$ of 0.008 among the three samples, suggesting a relatively steep slope of particle size distribution, relatively low bulk particle refractive index, or both [17]. This sample also shows the least steep near-forward scattering pattern (Figure 11c), which suggests a higher proportion of small particles relative to larger particles compared with the two other samples. Finally, the most turbid and least organic sample, sample C, exhibits an enhanced proportion of backscattering with the highest $\widetilde{b}_{bp}^{LISST}$ of 0.027. While this result may suggest a relatively high bulk particle refractive index [17] consistent with the lowest POC/SPM ratio among the three samples, the additional influence of particle size distribution cannot be ruled out. Sample C shows steep near-forward scattering pattern (Figure 11c), which is typically indicative of an increased proportion of large particles relative to small particles. Note that sample C also has the lowest $\beta_p^{LISST}(45°)/\beta_p^{LISST}(135°)$ ratio of 12, which indicates a higher degree of symmetry in the angular pattern of scattering about 90°, which is consistent with the relatively high value of $\widetilde{b}_{bp}^{LISST}$ for this sample. In addition, sample C has the lowest $DoLP_{p,max}^{LISST}$ of 0.58.

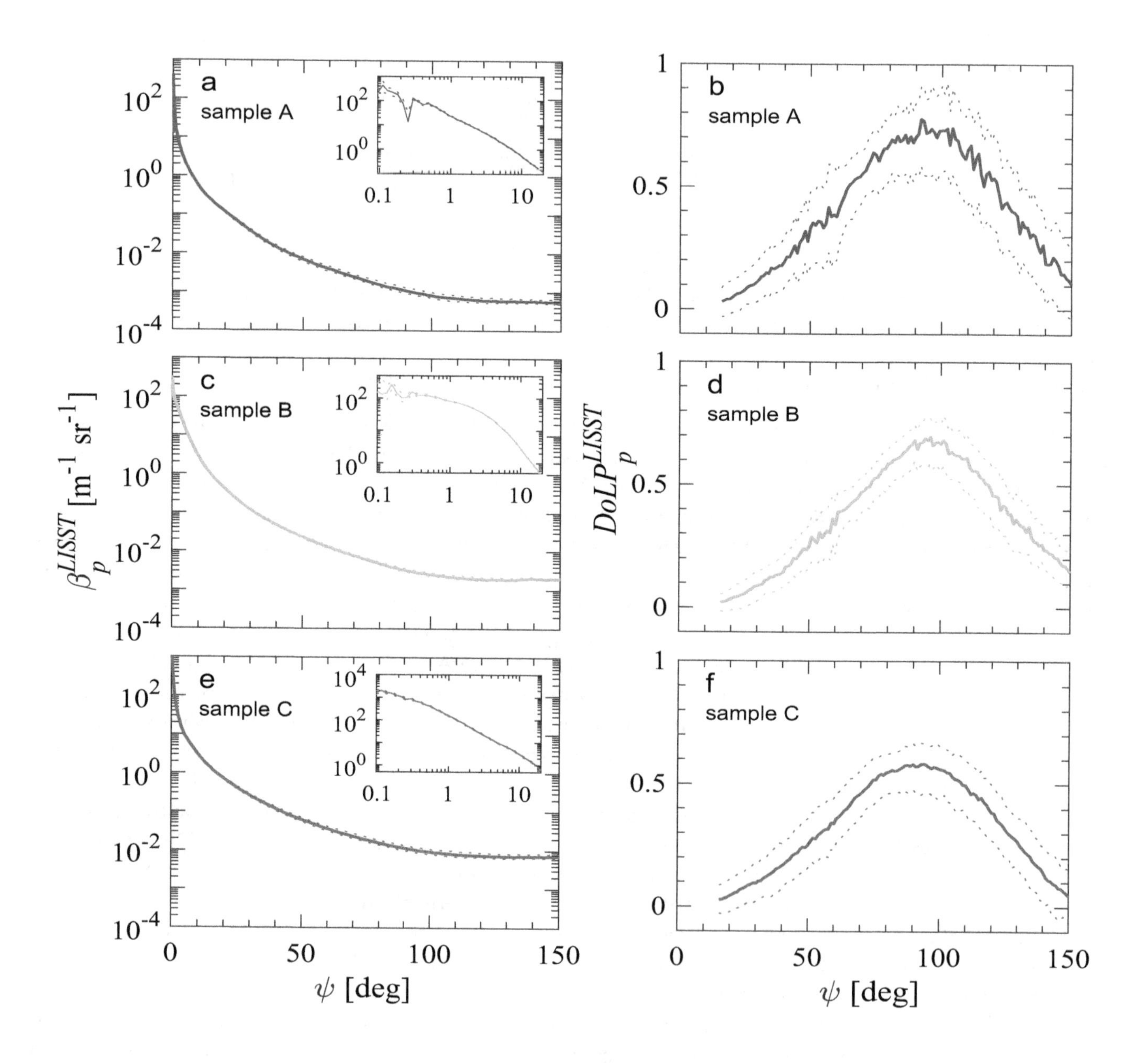

Figure 11. Measurements of β_p^{LISST} and $DoLP_p^{LISST}$ obtained with the LISST-VSF on natural seawater samples from the San Diego region representing (**a**,**b**) subsurface offshore waters, (**c**,**d**) SIO Pier, and (**e**,**f**) San Diego River Estuary. Solid lines represent median values while dotted lines indicate the 10th and 90th percentiles obtained from a series of measurements on each sample. Insets in (**a**,**c**,**d**) display greater detail on the near-forward scattering range.

Table 3. General information on particle characteristics and median values of optical quantities derived from LISST-VSF measurements for the three example natural seawater samples depicted in Figure 11. Values of selected optical quantities estimated from the measurements of Petzold [38] are shown for comparison.

Sample ID	Chla [mg m^{-3}]	SPM [g m^{-3}]	POC/SPM [g/g]	b_p [m−1]	$\widetilde{b}_{bp}$ [dim]	$\frac{\beta_p(45°)}{\beta_p(135°)}$	$DoLP_{p,max}$ [dim]	ψ_{max} [deg]
A	0.75	0.36	0.43	0.36	0.012	16.5	0.77	92
B	2.49	1.13	0.47	1.75	0.008	17.1	0.69	96
C	1.21	3.18	0.14	2.23	0.022	12.0	0.58	94
Petzold Measurements								
Clear				0.03	0.015	18.0		
Coastal				0.19	0.009	17.8		
Turbid				1.74	0.020	12.2		

For comparison, Table 3 also includes the values for the selected optical parameters estimated from measurements reported by Petzold [38] for clear ocean waters (off Bahamas), coastal waters (San Diego coastal region), and turbid waters (San Diego Harbor). These measurements span a generally similar range of scattering angles as the LISST-VSF (10–180° in 5° increments), but are based on a spectrally broader incident beam (75 nm full width half maximum) centered at 514 nm. The estimates of particulate volume scattering function β_p from Petzold's measurements were obtained by subtracting pure seawater contribution β_w from the measured total β. The β_w value was calculated assuming a water temperature of 15 °C and salinity of 33 PSU [77]. Although Petzold's data include measurements made in clearer waters compared with our samples, the range of values for the dimensionless parameters associated with the shape of angular scattering pattern, $\widetilde{b}_{bp}^{LISST}$ and $\beta_p^{LISST}(45°)/\beta_p^{LISST}(135°)$, is very similar. Specifically, our data for the offshore sample A are similar to Petzold's data from clear waters, sample B from the SIO Pier aligns with Petzold's data from coastal San Diego waters, and sample C from San Diego River Estuary with Petzold's data from the San Diego Harbor.

We note that the dotted lines in Figure 11 reflect some variations in $\beta_p^{LISST}(\psi)$ and $DoLP_p^{LISST}(\psi)$ between the individual measurements within a given set of measurements for each sample. For example, on the basis of the collection of 200 to 400 measurements for each of the two linear polarization states of the incident beam and the scattered light, the coefficient of variation at ψ = 90° was 14%, 8%, and 13% for β_p^{LISST} and 20%, 13%, and 15% for $DoLP_p^{LISST}$ for samples A, B, and C, respectively. These variations between the individual measurements that have been taken in rapid succession do not necessarily reflect the measurement precision, as they can also be associated with actual variations in the sample, for example the fluctuations in the presence of relatively rare large particles within the interrogated sample volume. Another important point is that the small negative values of $DoLP_p^{LISST}$ observed for some individual measurements at forward scattering angles <30° (see the 10th percentile dotted lines in Figure 11b,d,f) are not necessarily an indication of measurement uncertainty because the negative values, especially in this angular range, are physically possible for certain types of particles [30,54,56,58].

Figure 12 depicts scatter plots of the relationships between the dimensionless optical parameters, $\widetilde{b}_{bp}^{LISST}$ and $DoLP_{p,max}^{LISST}$, and the dimensionless particulate compositional properties, POC/SPM and Chla/SPM, for all 17 samples examined in this study. The overall range of POC/SPM in our dataset is 0.04 to 0.6. The presented data have been divided into three groups according to the values of POC/SPM as follows: the least organic-dominated (or the most mineral-dominated) data with POC/SPM < 0.15; the most organic-dominated data with POC/SPM > 0.3, and the intermediate data with 0.15 ≤ POC/SPM ≤ 0.3. The selected boundary values of POC/SPM for discriminating between the organic-dominated and mineral-dominated groups of data differ from those used in our previous studies [65,78], but appear to adequately reflect the patterns in the present data. In particular, the most mineral-dominated samples with POC/SPM < 0.15 form a clear cluster of data points with the highest $\widetilde{b}_{bp}^{LISST}$ (Figure 12a) and the lowest Chla/SPM (Figure 12b,d). We also note that no data were collected for POC/SPM between 0.15 and 0.2, so we will refer to all data with POC/SPM > 0.2 as highly organic because they all represent highly significant or dominant role of organic particles.

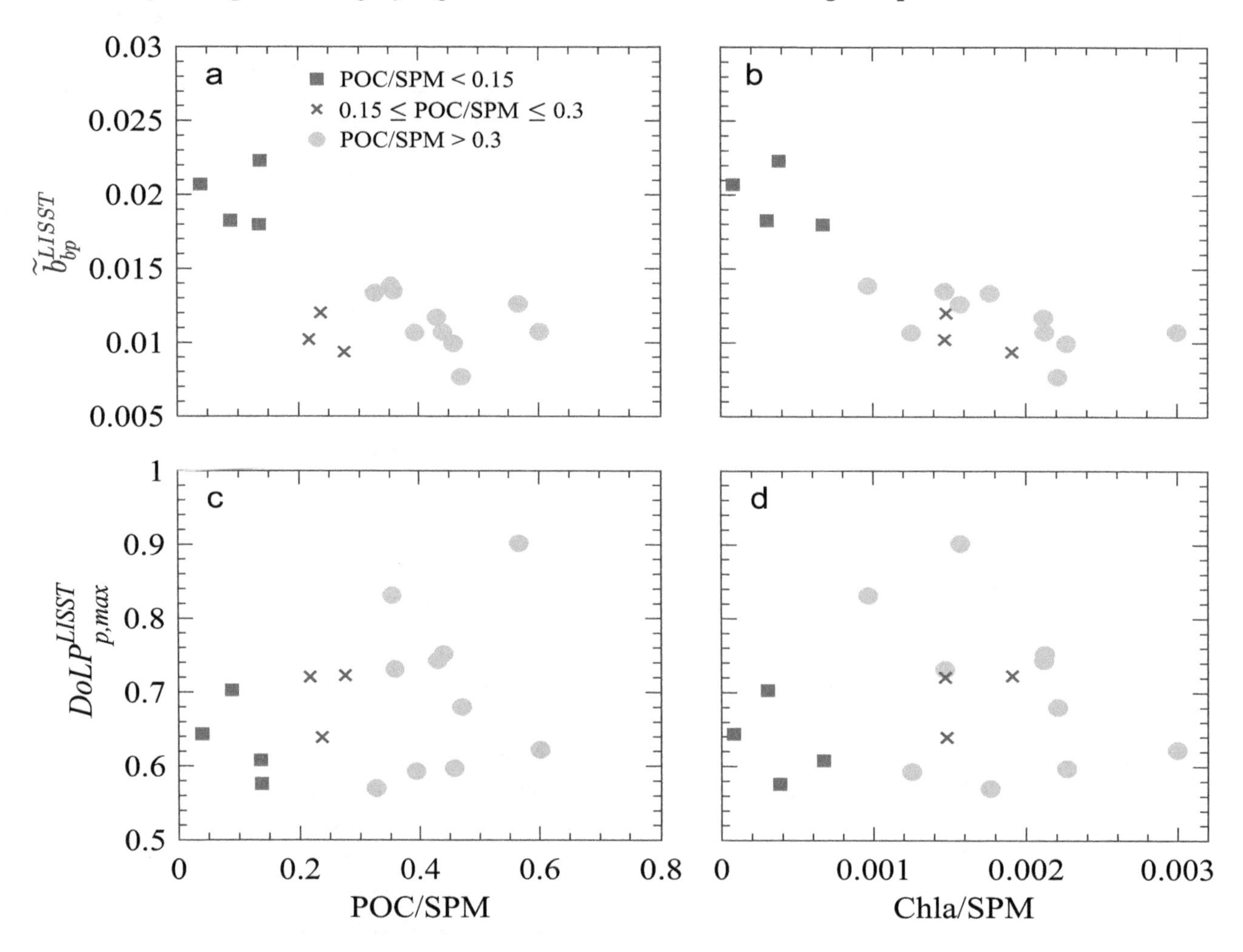

Figure 12. LISST-VSF measurements of (**a**,**b**) the particulate backscattering ratio, $\widetilde{b}_{bp}^{LISST}$, and (**c**,**d**) the maximum value of the degree of linear polarization of scattered light, $DoLP_{p,max}^{LISST}$, as a function of the POC/SPM or Chla/SPM ratio. The data are divided into three groups defined by the range of POC/SPM as indicated in the legend.

The scatter plot for the data of $\widetilde{b}_{bp}^{LISST}$ vs. POC/SPM suggests the presence of a relational trend with significant negative correlation between the variables (the correlation coefficient $R = -0.73$). While the $\widetilde{b}_{bp}^{LISST}$ values are clearly highest for POC/SPM < 0.15, the organic-dominated samples have consistently lower $\widetilde{b}_{bp}^{LISST}$. This result is consistent with the notion that mineral-dominated particulate assemblages with relatively high bulk refractive index of particles tend to have higher backscattering ratio $\widetilde{b}_{bp}$ compared with organic-dominated assemblages with lower refractive index [17]. However, we also note that highly organic samples with POC/SPM > 0.2 show no clear relationship and essentially no correlation between $\widetilde{b}_{bp}^{LISST}$ and POC/SPM ($R = -0.11$). This result may be attributable to the effect of other particle characteristics on $\widetilde{b}_{bp}^{LISST}$, such as variations in refractive index of particles associated with changes in the composition of particulate organic matter, variations in particle size distribution, or both. The scatter plot of $\widetilde{b}_{bp}^{LISST}$ vs. Chla/SPM (Figure 12b) provides interesting insight into this question, as this relationship is significantly better compared with $\widetilde{b}_{bp}^{LISST}$ vs. POC/SPM. Whereas the correlation between $\widetilde{b}_{bp}^{LISST}$ and Chla/SPM for all data is strong ($R = -0.85$), the subset of data for highly organic samples (POC/SPM > 0.2) has also a relatively high correlation coefficient of -0.51. This is an important result, suggesting that for particulate assemblages with high organic content, the backscattering ratio $\widetilde{b}_{bp}$ tends to decrease with increasing proportion of phytoplankton in the particulate assemblage. It is likely that the relationship in Figure 12b is largely driven by a decrease in the bulk particle refractive index with increasing proportion of phytoplankton in the particulate assemblage. Because this trend also holds for the subset of highly organic samples, it may indicate that live phytoplankton cells have generally lower refractive index than non-living organic particles.

In contrast to $\widetilde{b}_{bp}^{LISST}$, the $DoLP_{p,max}^{LISST}$ data show no clear relational trend and very weak correlation with POC/SPM (R = 0.31), indicating that the maximum degree of linear polarization does not provide a useful optical signature for the organic vs. inorganic content of particulate assemblages in our dataset (Figure 12c). A similar result with no correlation (R = 0.07) is observed for $DoLP_{p,max}^{LISST}$ vs. Chla/SPM, indicating that varying proportion of phytoplankton in total particulate assemblage has no discernible systematic effect on the maximum degree of linear polarization (Figure 12d). We also determined that there is no significant correlation between $DoLP_{p,max}^{LISST}$ and $\widetilde{b}_{bp}^{LISST}$ in our dataset ($R = -0.22$), as well as between ψ_{max} and POC/SPM or Chla/SPM ($R = -0.09$ and 0.06, respectively). It is also of interest to note that the range of our $DoLP_{p,max}^{LISST}$ data is generally consistent with the range of values reported in literature for natural seawater samples, although the reported range in some earlier studies extends to somewhat lower values, as low as about 0.4 [15,39–42,79,80].

The assessment of potential presence of systematic effects of particle size distribution (PSD) on $\widetilde{b}_{bp}^{LISST}$ and $DoLP_{p,max}^{LISST}$ is presented in Figure 13. In this assessment, we use two PSD metrics: the 90th percentile diameter, D_V^{90}, derived from the particle volume distribution, and the power function slope, ζ, derived from the particle number distribution. We also tested other percentile-based diameters such as the median D_V^{50} but no improvements in the examined relationships were observed. Figure 13a,b shows no trend in the data of $\widetilde{b}_{bp}^{LISST}$ associated with variations in the particle size metrics, even though these metrics vary over a significant dynamic range. This is the case for the entire dataset as well as a subset of highly organic samples with POC/SPM > 0.2, which supports the interpretation of results presented in Figure 12b in terms of the role of refractive index. The data of $DoLP_{p,max}^{LISST}$ vs. D_V^{90} show the potential for the presence of a relational trend (Figure 13c). Although the scatter in these data points is significant and correlation is weak ($R = -0.47$), the lowest values of $DoLP_{p,max}^{LISST}$ tend to occur along with the highest values of D_V^{90}. This result indicates that the decrease in the maximum degree of linear polarization tends to be associated with particulate assemblages exhibiting a higher proportion of large-sized particles. The potential usefulness of the relationship between the degree of linear polarization and particle size has been proposed for the first time in 1930 [81], and the trend observed in our data is consistent with those early results.

4. Concluding Remarks

Our laboratory measurements combined with Mie scattering calculations for samples of standard polystyrene beads illustrate the value of such an approach for evaluating the calibration and performance of light scattering instruments. For the specific version of LISST-VSF instrument and data processing code used in our study, we determined the calibration correction functions for improved determinations of the particulate volume scattering function $\beta_p(\psi)$ and the degree of linear polarization $DoLP_p(\psi)$. The required correction was found to be particularly significant for $\beta_p(\psi)$ (a correction factor of ~1.7 to 1.9). The improved determinations of $\beta_p(\psi)$ and $DoLP_p(\psi)$ were validated with measurements on independent samples, and also using another independently-calibrated light scattering instrument, DAWN-EOS. Although the correction functions developed in this study are applicable only to the specific version of LISST-VSF instrument and the data processing code used in this study, our results emphasize a general need for evaluating the performance of light scattering instruments and minimizing the associated uncertainties in quantitative determinations from measurements.

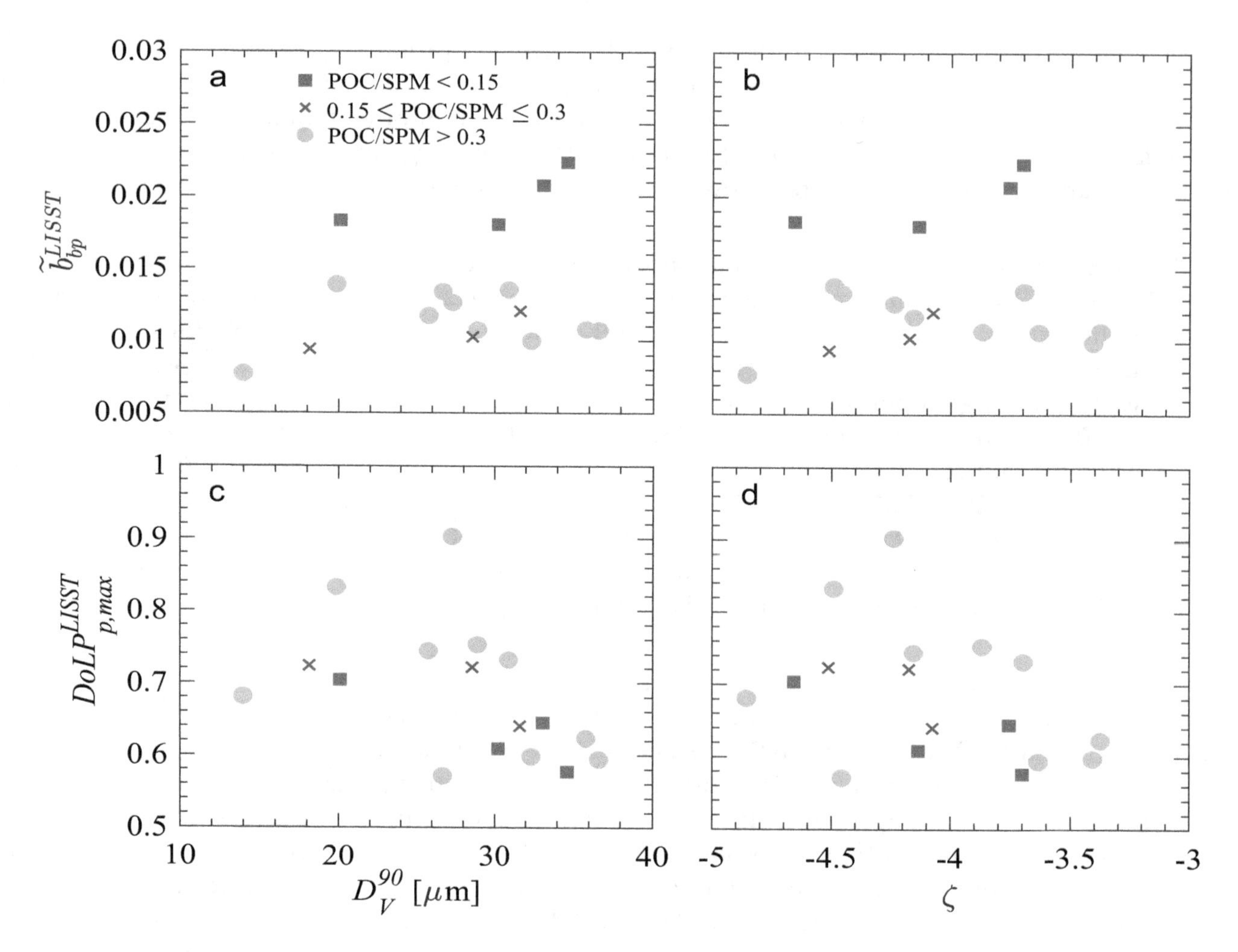

Figure 13. Similar to Figure 12, but with optical quantities shown as a function of the particle size metrics (**a**,**c**) D_V^{90}, representing the diameter corresponding to the 90th percentile of the particle volume distribution, and (**b**,**d**) ζ, the power law slope of the particle number distribution.

The improved protocol for measurements of light scattering with our LISST-VSF instrument was applied to measurements taken on 17 natural seawater samples from coastal and offshore marine environments characterized by contrasting assemblages of suspended particles. The particulate volume scattering function, degree of linear polarization, and backscattering ratio were determined from LISST-VSF measurements. For our dataset, these light scattering properties exhibit significant variations related to a broad range of measured particle properties characterizing the organic vs. inorganic composition and size distribution of particulate assemblages. For example, we observed negative relational trends between the particulate backscattering ratio and the increasing proportions of organic particles or phytoplankton in the total particulate assemblage. These proportions were parameterized in terms of the measured ratio of particulate organic carbon (POC) or chlorophyll-*a* (Chla) concentration to the total dry mass concentration of suspended particulate matter (SPM). The observed trends can be useful in the development of optical approaches for characterizing the composition of particulate assemblages. Our results also suggest a potential trend between the maximum degree of linear polarization of light scattered by particles and particle size metrics. Specifically, the decrease in the maximum degree of linear polarization observed at scattering angles close to 90° tends to be associated with particulate assemblages exhibiting a higher proportion of large-sized particles.

Earlier theoretical studies have shown that changes in the angular shape and the maximum value of the degree of linear polarization depend on particle refractive index and size distribution [82,83]; however, the experimental data of the degree of linear polarization of scattered light for natural marine particle assemblages are very scarce. Our results provide a contribution to filling this gap. This type of data can also be useful for improving an understanding of the polarization properties of marine light fields including polarization of water-leaving radiance and advancing related applications, including remote sensing applications [84–90]. The various potential applications of angular light scattering measurements, including the polarization effects associated with light scattering by marine particles, call for further efforts in this research area.

Author Contributions: All authors made intellectual contributions to the study design and participated in the collection of experimental data. D.K. and R.A.R. processed the data and conducted Mie scattering calculations, and D.K. analyzed the results. D.K. and D.S. wrote the paper, and all authors reviewed and provided final edits in the manuscript.

Acknowledgments: This work was supported by NASA Terrestrial Hydrology and Ocean Biology and Biogeochemistry Programs (Grant #NNX13AN72G) and the NASA Earth and Space Science Fellowship Program (Grant #NNX14AK93H). We acknowledge Edward R. Blocker, Eric Chen, and Linhai Li for assistance during measurements and processing of data. We also thank Xiaodong Zhang for providing computer code for backscattering extrapolation method and related discussions, and Wayne Slade for discussions about LISST-VSF instrument and processing code. We thank Hubert Loisel and three anonymous reviewers for valuable comments on the manuscript.

References

1. Kattawar, G.W.; Adams, C.N. Stokes vector calculations of the submarine light field in an atmosphere–ocean with scattering according to a Rayleigh phase matrix: Effect of interface refractive index on radiance and polarization. *Limnol. Oceanogr.* **1989**, *34*, 1453–1472. [CrossRef]
2. Mobley, C.D. *Light and Water: Radiative Transfer in Natural Waters*; Academic Press: San Diego, CA, USA, 1994.
3. Mobley, C.D.; Sundman, L.K.; Boss, E. Phase function effects on oceanic light fields. *Appl. Opt.* **2002**, *41*, 1035–1050. [CrossRef] [PubMed]
4. Kattawar, G.W.; Yang, P.; You, Y.; Bi, L.; Xie, Y.; Huang, X.; Hioki, S. Polarization of light in the atmosphere and ocean. In *Light Scattering Reviews 10*; Kokhanovsky, A.A., Ed.; Springer: Berlin, Germany, 2016; pp. 3–39.
5. Babin, M.; Morel, A.; Fournier-Sicre, V.; Fell, F.; Stramski, D. Light scattering properties of marine particles in coastal and open ocean waters as related to the particle mass concentration. *Limnol. Oceanogr.* **2003**, *48*, 843–859. [CrossRef]
6. Balch, W.M.; Gordon, R.G.; Bowler, B.C.; Drapeau, D.T.; Booth, E.S. Calcium carbonate measurements in the surface global ocean based on moderate-resolution imaging spectroradiometer data. *J. Geophys. Res.* **2005**, *110*, C07001. [CrossRef]
7. Stramski, D.; Reynolds, R.A.; Babin, M.; Kaczmarek, S.; Lewis, M.R.; Röttgers, R.; Sciandra, A.; Stramska, M.; Twardowski, M.S.; Franz, B.A.; et al. Relationships between the surface concentration of particulate organic carbon and optical properties in the eastern South Pacific and eastern Atlantic Oceans. *Biogeosciences* **2008**, *5*, 171–201. [CrossRef]
8. Bale, A.J.; Morris, A.W. In situ measurement of particle size in estuarine waters. *Estuar. Coast. Shelf Sci.* **1987**, *24*, 253–263. [CrossRef]
9. Agrawal, Y.C.; Pottsmith, H.C. Instruments for particle size and settling velocity observations in sediment transport. *Mar. Geol.* **2000**, *168*, 89–114. [CrossRef]
10. Agrawal, Y.C.; Whitmire, A.; Mikkelsen, O.A.; Pottsmith, H.C. Light scattering by random shaped particles and consequences on measuring suspended sediments by Laser Diffraction. *J. Geophys. Res.* **2008**, *113*, C04023. [CrossRef]
11. Reynolds, R.A.; Stramski, D.; Wright, V.M.; Woźniak, S.B. Measurements and characterization of particle size distributions in coastal waters. *J. Geophys. Res.* **2010**, *115*, C08024. [CrossRef]
12. Wyatt, P.J.; Villalpando, D.N. High-precision measurement of submicrometer particle size distributions. *Langmuir* **1997**, *13*, 3913–3914. [CrossRef]
13. Wyatt, P.J. Submicrometer particle sizing by multiangle light scattering following fractionation. *J. Colloid Interface Sci.* **1998**, *197*, 9–20. [CrossRef] [PubMed]

14. Uitz, J.; Stramski, D.; Baudoux, A.C.; Reynolds, R.A.; Wright, V.M.; Dubranna, J.; Azam, F. Variations in the optical properties of a particle suspension associated with viral infection of marine bacteria. *Limnol. Oceanogr.* **2010**, *55*, 2317–2330. [CrossRef]
15. Morel, A. Diffusion de la lumière par les eaux de mer: Resultats expérimentaux et approche théorique. In *Optics of the Sea*; North Atlantic Treaty Organization AGARD Lecture Series, No. 61; Technical Editing and Reproduction Ltd: London, UK, 1973; pp. 3.1.1–3.1.76.
16. Ackleson, S.G.; Spinrad, R.W. Size and refractive index of individual marine particulates: A flow cytometric approach. *Appl. Opt.* **1988**, *27*, 1270–1277. [CrossRef] [PubMed]
17. Twardowski, M.S.; Boss, E.; Macdonald, J.B.; Pegau, W.S.; Barnard, A.H.; Zaneveld, J.R.V. A model for estimating bulk refractive index from the optical backscattering ratio and the implications for understanding particle composition in case I and case II waters. *J. Geophys. Res.* **2001**, *106*, 14129–14142. [CrossRef]
18. Sullivan, J.M.; Twardowski, M.S.; Donaghay, P.L.; Freeman, S.A. Use of optical scattering to discriminate particle types in coastal waters. *Appl. Opt.* **2005**, *44*, 1667–1680. [CrossRef] [PubMed]
19. Zhang, X.; Huot, Y.; Gray, D.J.; Weidemann, A.; Rhea, W.J. Biogeochemical origins of particles obtained from the inversion of the volume scattering function and spectral absorption in coastal waters. *Biogeosciences* **2013**, *10*, 6029–6043. [CrossRef]
20. Bohren, C.F.; Huffman, D.R. *Absorption and Scattering of Light by Small Particles*; Wiley: New York, NY, USA, 1983.
21. Bickel, W.S.; Bailey, W.M. Stokes vectors, Mueller matrices, and polarized scattered light. *Am. J. Phys.* **1985**, *53*, 468–478. [CrossRef]
22. Kattawar, G.W. Polarization of light in the ocean. In *Ocean Optics*; Spinrad, R.W., Carder, K.L., Perry, M.J., Eds.; Oxford University Press: New York, NY, USA, 1994; p. 202.
23. Jonasz, M.; Fournier, G.R. *Light Scattering by Particles in Water: Theoretical and Experimental Foundations*; Academic Press: San Diego, CA, USA, 2007.
24. Fry, E.S; Voss, K.J. Measurement of the Mueller matrix for phytoplankton. *Limnol. Oceanogr.* **1985**, *30*, 1322–1326. [CrossRef]
25. Quinby-Hunt, M.S.; Hunt, A.J.; Lofftus, K.; Shapiro, D. Polarized-light scattering studies of marine Chlorella. *Limnol. Oceanogr.* **1989**, *34*, 1587–1600. [CrossRef]
26. Wyatt, P.J.; Jackson, C. Discrimination of phytoplankton via light-scattering properties. *Limnol. Oceanogr.* **1989**, *34*, 96–112. [CrossRef]
27. Shapiro, D.B.; Quinby-Hunt, M.S.; Hunt, A.J. Origin of the induced circular-polarization in the light scattered from a dinoflagellate. In *Ocean Optics X*; Spinrad, R.W., Ed.; SPIE: Bellingham, WA, USA, 1990; Volume 1302, pp. 281–289.
28. Witkowski, K.; Wolinski, L.; Turzynski, Z.; Gedziorowska, D.; Zielinski, A. The investigation of kinetic growth of Chlorella vulgaris cells by the method of integral and dynamic light-scattering. *Limnol. Oceanogr.* **1993**, *38*, 1365–1372. [CrossRef]
29. Volten, H.; De Haan, J.F.; Hovenier, J.W.; Schreurs, R.; Vassen, W.; Dekker, A.G.; Hoogenboom, H.J.; Charlton, F.; Wouts, R. Laboratory measurements of angular distributions of light scattered by phytoplankton and silt. *Limnol. Oceanogr.* **1998**, *43*, 1180–1197. [CrossRef]
30. Volten, H.; Muñoz, O.; Rol, E.; Haan, J.D.; Vassen, W.; Hovenier, J.W.; Nousiainen, T. Scattering matrices of mineral aerosol particles at 441.6 nm and 632.8 nm. *J. Geophys. Res.* **2001**, *106*, 17375–17401. [CrossRef]
31. Volten, H.; Muñoz, O.; Hovenier, J.W.; Waters, L.B.F.M. An update of the Amsterdam light scattering database. *J. Quant. Spectrosc. Radiat. Transf.* **2006**, *100*, 437–443. [CrossRef]
32. Svensen, Ø.; Stamnes, J.J.; Kildemo, M.; Aas, L.M.S.; Erga, S.R.; Frette, Ø. Mueller matrix measurements of algae with different shape and size distributions. *Appl. Opt.* **2011**, *50*, 5149–5157. [CrossRef] [PubMed]
33. Muñoz, O.; Moreno, F.; Guirado, D.; Dabrowska, D.D.; Volten, H.; Hovenier, J.W. The Amsterdam–Granada light scattering database. *J. Quant. Spectrosc. Radiat. Transf.* **2012**, *113*, 565–574. [CrossRef]
34. Liu, J.P.; Kattawar, G.W. Detection of dinoflagellates by the light scattering properties of the chiral structure of their chromosomes. *J. Quant. Spectrosc. Radiat. Transf.* **2013**, *131*, 24–33. [CrossRef]
35. Stramski, D.; Boss, E.; Bogucki, D.; Voss, K.J. The role of seawater constituents in light backscattering in the ocean. *Prog. Oceanogr.* **2004**, *61*, 27–56. [CrossRef]

36. Hovenier, J.W.; Volten, H.; Muñoz, O.; Van der Zande, W.J.; Waters, L.B.F.M. Laboratory studies of scattering matrices for randomly oriented particles: Potentials, problems, and perspectives. *J. Quant. Spectrosc. Radiat. Transf.* **2002**, *79*, 741–755. [CrossRef]
37. Kokhanovsky, A.A. Parameterization of the Mueller matrix of oceanic waters. *J. Geophys. Res.* **2003**, *108*, 3175. [CrossRef]
38. Petzold, T.J. *Volume Scattering Functions for Selected Ocean Waters*; SIO Ref. 72–78, Scripps Institution of Oceanography Visibility Lab; University of California: San Diego, CA, USA, 1972.
39. Beardsley, G.F. Mueller scattering matrix of sea water. *J. Opt. Soc. Am.* **1968**, *58*, 52–57. [CrossRef]
40. Kadyshevich, Y.A.; Lyubovtseva, Y.S.; Rozenberg, G.V. Light-scattering matrices of Pacific and Atlantic ocean waters. *Izv. Acad. Sci. USSR Atmos. Ocean. Phys.* **1976**, *12*, 106–111.
41. Kadyshevich, Y.A. Light-scattering matrices of inshore waters of the Baltic Sea. *Izv. Acad. Sci. USSR Atmos. Ocean. Phys.* **1977**, *13*, 77–78.
42. Voss, K.J.; Fry, E.S. Measurement of the Mueller matrix for ocean water. *Appl. Opt.* **1984**, *23*, 4427–4439. [CrossRef] [PubMed]
43. Lee, M.; Lewis, M. A new method for the measurement of the optical volume scattering function in the upper ocean. *J. Atmos. Ocean. Technol.* **2003**, *20*, 563–572. [CrossRef]
44. Sullivan, J.M.; Twardowski, M.S. Angular shape of the oceanic particulate volume scattering function in the backward direction. *Appl. Opt.* **2009**, *48*, 6811–6819. [CrossRef] [PubMed]
45. Tan, H.; Doerffer, R.; Oishi, T.; Tanaka, A. A new approach to measure the volume scattering function. *Opt. Express* **2013**, *21*, 18697–18711. [CrossRef] [PubMed]
46. Chami, M.; Thirouard, A.; Harmel, T. POLVSM (Polarized Volume Scattering Meter) instrument: An innovative device to measure the directional and polarized scattering properties of hydrosols. *Opt. Express* **2014**, *22*, 26403–26428. [CrossRef] [PubMed]
47. Zhang, X.; Lewis, M.; Lee, M.; Johnson, B.; Korotaev, G. The volume scattering function of natural bubble populations. *Limnol. Oceanogr.* **2002**, *47*, 1273–1282. [CrossRef]
48. Twardowski, M.S.; Zhang, X.; Vagle, S.; Sullivan, J.; Freeman, S.; Czerski, H.; You, Y.; Bi, L.; Kattawar, G. The optical volume scattering function in a surf zone inverted to derive sediment and bubble particle subpopulations. *J. Geophys. Res.* **2012**, *117*, C00H17. [CrossRef]
49. Slade, W.H.; Agrawal, Y.C.; Mikkelsen, O.A. Comparison of measured and theoretical scattering and polarization properties of narrow size range irregular sediment particles. In *Oceans San Diego*; IEEE: San Diego, CA, USA, 2013; pp. 1–6.
50. Slade, W.H.; Boss, E.S. Calibrated near-forward volume scattering function obtained from the LISST particle sizer. *Opt. Express* **2006**, *14*, 3602–3615. [CrossRef] [PubMed]
51. Van de Hulst, H.C. *Light Scattering by Small Particles*; Dover Publications: New York, NY, USA, 1981.
52. McCartney, E.J. *Optics of the Atmosphere: Scattering by Molecules and Particles*; Wiley: New York, NY, USA, 1976.
53. Mishchenko, M.I.; Travis, L.D. Light scattering by polydisperse, rotationally symmetric nonspherical particles: Linear polarization. *J. Quant. Spectrosc. Radiat. Transf.* **1994**, *51*, 759–778. [CrossRef]
54. Yanamandra-Fisher, P.A.; Hanner, M.S. Optical properties of nonspherical particles of size comparable to the wavelength of light: Application to comet dust. *Icarus* **1999**, *138*, 107–128. [CrossRef]
55. Petrova, E.V.; Jockers, K.; Kiselev, N.N. Light scattering by aggregates with sizes comparable to the wavelength: An application to cometary dust. *Icarus* **2000**, *148*, 526–536. [CrossRef]
56. Muñoz, O.; Volten, H.; Hovenier, J.W.; Min, M.; Shkuratov, Y.G.; Jalava, J.P.; van der Zande, W.J.; Waters, L.B.F.M. Experimental and computational study of light scattering by irregular particles with extreme refractive indices: Hematite and rutile. *Astron. Astrophys.* **2006**, *446*, 525–535. [CrossRef]
57. Muinonen, K.; Zubko, E.; Tyynelä, J.; Shkuratov, Y.G.; Videen, G. Light scattering by Gaussian random particles with discrete-dipole approximation. *J. Quant. Spectrosc. Radiat. Transf.* **2007**, *106*, 360–377. [CrossRef]
58. Zubko, E. Light scattering by irregularly shaped particles with sizes comparable to the wavelength. In *Light Scattering Reviews 6*; Kokhanovsky, A.A., Ed.; Springer: Berlin, Germany, 2012; pp. 39–74.
59. Agrawal, Y.C. The optical volume scattering function: Temporal and vertical variability in the water column off the New Jersey coast. *Limnol. Oceanogr.* **2005**, *50*, 1787–1794. [CrossRef]
60. Agrawal, Y.C.; Mikkelsen, O.A. Empirical forward scattering phase functions from 0.08 to 16 deg. for randomly shaped terrigenous 1–21 μm sediment grains. *Opt. Express* **2009**, *17*, 8805–8814. [CrossRef] [PubMed]

61. Babin, M.; Stramski, D.; Reynolds, R.A.; Wright, V.M.; Leymarie, E. Determination of the volume scattering function of aqueous particle suspensions with a laboratory multi-angle light scattering instrument. *Appl. Opt.* **2012**, *51*, 3853–3873. [CrossRef] [PubMed]
62. Heller, W.; Tabibian, R.M. Experimental investigations on the light scattering of colloidal spheres. II. Sources of error in turbidity measurements. *J. Colloid Sci.* **1957**, *12*, 25–39. [CrossRef]
63. Bateman, J.B.; Weneck, E.J.; Eshler, D.C. Determination of particle size and concentration from spectrophotometric transmission. *J. Colloid Sci.* **1959**, *14*, 308–329. [CrossRef]
64. Stramski, D.; Babin, M.; Woźniak, S.B. Variations in the optical properties of terrigenous mineral-rich particulate matter suspended in seawater. *Limnol. Oceanogr.* **2007**, *52*, 2418–2433. [CrossRef]
65. Woźniak, S.B.; Stramski, D.; Stramska, M.; Reynolds, R.A.; Wright, V.M.; Miksic, E.Y.; Cieplak, A.M. Optical variability of seawater in relation to particle concentration, composition, and size distribution in the nearshore marine environment at Imperial Beach, California. *J. Geophys. Res.* **2010**, *115*, C08027. [CrossRef]
66. Bohren, C.F. Multiple scattering of light and some of its observable consequences. *Am. J. Phys.* **1987**, *55*, 524–533. [CrossRef]
67. *LISST-VSF Multi-Angle Polarized Light Scattering Meter: User'S Manual Revision A*; Sequoia Scientific: Bellevue, WA, USA.
68. Ma, X.; Lu, J.Q.; Brock, R.S.; Jacobs, K.M.; Yang, P.; Hu, X.H. Determination of complex refractive index of polystyrene microspheres from 370 to 1610 nm. *Phys. Med. Biol.* **2003**, *48*, 4165–4172. [CrossRef] [PubMed]
69. Morel, A.; Bricaud, A. Inherent optical properties of algal cells including picoplankton: Theoretical and experimental results. *Can. Bull. Fish. Aquat. Sci.* **1986**, *214*, 521–559.
70. Van der Linde, D.W. Protocol for determination of total suspended matter in oceans and coastal zones. *JRC Tech. Note I* **1998**, *98*, 182.
71. Parsons, T.R.; Maita, Y.; Lalli, C.M. *A Manual of Chemical and Biological Methods for Seawater Analysis*; Elsevier: New York, NY, USA, 1984.
72. Knap, A.; Michaels, A.; Close, A.; Ducklow, H.; Dickson, A. *Protocols for the Joint Global Ocean Flux Study (JGOFS) Core Measurements*; UNESCO: Paris, France, 1994.
73. Ritchie, R.J. Universal chlorophyll equations for estimating chlorophylls a, b, c, and d and total chlorophylls in natural assemblages of photosynthetic organisms using acetone, methanol, or ethanol solvents. *Photosynthetica* **2008**, *46*, 115–126. [CrossRef]
74. Bader, H. The hyperbolic distribution of particle sizes. *J. Geophys. Res.* **1970**, *75*, 2822–2830. [CrossRef]
75. Beardsley, G.F., Jr.; Zaneveld, J.R.V. Theoretical dependence of the near-asymptotic apparent optical properties on the inherent optical properties of sea water. *J. Opt. Soc. Am.* **1969**, *58*, 373–377. [CrossRef]
76. Zhang, X.; Fournier, G.R.; Gray, D.J. Interpretation of scattering by oceanic particles around 120 degrees and its implication in ocean color studies. *Opt. Express* **2017**, *25*, A191–A199. [CrossRef] [PubMed]
77. Zhang, X.; Hu, L.; He, M. Scattering by pure seawater: Effect of salinity. *Opt. Express* **2009**, *17*, 5698–5710. [CrossRef] [PubMed]
78. Reynolds, R.A.; Stramski, D.; Neukermans, G. Optical backscattering by particles in Arctic seawater and relationships to particle mass concentration, size distribution, and bulk composition. *Limnol. Oceanogr.* **2016**, *61*, 1869–1890. [CrossRef]
79. Ivanoff, A. Optical method of investigation of the oceans: The p-*ß* diagram. *J. Opt. Soc. Am.* **1959**, *49*, 103–104. [CrossRef]
80. Ivanoff, A.; Jerlov, N.; Waterman, T.H. A comparative study of irradiance, beam transmittance and scattering in the sea near Bermuda. *Limnol. Oceanogr.* **1961**, *6*, 129–148. [CrossRef]
81. Hatch, T.; Choate, S.P. Measurement of polarization of the Tyndall beam of aqueous suspension as an aid in determining particle size. *J. Franklin Inst.* **1930**, *210*, 793–804. [CrossRef]
82. Chami, M.; Santer, R.; Dilligeard, E. Radiative transfer model for the computation of radiance and polarization in an ocean–atmosphere system: Polarization properties of suspended matter for remote sensing. *Appl. Opt.* **2001**, *40*, 2398–2416. [CrossRef] [PubMed]
83. Lotsberg, J.K.; Stamnes, J.J. Impact of particulate oceanic composition on the radiance and polarization of underwater and backscattered light. *Opt. Express* **2010**, *18*, 10432–10445. [CrossRef] [PubMed]
84. Waterman, T.H. Polarization patterns in submarine illumination. *Science* **1954**, *120*, 927–932. [CrossRef] [PubMed]

85. Ivanoff, A. Polarization measurements in the sea. In *Optical Aspects of Oceanography*; Jerlov, N.G., Steeman-Nielsen, E., Eds.; Academic Press: London, UK; New York, NY, USA, 1974; pp. 151–175.
86. Chami, M. Importance of the polarization in the retrieval of oceanic constituents from the remote sensing reflectance. *J. Geophys. Res.* **2007**, *112*, C05026. [CrossRef]
87. Loisel, H.; Duforet, L.; Dessailly, D.; Chami, M.; Dubuisson, P. Investigation of the variations in the water leaving polarized reflectance from the POLDER satellite data over two biogeochemical contrasted oceanic areas. *Opt. Express* **2008**, *16*, 12905–12918. [CrossRef] [PubMed]
88. Tonizzo, A.; Gilerson, A.; Harmel, T.; Ibrahim, A.; Chowdhary, J.; Gross, B.; Ahmed, S. Estimating particle composition and size distribution from polarized water-leaving radiance. *Appl. Opt.* **2011**, *50*, 5047–5058. [CrossRef]
89. Ibrahim, A.; Gilerson, A.; Chowdhary, J.; Ahmed, S. Retrieval of macro- and micro-physical properties of oceanic hydrosols from polarimetric observations. *Rem. Sens. Environ.* **2016**, *186*, 548–566. [CrossRef]
90. Zhai, P.W.; Knobelspiesse, K.; Ibrahim, A.; Franz, B.A.; Hu, Y.; Gao, M.; Frouin, R. Water-leaving contribution to polarized radiation field over ocean. *Opt. Express* **2017**, *25*, A689–A708. [CrossRef] [PubMed]

5

Progress in Forward-Inverse Modeling Based on Radiative Transfer Tools for Coupled Atmosphere-Snow/Ice-Ocean Systems: A Review and Description of the AccuRT Model

Knut Stamnes [1,*], Børge Hamre [2,†], Snorre Stamnes [3,†], Nan Chen [1,†], Yongzhen Fan [1,†], Wei Li [1,†], Zhenyi Lin [1,†] and Jakob Stamnes [2,†]

1 Stevens Institute of Technology, Hoboken, NJ 07030, USA; nchen@stevens.edu (N.C.); yfan3021@gmail.com (Y.F.); wli4@stevens.edu (W.L.); lzhenyi@stevens.edu (Z.L.)
2 University of Bergen, 5020 Bergen, Norway; Borge.Hamre@uib.no (B.H.); Jakob.Stamnes@uib.no (J.S.)
3 NASA LARC, Hampton, VA 23681, USA; snorre.a.stamnes@nasa.gov
* Correspondence: Knut.Stamnes@stevens.edu
† These authors contributed equally to this work.

Abstract: A tutorial review is provided of forward and inverse radiative transfer in coupled atmosphere-snow/ice-water systems. The coupled system is assumed to consist of two adjacent horizontal slabs separated by an interface across which the refractive index changes abruptly from its value in air to that in ice/water. A comprehensive review is provided of the inherent optical properties of air and water (including snow and ice). The radiative transfer equation for unpolarized as well as polarized radiation is described and solutions are outlined. Several examples of how to formulate and solve inverse problems encountered in environmental optics involving coupled atmosphere-water systems are discussed in some detail to illustrate how the solutions to the radiative transfer equation can be used as a forward model to solve practical inverse problems.

Keywords: vector radiative transfer; polarization; coupled systems; atmosphere; ocean; forward modeling; inverse problems

1. Introduction

Reliable, accurate, and efficient modeling of electromagnetic radiation transport in turbid media has important applications in studies of Earth's climate by remote sensing. For example, such modeling is needed to develop forward-inverse methods used to quantify types and concentrations of aerosol and cloud particles in the atmosphere, as well as dissolved organic and particulate biogeochemical matter in lakes, rivers, coastal water, and open-ocean water, and to simulate the performance of remote sensing detectors deployed on aircraft, balloons, and satellites. Accurate radiative transfer (RT) modeling is also required to compute irradiances and scalar irradiances that are used to compute warming/cooling and photolysis rates in the atmosphere, solar energy deposition in the cryosphere including frozen fresh water (lakes and rivers), sea ice, and glaciers, as well as primary production rates in the water. Finally, RT modeling is needed to compute the Stokes vector describing the polarization state of the radiation field, which is desired in many remote sensing applications.

Accurate, efficient, and easy-to-use radiative transfer (RT) simulation tools are important because they (i) can be used to generate irradiances as well as total and polarized radiances (including degree of polarization) at any location and direction; (ii) will provide accurate results for given input parameters and specified inherent optical properties (IOPs); (iii) will lead to significant progress in research areas such as remote sensing algorithm development, climate research, and other atmospheric and hydrologic applications.

Available tools for atmospheric applications include: (i) SBDART [1], Streamer [2], and LibRadtran (www.libradtran.org), which all apply to the atmosphere only; there is no coupling to an underlying surface consisting of e.g., solid (snow/ice) or liquid water; (ii) Hydrolight, which applies to water only, provides water-leaving radiance, but not top-of-the-atmosphere (TOA) radiance; there is no coupling to the atmosphere (assumed to be a boundary condition). To remedy this situation a new Accurate Radiative Transfer (AccuRT) tool was developed to facilitate well-tested and robust RT simulations in coupled systems consisting of two slabs with different refractive indices. Please note that we here use the word "water" generically to describe the solid phase (i.e., snow and ice) as well as the liquid phase. The AccuRT tool accounts for reflection and transmission at the interface between the two slabs, and allows each slab to be divided into a sufficient number of layers to resolve the variation in the IOPs with depth.

Notation

Radiative transfer practitioners in the atmosphere, liquid water (ocean, lakes, rivers) and cryosphere (snow/ice) communities use different nomenclatures and terminologies. This situation can be confusing and frustrating to students and researchers addressing interdisciplinary problems in environmental optics. In this paper, we will adopt the notation of [3]. Letting z be the vertical position in the plane-parallel medium under consideration, and letting Θ be the scattering angle, we will denote the:

1. the absorption coefficient [m^{-1}] by the letter $\alpha(z)$;
2. the scattering coefficient [m^{-1}] by the letter $\beta(z)$;
3. the extinction coefficient [m^{-1}] by the letter $\gamma(z) = \alpha(z) + \beta(z)$;
4. the single-scattering albedo by $\varpi(z) = \beta(z)/(\alpha(z) + \beta(z))$;
5. the volume scattering function [$\mathrm{m}^{-1}\mathrm{sr}^{-1}$] by $\mathrm{vsf}(z, \cos\Theta, \phi)$ and the related scattering phase function (dimensionless) by $p(z, \cos\Theta, \phi)$;
6. the scattering phase matrix (dimensionless) by $\mathbf{P}_S(\Theta)$ in the Stokes vector representation $\mathbf{I}_S = [I, Q, U, V]^T$ and by $\mathbf{P}(\Theta)$ in the Stokes vector representation $\mathbf{I} = [I_{\parallel}, I_{\perp}, U, V]^T$.

The corresponding notation used in the Ocean Optics community is a instead of α, b instead of β, and c instead of γ. Since α, β, and γ are the three first letters in the Greek alphabet it should be easy to recall the connection with a, b, and c.

This tutorial review is organized as follows. In Section 2 the input parameters needed to describe the coupled system are specified. Then in Section 3 we describe the inherent optical properties (IOPs) of the two adjacent coupled slabs, consisting of absorption and scattering coefficients as well as scattering phase functions. In addition, one needs the scattering phase matrix for polarized radiation. For unpolarized radiation only one element of this matrix, namely the scattering phase function, is required. These IOPs appear in the radiative transfer equation (RTE) described in Section 4, where we also review how to formulate and solve the RTE for unpolarized as well as polarized radiation. In Section 5 we provide several examples of how the solution of the forward problem discussed in Section 4 can be used to solve the corresponding inverse problem. In the forward problem, the IOPs are assumed to be known, so that the solution of the RTE provides the total (and polarized) radiances. To solve the inverse problem we ask: given the measured total (and/or polarized) radiances can we determine the IOPs? The inverse problem is generally much more difficult to solve than the forward problem. It can be formulated as a classical, nonlinear minimization problem, which can be solved in an iterative manner. We will also demonstrate how neural networks can be used to help tackle the inverse problem in a reliable and efficient manner. Finally, in Section 6 we briefly discuss some remaining problems in Ocean Optics, while in Section 7 we provide a brief summary.

2. Input and Output Parameters for the Forward Radiative Transfer Problem

2.1. Input Parameters

The following input parameters must be specified (i) the physical properties of each of the two slabs that constitute the coupled system, (ii) the radiative energy input at the top of the upper slab (top-of-the-atmosphere, TOA), and (iii) the boundary conditions at the bottom of the lower slab (water bottom). Each of the two slabs is assumed to be a plane-parallel, vertically stratified structure in which the scattering and absorption properties, i.e., the IOPs are defined in Section 3. To resolve changes in the IOPs as a function of vertical position z, each slab can be divided into several adjacent layers such that the IOPs are constant within each layer, but allowed to vary from one layer to the next. The impact of a wind-roughened air-water interface is described in Section 5.3.

To specify the IOPs, we will use the concept of materials, which are radiatively significant constituents in the atmosphere-water system. Examples of such materials are atmospheric gases, aerosols, clouds, snow, ice, pure water and water impurities. These materials can be designed to account for the wavelength dependence of the IOPs so that all one needs to do is to decide which of them to include in each layer. More specifically the following input parameters must be specified: (i) solar beam irradiance [$W{\cdot}m^{-2}$]; (ii) wavelength range, number of center wavelengths and widths [nm]; (iii) solar zenith angle(s) in degrees; (iv) the number of "discrete ordinate streams" used to solve the radiative transfer equation (RTE) as described in Section 4; (v) IOPs of materials used. Specifications required for the upper slab include (i) layer boundaries, ground-level altitude (sea-level is default); (ii) atmospheric type; (iii) aerosol particle types (e.g., a bimodal lognormal volume distribution) for each layer as described in Section 3.4.2; (iv) cloud particles as described in Section 3.4.3; (v) snow particles as described in Section 3.5. Specifications required for the lower slab include (i) layer boundaries; (ii) ice material as described in Section 3.5; (iii) refractive index as a function of wavelength in the lower slab (water)—set to "one" in the upper slab (atmosphere); (iv) water impurities; (v) water bottom albedo.

2.2. Output Parameters

Once the input parameters above have been specified, the solution of the RTE as described in Section 4, will provide two types of output, namely, (i) irradiances and mean intensities (scalar irradiances in Ocean Optics terminology) at specified vertical positions in the coupled system; (ii) total and polarized radiances in desired directions at specified vertical positions in the coupled system.

3. Inherent Optical Properties (IOPs)

3.1. General Definitions

An *inherent optical property* (IOP) depends only on the medium itself, and is independent of the ambient light field within the medium [4]. An *apparent optical property* (AOP) depends also on the illumination, i.e., on light propagating in particular directions inside and outside the medium (Apparent optical properties (1) depend both on the medium (the IOPs) and on the geometric (directional) structure of the polarized radiance distribution, and (2) display enough regular features and stability to be useful descriptors of a water body [4]. Hence, a radiance or an irradiance would satisfy only the first part of the definition, while a radiance reflectance or irradiance reflectance, obtained by division of the radiance or the upward irradiance by the downward irradiance, would satisfy also the second part of the definition.).

Two important IOPs are the absorption coefficient $\alpha(z)$ and the scattering coefficient $\beta(z)$ defined as [3]

$$\alpha(z) = \frac{1}{I^i}\left(\frac{dI^{\alpha}}{dz}\right), \qquad \beta(z) = \frac{1}{I^i}\left(\frac{dI^{\beta}}{dz}\right) \qquad [m^{-1}]. \tag{1}$$

Here I^i is the incident radiance entering a volume element $dV = dAdz$ of the medium of cross sectional area dA and thickness dz, and $dI^\alpha > 0$ and $dI^\beta > 0$, respectively, are the radiances that are absorbed and scattered in all directions as the light propagates the distance dz along the direction of the incident light. The extinction coefficient is given by $\gamma(z) = \alpha(z) + \beta(z)$, and the single-scattering albedo is defined as $\varpi(z) \equiv \beta(z)/\gamma(z)$.

The angular distribution of the scattered light is given in terms of the *volume scattering function* (vsf), which is defined as

$$\text{vsf}(z, \hat{\Omega}', \hat{\Omega}) = \frac{1}{I^i}\frac{d^2 I^\beta}{dz d\omega} = \frac{1}{I^i}\frac{d}{dz}\left(\frac{dI^\beta}{d\omega}\right) \qquad [\text{m}^{-1}\,\text{sr}^{-1}]. \tag{2}$$

Here $\hat{\Omega}'$ and $\hat{\Omega}$ are unit vectors, and $d^2 I^\beta$ is the radiance scattered from an incident direction $\hat{\Omega}'$ into a cone of solid angle $d\omega$ around the direction $\hat{\Omega}$ as the light propagates the distance dz along the direction $\hat{\Omega}'$ (see Figure 1). The plane spanned by $\hat{\Omega}'$ and $\hat{\Omega}$ is called the *scattering plane*, and the *scattering angle* Θ is given by $\cos\Theta = \hat{\Omega}' \cdot \hat{\Omega}$. Integration of Equation (2) over all scattering directions yields

$$\begin{aligned}\beta(z) &= \frac{1}{I^i}\frac{d}{dz}\int_{4\pi}\left(\frac{dI^\beta}{d\omega}\right)d\omega = \frac{1}{I^i}\left(\frac{dI^\beta}{dz}\right)\\ &= \int_{4\pi}\text{vsf}(z, \hat{\Omega}', \hat{\Omega})d\omega = \int_0^{2\pi}\int_0^{\pi}\text{vsf}(z, \cos\Theta, \phi)\sin\Theta d\Theta d\phi \qquad [\text{m}^{-1}]\end{aligned} \tag{3}$$

where Θ and ϕ are respectively the polar angle and the azimuth angle in a spherical coordinate system in which the polar axis is along $\hat{\Omega}'$. As indicated in Equation (3), the volume scattering function [$\text{vsf}(z, \cos\Theta, \phi)$] is generally a function of both Θ and ϕ, but for randomly oriented scatterers one may assume that the scattering potential is spherically symmetric implying that there is no dependence on azimuth, so that $\text{vsf} = \text{vsf}(z, \cos\Theta)$. Then one finds, with $x = \cos\Theta$

$$\beta(z) = 2\pi\int_0^{\pi}\text{vsf}(z, \cos\Theta)\sin\Theta d\Theta = 2\pi\int_{-1}^{1}\text{vsf}(z, x)dx \qquad [\text{m}^{-1}]. \tag{4}$$

A normalized vsf, denoted by $p(z, \cos\Theta)$ and referred to hereafter as the *scattering phase function*, may be defined as follows

$$p(z, \cos\Theta) = 4\pi\frac{\text{vsf}(z, \cos\Theta)}{\int_{4\pi}\text{vsf}(z, \cos\Theta)d\omega} = \frac{\text{vsf}(z, x)}{\frac{1}{2}\int_{-1}^{1}\text{vsf}(z, x)dx} \tag{5}$$

so that

$$\frac{1}{4\pi}\int_{4\pi}p(z, \cos\Theta)d\omega = \frac{1}{2}\int_{-1}^{1}p(z, x)dx = 1. \tag{6}$$

The scattering phase function has the following physical interpretation. Given that a scattering event has occurred, $p(z, \cos\Theta)d\omega/4\pi$ is the probability that a light beam traveling in the direction $\hat{\Omega}'$ is scattered into a cone of solid angle $d\omega$ around the direction $\hat{\Omega}$.

The scattering phase function [$p(z, \cos\Theta)$] describes the angular distribution of the scattered light, while the scattering coefficient $\beta(z)$ describes the total amount of scattered light integrated over all scattering directions. A convenient measure of the "shape" of the scattering phase function is the average over all scattering directions (weighted by $p(z, \cos\Theta)$) of the cosine of the scattering angle Θ, i.e.,

$$\begin{aligned}g(z) = \langle\cos\Theta\rangle &= \frac{1}{4\pi}\int_{4\pi}p(z, \cos\Theta)\,\cos\Theta\,d\omega\\ &= \frac{1}{2}\int_0^{\pi}p(z, \cos\Theta)\,\cos\Theta\,\sin\Theta d\Theta = \frac{1}{2}\int_{-1}^{1}p(z, x)x dx.\end{aligned} \tag{7}$$

The average cosine $g(z)$ is called the *asymmetry factor* of the scattering phase function. Equation (7) yields complete forward scattering if $g = 1$, complete backward scattering if $g = -1$, and $g = 0$ if $p(z, \cos\Theta)$ is symmetric about $\Theta = 90°$. Thus, isotropic scattering also gives $g = 0$. Similarly, the probability of scattering into the backward hemisphere, is given by the backscattering ratio (or backscatter fraction) b, defined as

$$b(z) = \frac{1}{2}\int_{\pi/2}^{\pi} p(z, \cos\Theta)\ \sin\Theta\, d\Theta = \frac{1}{2}\int_0^1 p(z, -x)dx. \tag{8}$$

The scattering phase function $p(z, \cos\Theta)$ depends on the refractive index as well as the size and shape of the scattering particles, and will thus depend on the physical situation and the practical application of interest. Two different scattering phase functions, which are useful in practical applications, are discussed below.

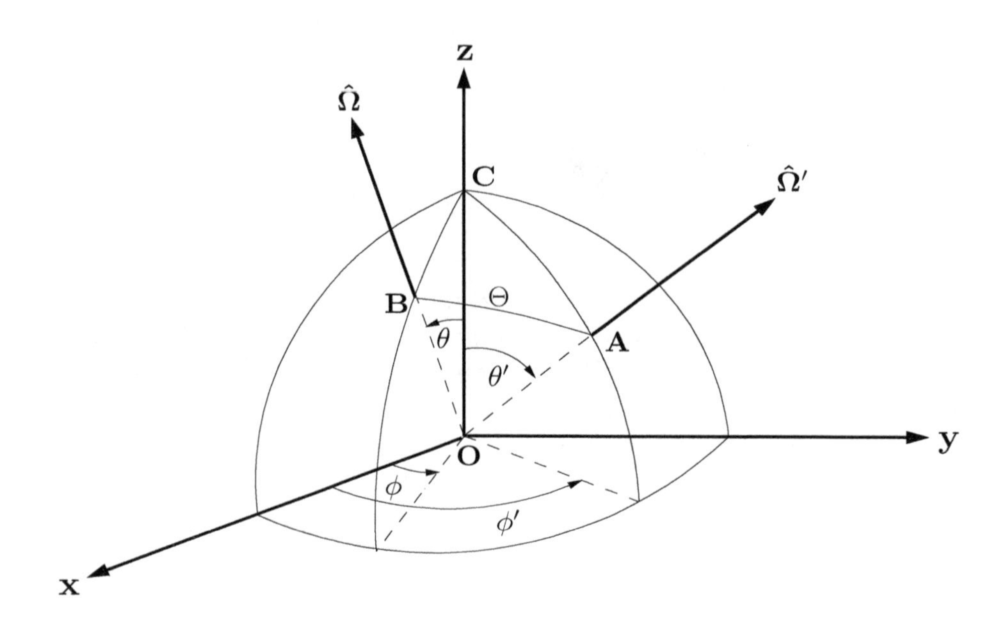

Figure 1. Coordinate system for scattering by a volume element at **O**. The points **C**, **A** and **B** are located on the unit sphere. The incident light beam with Stokes vector $\mathbf{I}_S^{\mathrm{inc}}$ is in direction $\mathbf{OA}(\theta', \phi')$ with unit vector $\hat{\mathbf{\Omega}}'$, the scattered beam with Stokes vector $\mathbf{I}_S^{\mathrm{sca}}$ is in direction $\mathbf{OB}(\theta, \phi)$ with unit vector $\hat{\mathbf{\Omega}}$ [5].

3.1.1. Rayleigh Scattering Phase Function

When the size d of the scatterers is small compared with the wavelength of light ($d < \frac{1}{10}\lambda$), the Rayleigh scattering phase function gives a good description of the angular distribution of the scattered light. The Rayleigh scattering phase function for unpolarized light is given by

$$p(\cos\Theta) = \frac{3}{3+f}(1 + f\cos^2\Theta) \tag{9}$$

where the parameter $f = \frac{1-\rho}{1+\rho}$, and ρ is the depolarization ratio, attributed to the anisotropy of the scatterer (molecule) [6–8]. Originally this scattering phase function was derived for light radiated by an electric dipole [9]. Since the Rayleigh scattering phase function is symmetric about $\Theta = 90°$, the asymmetry factor is $g = 0$. If the Rayleigh scattering phase function is expanded in Legendre polynomials, the expansion coefficients χ_ℓ [see Equation (117) below] are simply given by $\chi_0 = 1$, $\chi_1 = 0$, $\chi_2 = \frac{2f}{5(3+f)}$, and $\chi_\ell = 0$ for $\ell > 2$.

Using $\rho = 0.0286$ for air at 500 nm [10], we get $f = \frac{1-\rho}{1+\rho} = 0.944$, and using $\rho = 0.039$ [11,12] for water, we get $f = \frac{1-\rho}{1+\rho} = 0.925$. Hence, for Rayleigh scattering, the scattering phase function moments become:

- $\chi_0 = 1, \chi_1 = 0, \chi_2 = 0.0957$ and $\chi_\ell = 0$ for $\ell > 2$ for air, and
- $\chi_0 = 1, \chi_1 = 0, \chi_2 = 0.0943$, and $\chi_\ell = 0$ for $\ell > 2$ for water.

3.1.2. Henyey-Greenstein Scattering Phase Function

Henyey and Greenstein [13] proposed the one-parameter scattering phase function given by (oppressing the dependence on the position z)

$$p(\cos\Theta) = \frac{1-g^2}{(1+g^2-2g\cos\Theta)^{3/2}} \tag{10}$$

where the parameter g is the asymmetry factor defined in Equation (7). The Henyey–Greenstein (HG) scattering phase function has no physical basis, but is very useful for describing a highly scattering medium, such as turbid water or sea ice, for which the actual scattering phase function is unknown. The HG scattering phase function is convenient for Monte Carlo simulations and other numerical calculations because it has an analytical form. In deterministic plane-parallel RT models it is also very convenient because the addition theorem of spherical harmonics can be used to expand the scattering phase function in a series of Legendre polynomials [3,14], as reviewed in Section 4.1.1. For the HG scattering phase function, the expansion coefficients χ_ℓ in this series [see Equation (117) below] are simply given by $\chi_\ell = g^\ell$, where $g = \chi_1$ is the asymmetry factor defined in Equation (7). The HG scattering phase function is useful for scatterers with sizes comparable to or larger than the wavelength of light. Although the HG scattering phase function is easy to use, it is not as realistic as the Fournier-Forand scattering phase function discussed in Section 3.6.1.

3.2. Scattering Phase Matrix

The theoretical development of vector radiative transfer theory may start with the Stokes vector representation $\mathbf{I} = [I_\parallel, I_\perp, U, V]^T$, where the superscript ^{T} denotes the transpose. In terms of the complex transverse electric field components of the radiation field $E_\parallel = |E_\parallel|e^{-i\epsilon_1}$ and $E_\perp = |E_\perp|e^{-i\epsilon_2}$, these Stokes vector components are given by:

$$\begin{aligned} I_\parallel &= E_\parallel E_\parallel^* \\ I_\perp &= E_\perp E_\perp^* \\ U &= 2|E_\parallel||E_\perp|\cos\delta \\ V &= 2|E_\parallel||E_\perp|\sin\delta \end{aligned} \tag{11}$$

where $\delta = \epsilon_1 - \epsilon_2$. The connection between this Stokes vector representation, $\mathbf{I} = [I_\parallel, I_\perp, U, V]^T$, and the more commonly used representation $\mathbf{I}_S = [I, Q, U, V]^T$, where $I = I_\parallel + I_\perp$ and $Q = I_\parallel - I_\perp$, is given by:

$$\mathbf{I}_S = \mathbf{D}\mathbf{I} \tag{12}$$

where

$$\mathbf{D} = \begin{pmatrix} 1 & 1 & 0 & 0 \\ 1 & -1 & 0 & 0 \\ 0 & 0 & 1 & 0 \\ 0 & 0 & 0 & 1 \end{pmatrix}, \quad \mathbf{D}^{-1} = \frac{1}{2}\begin{pmatrix} 1 & 1 & 0 & 0 \\ 1 & -1 & 0 & 0 \\ 0 & 0 & 2 & 0 \\ 0 & 0 & 0 & 2 \end{pmatrix}. \tag{13}$$

The degree of polarization is defined as

$$p = [Q^2 + U^2 + V^2]^{1/2}/I \tag{14}$$

so that $0 \leq p \leq 1$, where $p = 1$ corresponds to completely polarized light and $p = 0$ to natural (unpolarized) light. The degree of circular polarization is defined as

$$p_c = V/I \tag{15}$$

the degree of linear polarization as

$$p_l = [Q^2 + U^2]^{1/2}/I \tag{16}$$

and alternatively, when $U = 0$ as

$$p_l = \frac{|Q|}{I} = \frac{|I_\perp - I_\parallel|}{I_\perp + I_\parallel}. \tag{17}$$

The transverse electric field vector $[E_\parallel, E_\perp]^T$ of the scattered field can be obtained in terms of the transverse field vector $[E_{\parallel 0}, E_{\perp 0}]^T$ of the incident field by a linear transformation:

$$\begin{pmatrix} E_\parallel \\ E_\perp \end{pmatrix} = \mathbf{A} \begin{pmatrix} E_{\parallel 0} \\ E_{\perp 0} \end{pmatrix}$$

where $\mathbf{A}$ is a 2×2 matrix, referred to as the amplitude scattering matrix, which includes a $1/r$ dependence of the scattered field. The corresponding linear transformation connecting the Stokes vectors of the incident and scattered fields in the scattering plane is called the Mueller matrix (in the case of a single scattering event). For scattering by a small volume containing an ensemble of particles, the ensemble-averaged Mueller matrix is referred to as the Stokes scattering matrix **F**. Finally, when transforming from the scattering plane to a fixed laboratory frame, the corresponding matrix is referred to as the scattering phase matrix **P**.

3.2.1. Stokes Vector Representation $\mathbf{I}_S = [I, Q, U, V]^T$

The scattering geometry is illustrated in Figure 1. The plane **AOB**, defined as the scattering plane, is spanned by the directions of propagation of the incident parallel beam with Stokes vector $\mathbf{I}_S^{\text{inc}}$ and the scattered parallel beam with Stokes vector $\mathbf{I}_S^{\text{sca}}$. Here the subscript S pertains to the Stokes vector representation $\mathbf{I}_S = [I, Q, U, V]^T$. The scattered radiation, represented by the Stokes vector $\mathbf{I}_S^{\text{sca}}$, is related to the incident radiation, represented by the Stokes vector $\mathbf{I}_S^{\text{inc}}$, by a 4×4 scattering matrix [see Equations (18) and (19) below] and two rotations are required to properly connect the two Stokes vectors as explained below. We describe the Stokes vector of the incident beam in terms of two unit vectors $\hat{\ell}'$ and $\hat{\mathbf{r}}'$, which are normal to one another and to the unit vector $\hat{\mathbf{\Omega}}' = \hat{\mathbf{r}}' \times \hat{\ell}'$ along the propagation direction of the incident beam. Similarly, we describe the Stokes vector of the scattered beam in terms of two unit vectors $\hat{\ell}$ and $\hat{\mathbf{r}}$, which are normal to one another and to the unit vector $\hat{\mathbf{\Omega}} = \hat{\mathbf{r}} \times \hat{\ell}$ along the propagation direction of the scattered beam. The unit vector $\hat{\ell}'$ is along the direction of $\mathbf{E}_\parallel'$ of the incident beam and lies in the meridian plane of that beam, which is defined as the plane **OAC** in Figure 1. Similarly, the unit vector $\hat{\ell}$ is along the direction of $\mathbf{E}_\parallel$ of the scattered beam and lies in the meridian plane of that beam, which is defined as the plane **OBC** in Figure 1. For the incident beam, the unit vector $\hat{\ell}'$, may be defined to be tangent at the point **A** to the unit circle passing through the points **A** and **C** in Figure 1. For the scattered beam, the unit vector $\hat{\ell}$ may be defined to be tangent at the point **B** to the unit circle passing through the points **B** and **C** in Figure 1. For either beam, its meridian plane acts as a plane of reference for the Stokes vector, so that the point **A** in Figure 1 is the starting point for the unit vector $\hat{\mathbf{\Omega}}' = \hat{\mathbf{r}}' \times \hat{\ell}'$ along the direction of propagation of the incident beam, and the point **B** in Figure 1 is the starting point for the unit vector $\hat{\mathbf{\Omega}} = \hat{\mathbf{r}} \times \hat{\ell}$ along the direction of propagation of the scattered beam.

As explained above, the Mueller matrix describes scattering by a single particle, and for scattering by a small volume of particles, the ensemble-averaged Mueller matrix is referred to as the Stokes scattering matrix $\mathbf{F}_S$. If any of the following conditions are fulfilled [15] (i) each particle in the volume element has a plane of symmetry, and the particles are randomly oriented, (ii) each volume element contains an equal number of particles and their mirror particles in random orientation, (iii) the particles are much smaller than the wavelength of the incident light, then the Stokes scattering matrix in the $\mathbf{I}_S = [I, Q, U, V]^T$ representation has the following form

$$\mathbf{F}_S(\Theta) = \begin{bmatrix} a_1(\Theta) & b_1(\Theta) & 0 & 0 \\ b_1(\Theta) & a_2(\Theta) & 0 & 0 \\ 0 & 0 & a_3(\Theta) & b_2(\Theta) \\ 0 & 0 & -b_2(\Theta) & a_4(\Theta) \end{bmatrix}. \tag{18}$$

Each of the six independent matrix elements in Equation (18) depends on the scattering angle Θ, and will in general also depend on the position in the medium. For spherical particles, the matrix in Equation (18) simplifies, since $a_1 = a_2$ and $a_3 = a_4$, so that only four independent elements remain.

As already mentioned, two rotations are required to connect the Stokes vector of the scattered radiation to that of the incident radiation. As illustrated in Figure 1, the first rotation is from the meridian plane **OAC**, associated with the Stokes vector $\mathbf{I}_S^{\mathrm{inc}}$, into the scattering plane **OAB**, whereas the second rotation is from the scattering plane **OAB** into the meridian plane **OBC**, associated with the Stokes vector $\mathbf{I}_S^{\mathrm{sca}}$. Hence, the Stokes vector for the scattered radiation is given by [16]

$$\mathbf{I}_S^{\mathrm{sca}} = \mathbf{R}_S(\pi - i_2)\mathbf{F}_S(\Theta)\mathbf{R}_S(-i_1)\mathbf{I}_S^{\mathrm{inc}} \equiv \mathbf{P}_S(\Theta)\mathbf{I}_S^{\mathrm{inc}}. \tag{19}$$

The matrix $\mathbf{R}_S$ is called the Stokes rotation matrix. It represents a rotation in the clockwise direction with respect to an observer looking into the direction of propagation, and can be written as $(0 \leq \omega \leq 2\pi)$

$$\mathbf{R}_S(\omega) = \begin{bmatrix} 1 & 0 & 0 & 0 \\ 0 & \cos(2\omega) & -\sin(2\omega) & 0 \\ 0 & \sin(2\omega) & \cos(2\omega) & 0 \\ 0 & 0 & 0 & 1 \end{bmatrix}. \tag{20}$$

Hence, according to Equation (19), the scattering phase matrix, which connects the Stokes vector of the scattered radiation to that of the incident radiation, is obtained from the Stokes scattering matrix $\mathbf{F}_S(\Theta)$ in Equation (18) by

$$\mathbf{P}_S(\theta', \phi'; \theta, \phi) = \mathbf{R}_S(\pi - i_2)\mathbf{F}_S(\Theta)\mathbf{R}_S(-i_1) = \mathbf{R}_S(-i_2)\mathbf{F}_S(\Theta)\mathbf{R}_S(-i_1) \tag{21}$$

where $\mathbf{R}_S$ is the rotation matrix described in Equation (20) [16], and $\mathbf{R}_S(\pi - i_2) = \mathbf{R}_S(-i_2)$ since the rotation matrix is periodic with a period π.

According to Equation (19) (see also Figure 1), the Stokes vector $\mathbf{I}_S^{\mathrm{inc}}$ of the incident parallel beam must be multiplied by the rotation matrix $\mathbf{R}_S(-i_1)$ before it is multiplied by the Stokes scattering matrix $\mathbf{F}_S(\Theta)$, whereafter it must be multiplied by the rotation matrix $\mathbf{R}_S(\pi - i_2)$. These matrix multiplications are carried out explicitly in some radiative transfer (RT) models including Monte Carlo simulations, while they are implicitly taken care of in other RT models such as the adding-doubling method [17] and the discrete ordinate method [18,19] which use the expansion of the scattering phase matrix in generalized spherical functions [20,21] as discussed in Section 3.2.3.

Carrying out the matrix multiplications in Equation (21) one finds:

$$\mathbf{P}_S(\Theta) = \begin{bmatrix} a_1 & b_1C_1 & -b_1S_1 & 0 \\ b_1C_2 & C_2a_2C_1 - S_2a_3S_1 & -C_2a_2S_1 - S_2a_3C_1 & -b_2S_2 \\ b_1S_2 & S_2a_2C_1 + C_2a_3S_1 & -S_2a_2S_1 + C_2a_3C_1 & -b_2C_2 \\ 0 & -b_2S_1 & -b_2C_1 & a_4 \end{bmatrix} \tag{22}$$

where $a_i = a_i(\Theta), i = 1, \ldots, 4$, $b_i = b_i(\Theta), i = 1, 2$, and

$$C_1 = \cos 2i_1, \quad C_2 = \cos 2i_2 \tag{23}$$

$$S_1 = \sin 2i_1, \quad S_2 = \sin 2i_2. \tag{24}$$

A comparison of Equations (18) and (22) shows that only the corner elements of $\mathbf{F}_S(\Theta)$ remain unchanged by the rotations of the reference planes. The (1,1)-element of the scattering phase matrix $\mathbf{P}_S(\Theta)$ (and of the Stokes scattering matrix $\mathbf{F}_S(\Theta)$) is the scattering phase function. Since also the (4,4)-element of the scattering phase matrix remains unchanged by the rotations, the state of circular polarization of the incident light does not affect the intensity of the scattered radiation after one scattering event.

To compute $\mathbf{P}_S(\theta', \phi'; \theta, \phi)$ given by Equation (21) we must relate the angles θ', ϕ', θ, and ϕ on the left side with the angles i_1, i_2, and Θ on the right side. Using spherical geometry, we may apply the cosine rule for Θ, θ, and θ' successively, in Figure 1, to obtain ($u = \cos\theta, u' = \cos\theta'$) [5]

$$\cos\Theta = uu' + (1-u^2)^{1/2}(1-u'^2)^{1/2}\cos(\phi' - \phi) \tag{25}$$

$$\cos i_1 = \frac{-u + u'\cos\Theta}{(1-u'^2)^{1/2}(1-\cos^2\Theta)^{1/2}} \tag{26}$$

$$\cos i_2 = \frac{-u' + u\cos\Theta}{(1-u^2)^{1/2}(1-\cos^2\Theta)^{1/2}}. \tag{27}$$

The trigonometric functions for the double angles can be obtained by using

$$\cos 2i = 2\cos^2 i - 1 \tag{28}$$

and

$$\sin 2i = 2\sin i \cos i \tag{29}$$

or

$$\sin 2i = \begin{cases} 2(1-\cos^2 i)^{1/2}\cos i & \text{if} \quad 0 < \phi' - \phi < \pi \\ -2(1-\cos^2 i)^{1/2}\cos i & \text{if} \quad \pi < \phi' - \phi < 2\pi \end{cases} \tag{30}$$

where i is i_1 or i_2. We now have all the information needed to compute the scattering phase matrix [see Equation (21)] as a function of the three variables u, u' and $\phi' - \phi$:

$$\mathbf{P}_S(u', u, \phi' - \phi) = \mathbf{R}_S(-i_2)\mathbf{F}_S(\Theta)\mathbf{R}_S(-i_1).$$

If there is no difference in azimuth (i.e. $\phi' - \phi = 0$), then the meridian planes of the incident and scattered beams in Figure 1 coincide with the scattering plane. Hence there is no need to rotate the reference planes ($\mathbf{R}(-i_2)$ and $\mathbf{R}(-i_1)$ both reduce to the identity matrix), so that

$$\mathbf{P}_S(u', u, 0) = \mathbf{P}_S(u', u, \pi) = \mathbf{F}_S(\Theta). \tag{31}$$

It follows from Equation (25) that the phase matrix is invariant to three basic changes in the polar angles u' and u and azimuthal angles ϕ' and ϕ which leave the scattering angle unaltered: (i) changing the signs of u and u' simultaneously: $\mathbf{P}_S(-u', -u, \phi' - \phi) = \mathbf{P}_S(u', u, \phi' - \phi)$, (ii) interchange of u and

u': $\mathbf{P}_S(u', u, \phi' - \phi) = \mathbf{P}_S(u, u', \phi' - \phi)$ (iii) interchange of ϕ and ϕ': $\mathbf{P}_S(u', u, \phi' - \phi) = \mathbf{P}_S(u', u, \phi - \phi')$. Also, if the b_2-element in Equation (22) is zero, the circular polarization component decouples from the other three components. Then, the Stokes parameter V is scattered independently of the others, according to the phase function $a_4(\Theta)$, and the remaining part of the scattering phase matrix referring to I, Q, and U becomes a 3×3 matrix:

$$\mathbf{P}_S(\Theta) = \begin{bmatrix} a_1 & b_1 C_1 & -b_1 S_1 \\ b_1 C_2 & C_2 a_2 C_1 - S_2 a_3 S_1 & -C_2 a_2 S_1 - S_2 a_3 C_1 \\ b_1 S_2 & S_2 a_2 C_1 + C_2 a_3 S_1 & -S_2 a_2 S_1 + C_2 a_3 C_1 \end{bmatrix}. \quad (32)$$

Finally, in a plane-parallel or slab geometry, there is no azimuth-dependence for light beams traveling in directions perpendicular to the slab (either up or down). Thus, if either the incident or the scattered beam travels in a perpendicular direction, we may use the meridian plane of the other beam as a reference plane for both beams. Since this plane coincides with the scattering plane, Equation (31) applies in this situation too.

3.2.2. Stokes Vector Representation $\mathbf{I} = [I_{||}, I_\perp, U, V]^T$

The Stokes vector $\mathbf{I} = [I_{||}, I_\perp, U, V]^T$ is related to $\mathbf{I}_S = [I, Q, U, V]^T$ by

$$\mathbf{I}_S = \mathbf{D}\mathbf{I} \quad (33)$$

where $\mathbf{D}$ is given by Equation (13), so that $I = I_{||} + I_\perp$, and $Q = I_{||} - I_\perp$. Denoting the Stokes vector obtained after a rotation by

$$\mathbf{I}'_S = \mathbf{R}_S(\omega)\mathbf{I}_S \quad (34)$$

we find

$$\mathbf{I}' = \mathbf{D}^{-1}\mathbf{I}'_S = \mathbf{D}^{-1}\mathbf{R}_S(\omega)\mathbf{I}_S = \mathbf{D}^{-1}\mathbf{R}_S(\omega)\mathbf{D}\,\mathbf{I} = \mathbf{R}(\omega)\mathbf{I}. \quad (35)$$

Hence, the rotation matrix for the Stokes vector in the representation $\mathbf{I} = [I_{||}, I_\perp, U, V]^T$ becomes:

$$\mathbf{R}(\omega) = \mathbf{D}^{-1}\mathbf{R}_S(\omega)\mathbf{D} = \begin{bmatrix} \cos^2\omega & \sin^2\omega & -\frac{1}{2}\sin(2\omega) & 0 \\ \sin^2\omega & \cos^2\omega & \frac{1}{2}\sin(2\omega) & 0 \\ \sin(2\omega) & -\sin(2\omega) & \cos(2\omega) & 0 \\ 0 & 0 & 0 & 1 \end{bmatrix}. \quad (36)$$

The scattering phase matrix $\mathbf{P}(\Theta)$ in the Stokes vector representation $\mathbf{I} = [I_{||}, I_\perp, U, V]^T$ is related to scattering phase matrix $\mathbf{P}_S(\Theta)$ in the Stokes vector representation $\mathbf{I}_S = [I, Q, U, V]^T$ by

$$\mathbf{P}(\Theta) = \mathbf{D}^{-1}\mathbf{P}_S(\Theta)\mathbf{D}. \quad (37)$$

Similarly, the Stokes scattering matrix $\mathbf{F}(\Theta)$ associated with the Stokes vector representation $\mathbf{I} = [I_{||}, I_\perp, U, V]^T$ is related to the Stokes scattering matrix $\mathbf{F}_S(\Theta)$ in Equation (18) by

$$\mathbf{F}(\Theta) = \mathbf{D}^{-1}\mathbf{F}_S(\Theta)\mathbf{D} = \begin{pmatrix} \frac{1}{2}(a_1 + a_2 + 2b_1) & \frac{1}{2}(a_1 - a_2) & 0 & 0 \\ \frac{1}{2}(a_1 - a_2) & \frac{1}{2}(a_1 + a_2 - 2b_1) & 0 & 0 \\ 0 & 0 & a_3 & b_2 \\ 0 & 0 & -b_2 & a_4 \end{pmatrix}. \quad (38)$$

For Rayleigh scattering with parameter $f = \frac{1-\rho}{1+\rho}$, where ρ is the depolarization factor defined in Equation (43), the Stokes scattering matrix in the Stokes vector representation $\mathbf{I}_S = [I, Q, U, V]^T$ is given by [16,22]

$$\mathbf{F}_S(\Theta) = \frac{3}{3+f}\begin{bmatrix} 1+f\cos^2\Theta & -f\sin^2\Theta & 0 & 0 \\ -f\sin^2\Theta & f(1+\cos^2\Theta) & 0 & 0 \\ 0 & 0 & 2f\cos\Theta & 0 \\ 0 & 0 & 0 & (3f-1)\cos\Theta \end{bmatrix}. \tag{39}$$

For the first scattering event of unpolarized light, only the (1,1)-element of Equation (39) matters, and leads to the scattering phase function given by Equation (9).

In the Stokes vector representation $\mathbf{I} = [I_{\|}, I_{\perp}, U, V]^T$, the corresponding Stokes scattering matrix for Rayleigh scattering becomes (using Equations (38) and (39) [16]):

$$\mathbf{F}(\Theta) = \frac{3}{2(1+2\zeta)}\begin{pmatrix} \cos^2\Theta + \zeta\sin^2\Theta & \zeta & 0 & 0 \\ \zeta & 1 & 0 & 0 \\ 0 & 0 & (1-\zeta)\cos\Theta & 0 \\ 0 & 0 & 0 & (1-3\zeta)\cos\Theta \end{pmatrix} \tag{40}$$

where $\zeta = \rho/(2-\rho) = \frac{1-f}{1+3f}$.

From Equation (40) we see that for an incident beam of natural unpolarized light given by $\mathbf{I}^{\mathrm{inc}} = [I_{\|}^{\mathrm{inc}}, I_{\perp}^{\mathrm{inc}}, U^{\mathrm{inc}}, V^{\mathrm{inc}}]^T = [\frac{1}{2}I^{\mathrm{inc}}, \frac{1}{2}I^{\mathrm{inc}}, 0, 0]^T$, the scattered intensities in the plane parallel and perpendicular to the scattering plane are obtained by carrying out the multiplication $\mathbf{I}^{\mathrm{sca}} = \mathbf{F}(\Theta)\mathbf{I}^{\mathrm{inc}}$:

$$I_{\|}^{\mathrm{sca}} \propto \frac{3}{4(1+2\zeta)}[2\zeta + (1-\zeta)\cos^2\Theta]I^{\mathrm{inc}} \tag{41}$$

$$I_{\perp}^{\mathrm{sca}} \propto \frac{3}{4(1+2\zeta)}[(1+\zeta)]I^{\mathrm{inc}}. \tag{42}$$

Thus, for unpolarized incident light, the scattered light at right angles ($\Theta = 90°$) to the direction of incidence defines the depolarization ratio:

$$\rho \equiv \left(\frac{I_{\|}^{\mathrm{sca}}}{I_{\perp}^{\mathrm{sca}}}\right)_{\Theta=90°} = \frac{2\zeta}{1+\zeta} \tag{43}$$

whereas the degree of linear polarization becomes [Equation (17)]:

$$p_l = \frac{I_{\perp} - I_{\|}}{I_{\perp} + I_{\|}} = \frac{(1-\zeta)(1-\cos^2\Theta)}{1+3\zeta+(1-\zeta)\cos^2\Theta} \to \frac{1-\zeta}{1+3\zeta} = \frac{1-\rho}{1+\rho} = f \quad \text{as} \quad \Theta \to 90°.$$

3.2.3. Generalized Spherical Functions—The "Greek Constants"

For unpolarized radiation, only the $a_1(\Theta)$ element of the Stokes scattering matrix Equation (18) is relevant, and this element is the scattering phase function given by Equation (5) in general, and by Equation (9) for Rayleigh scattering. As discussed below, the scattering phase function can be expanded in Legendre polynomials (see Equation (116)), enabling one to express it as a Fourier cosine series (see Equation (115)).

In a similar manner, the scattering phase matrix can be expanded in generalized spherical functions. In the Stokes vector representation $\mathbf{I}_S = [I, Q, U, V]^T$, the scattering phase matrix is $\mathbf{P}_S(\Theta) = \mathbf{P}_S(u', u; \phi' - \phi)$ with $u = \cos\theta$, θ being the polar angle after scattering, and $u' = \cos\theta'$, θ' being the polar angle prior to scattering. Similarly, ϕ and ϕ' are the azimuth angles after and prior to scattering, respectively. To accomplish the expansion in generalized spherical functions, the scattering phase matrix is first expanded in a $(M+1)$-term Fourier series in the azimuth angle difference $(\Delta\phi = \phi' - \phi)$:

$$\mathbf{P}_S(u',u;\Delta\phi) = \sum_{m=0}^{M} \{\mathbf{P}_c^m(u',u)\cos m(\Delta\phi) + \mathbf{P}_s^m(u',u)\sin m(\Delta\phi)\} \tag{44}$$

where $\mathbf{P}_c^m(u',u)$ and $\mathbf{P}_s^m(u',u)$ are the coefficient matrices of the cosine and sine terms, respectively, of the Fourier series.

An addition theorem for the generalized spherical functions can be used to express the Fourier expansion coefficient matrices directly in terms of the expansion coefficients of the Stokes scattering matrix $\mathbf{F}_S(\Theta)$ [see Equation (18)] as follows [20,21,23]:

$$\mathbf{P}_c^m(u',u) = \mathbf{A}^m(u',u) + \mathbf{\Delta}_{3,4}\mathbf{A}^m(u',u)\mathbf{\Delta}_{3,4} \tag{45}$$

$$\mathbf{P}_s^m(u',u) = \mathbf{A}^m(u',u)\mathbf{\Delta}_{3,4} - \mathbf{\Delta}_{3,4}\mathbf{A}^m(u',u) \tag{46}$$

where $\mathbf{\Delta}_3 = \text{diag}(1,1,-1,1)$. The matrix $\mathbf{A}^m(u',u)$ is given by:

$$\mathbf{A}^m(u',u) = \sum_{\ell=m}^{M} \mathbf{P}_\ell^m(u)\mathbf{\Lambda}_\ell\mathbf{P}_\ell^m(u') \tag{47}$$

where

$$\mathbf{\Lambda}_\ell = \begin{pmatrix} \alpha_{1,\ell} & \beta_{1,\ell} & 0 & 0 \\ \beta_{1,\ell} & \alpha_{2,\ell} & 0 & 0 \\ 0 & 0 & \alpha_{3,\ell} & \beta_{2,\ell} \\ 0 & 0 & -\beta_{2,\ell} & \alpha_{4,\ell} \end{pmatrix} \tag{48}$$

and

$$a_1(\Theta) = \sum_{\ell=0}^{M} \alpha_{1,\ell} P_\ell^{0,0}(\cos\Theta) \tag{49}$$

$$a_2(\Theta) + a_3(\Theta) = \sum_{\ell=2}^{M} (\alpha_{2,\ell} + \alpha_{3,\ell}) P_\ell^{2,2}(\cos\Theta) \tag{50}$$

$$a_2(\Theta) - a_3(\Theta) = \sum_{\ell=2}^{M} (\alpha_{2,\ell} - \alpha_{3,\ell}) P_\ell^{2,-2}(\cos\Theta) \tag{51}$$

$$a_4(\Theta) = \sum_{\ell=0}^{M} \alpha_{4,\ell} P_\ell^{0,0}(\cos\Theta) \tag{52}$$

$$b_1(\Theta) = \sum_{\ell=2}^{M} \beta_{1,\ell} P_\ell^{0,2}(\cos\Theta) \tag{53}$$

$$b_2(\Theta) = \sum_{\ell=2}^{M} \beta_{2,\ell} P_\ell^{0,2}(\cos\Theta). \tag{54}$$

Here the so-called "Greek constants" $\alpha_{j,\ell}$ and $\beta_{j,\ell}$ are expansion coefficients, and $a_j(\Theta)$ and $b_j(\Theta)$ are the elements of the Stokes scattering matrix $\mathbf{F}_S(\Theta)$ in Equation (18). An example of Greek constants for Rayleigh scattering is provided in Table 1 (see [24]):

Table 1. Expansion Coefficients for Rayleigh Scattering.

ℓ	$\alpha_{1,\ell}$	$\alpha_{2,\ell}$	$\alpha_{3,\ell}$	$\alpha_{4,\ell}$	$\beta_{1,\ell}$	$\beta_{2,\ell}$
0	1	0	0	0	0	0
1	0	0	0	$3d/2$	0	0
2	$c/2$	$3c$	0	0	$\sqrt{3/2}c$	0

where

$$c = \frac{2(1-\rho)}{2+\rho} \qquad d = \frac{2(1-2\rho)}{2+\rho}$$

and ρ is the depolarization ratio given by Equation (43).

The matrix $\mathbf{P}_\ell^m(u)$ occurring in Equation (47) is defined as:

$$\mathbf{P}_\ell^m(u) = \begin{pmatrix} P_\ell^{m,0}(u) & 0 & 0 & 0 \\ 0 & P_\ell^{m,+}(u) & P_\ell^{m,-}(u) & 0 \\ 0 & P_\ell^{m,-}(u) & P_\ell^{m,+}(u) & 0 \\ 0 & 0 & 0 & P_\ell^{m,0}(u) \end{pmatrix} \tag{55}$$

where

$$P_\ell^{m,\pm}(u) = \frac{1}{2}[P_\ell^{m,-2}(u) \pm P_\ell^{m,2}(u)] \tag{56}$$

and the functions $P_\ell^{m,0}(u)$ and $P_\ell^{m,\pm 2}(u)$ are the generalized spherical functions. More details about these functions and how they are computed are available in Appendix B of the book by Hovenier et al. [5].

We note that in the scalar (unpolarized) case all the components of the Stokes scattering matrix $\mathbf{F}_S(\Theta)$ [see Equation (18)] are zero except for $a_1(\Theta)$, and:

$$a_1(\Theta) = \sum_{\ell=0}^{M} \alpha_{1,\ell}(\tau) P_\ell^{0,0}(\cos\Theta) \equiv p(\tau, \cos\Theta) \approx \sum_{\ell=0}^{M} (2\ell+1)\chi_\ell(\tau) P_\ell(\cos\Theta) \tag{57}$$

since $P_\ell^{0,0}(\cos\Theta) \equiv P_\ell(\cos\Theta)$, where $P_\ell(\cos\Theta)$ is the Legendre polynomial of order ℓ, and $\alpha_{1,\ell}(\tau) \equiv (2\ell+1)\chi_\ell(\tau)$. Note also that the expansion coefficients given above are for the scattering phase matrix $\mathbf{P}_S(\Theta)$, which relates the incident and scattered Stokes vectors in the representation $\mathbf{I}_S = [I, Q, U, V]^T$.

3.3. *IOPs for a Size Distribution of Particles*

Particles encountered in nature consist of a variety of chemical compositions, sizes, and shapes. The chemical composition determines the refractive index of the particle, and unless the composition is the same throughout the particle, the refractive index will depend on location inside the particle. The computation of IOPs for such a collection of particles requires solutions of Maxwell's equations for electromagnetic radiation interacting with an inhomogeneous, non-spherical particle of a given size. Then one needs to integrate over size and shape for particles of a given chemical composition, and finally average over the particle composition. To avoid having to deal with this complexity it is frequently assumed that the particles are homogeneous with a constant refractive index, and that the shape can be taken to be spherical. Even with these assumptions, one still needs to deal with the variety of particles sizes encountered in nature.

For a spherical particle with a specified radius and refractive index, Mie theory (and its numerical implementation) may be used to generate IOPs for a single particle. Thus, if we have computed the IOPs for a single spherical particle with specified refractive index and a given size, we may compute the absorption and scattering coefficients and the scattering phase function for a polydispersion of particles by integrating over the particle size distribution (PSD):

$$\alpha_p(\lambda) = \int_{r_{min}}^{r_{max}} \alpha_n(\lambda, r) n(r) dr = \int_{r_{min}}^{r_{max}} \pi r^2 Q'_\alpha(\lambda, r) n(r) dr \tag{58}$$

$$\beta_p(\lambda) = \int_{r_{min}}^{r_{max}} \beta_n(\lambda, r) n(r) dr = \int_{r_{min}}^{r_{max}} \pi r^2 Q'_\beta(\lambda, r) n(r) dr \tag{59}$$

$$p_p(\lambda, \Theta) = \frac{\int_{r_{min}}^{r_{max}} p(\lambda, \Theta, r) n(r) dr}{\int_{r_{min}}^{r_{max}} n(r) dr} \tag{60}$$

where $n(r)$ is the PSD and $\alpha_n(\lambda,r)$, $\beta_n(\lambda,r)$, and $p(\lambda,\Theta,r)$ are the absorption cross section, the scattering cross section, and the scattering phase function per particle of radius r. The absorption or scattering "efficiency", $Q'_\alpha(r)$ or $Q'_\beta(r)$, is defined as the ratio of the absorption or scattering cross section for a spherical particle of radius r to the geometrical cross section πr^2. The scattering phase function $p(\lambda,\Theta,r)$ in Equation (60) is the $a_1(\Theta)$ element of the Stokes scattering matrix [Equation (18)]. Since a Mie code can be used to compute all elements of the Stokes scattering matrix in Equation (18), we may use an expression analogous to Equation (60) to carry out the integration over the PSD for each of the matrix elements.

IOPs for a Mixture of Different Particle Types

Consider a particle mixture consisting of a total of N particles per unit volume in a layer of thickness Δz, and let $N = \sum_i n_i$ and $f_i = n_i/N$, where n_i is the concentration and f_i the fraction of homogeneous particles (with fixed chemical composition or refractive index) of type labeled i. To compute IOPs for the mixture of particles, we define $\beta_{n,i}$ = scattering cross section, $\alpha_{n,i}$ = absorption cross section, $\gamma_{n,i} = \beta_{n,i} + \alpha_{n,i}$ = extinction cross section, and $\varpi_i = \beta_{n,i}n_i/\gamma_{n,i}n_i = \beta_i/\gamma_i$ = single-scattering albedo, where the subscript i stands for particle type. Weighting by number concentration may be used to create IOPs for the particle mixture. Thus, by combining the absorption and scattering cross sections, and the moments of the scattering phase matrix elements, one obtains the following IOPs for the mixture (subscript m stands for mixture):

$$\Delta\tau_m = \Delta z \sum_i n_i \gamma_{n,i} = \Delta z \sum_i \gamma_i = \gamma_m \Delta z \tag{61}$$

$$\varpi_m = \frac{\beta_m}{\gamma_m} = \frac{\sum_i \beta_{n,i} n_i}{\sum_i \gamma_{n,i} n_i} = \frac{\sum_i \beta_i}{\sum_i \gamma_i} = \frac{\sum_i \varpi_i \gamma_{n,i} f_i}{\sum_i \gamma_{n,i} f_i} \tag{62}$$

$$\chi_{m,\ell} = \frac{\sum_i \beta_{n,i} n_i \chi_{i,\ell}}{\sum_i \beta_{n,i} n_i} = N\frac{\sum_i \beta_{n,i} f_i \chi_{i,\ell}}{\beta_m} = \frac{\sum_i f_i \varpi_i \gamma_{n,i} \chi_{i,\ell}}{\sum_i f_i \beta_{n,i}} \tag{63}$$

where $\Delta\tau_m$ = layer optical depth; β_m = total scattering coefficient; γ_m = total extinction coefficient; ϖ_m = single-scattering albedo; and $\chi_{m,\ell}$ = scattering phase function expansion coefficient for the particle mixture. A mixing rule similar to Equation (63) may be used for each element of the scattering phase matrix.

3.4. Atmosphere IOPs

The stratified vertical structure of the bulk properties of an atmosphere is a consequence of hydrostatic balance. By equating pressure forces and gravitational forces and invoking the ideal gas law, one may derive the barometric law for the pressure $p(z)$ as function of altitude z above the surface z_0 [14]:

$$p(z) = p(z_0)\exp\left[-\int_{z_0}^{z} dz'/H(z')\right] \tag{64}$$

where $H(z) = kT(z)/\bar{M}g$ is the atmospheric scale height, $\bar{M}$ is the mean molecular weight, k is is Boltzmann's constant, g is the acceleration due to gravity, and $T(z)$ is the temperature. The ideal gas law allows one to write similar expressions for the bulk density $\rho(z)$ and the bulk concentration $n(z)$. Clearly, from a knowledge of the surface pressure $p(z_0)$ and the variation of the scale height $H(z)$ with height z, Equation (64) allows us to determine the bulk gas properties at any height. Equation (64) applies to well-mixed gases, but not to short-lived species such as ozone, which is chemically created and destroyed, or water, which undergoes phase changes on short time scales.

3.4.1. Gases in the Earth's Atmosphere

The total number of air molecules in a 1 m^2 wide vertical column extending from sea level to the top of the atmosphere is about 2.15×10^{29}. In comparison, the total amount of ozone (a trace gas) in the same vertical column is about 1.0×10^{23}. Anderson et al. [25] compiled six model atmospheres including, (i) the US Standard atmosphere 1976, (ii) tropical, (iii) midlatitude summer, (iv) midlatitude winter, (v) subarctic summer, and (vi) subarctic winter (see Appendix U of [14] for a numerical tabulation of these models). These atmospheric models contain profiles of temperature, pressure, and the concentrations of the main atmospheric constituents, molecular nitrogen (N_2) and molecular oxygen (O_2). In addition, they contain profiles of the concentrations of several trace gases including water vapor (H_2O), ozone O_3, carbon dioxide (CO_2), methane (CH_4), nitrous oxide (N_2O), and the chlorofluoromethanes (CFCs) in the Earth's atmosphere.

The clear atmosphere (no clouds or aerosols) molecular (Rayleigh) scattering coefficient can be expressed as

$$\sigma_{\rm Ray}(\lambda, z) \equiv \sigma_{\rm Ray,n} n(z) = \frac{32\pi^3(m_{\rm r} - 1)^2}{3\lambda^4 n(z)} \qquad [\mathrm{m}^{-1}] \tag{65}$$

where $n(z)$ is the bulk air concentration (see Equation (64)), $m_{\rm r}$ is the real part of the refractive index, and $\sigma_{\rm Ray,n}$ is the Rayleigh scattering cross section. Please note that since $m_{\rm r}$ depends on wavelength, the Rayleigh scattering coefficient does not have an exact λ^{-4} dependence. For air, a convenient numerical formula for the Rayleigh scattering cross section (accurate to 0.3%) is given by [14]

$$\sigma_{\rm Ray,n} = \lambda^{-4} \sum_{i=0}^{3} a_i \lambda^{-2i} \times 10^{-28} \quad [\mathrm{cm}^2] \quad (0.205 < \lambda < 1.05\ \mu\mathrm{m})$$

where the coefficients are $a_0 = 3.9729066$, $a_1 = 4.6547659 \times 10^{-2}$, $a_2 = 4.5055995 \times 10^{-4}$, and $a_3 = 2.3229848 \times 10^{-5}$. The scattering phase function is given by Equation (9).

Computer codes like MODTRAN [26,27] have been developed to provide atmospheric transmittance and thereby absorption coefficients for all important atmospheric trace gases for a large variety of atmospheric conditions. In AccuRT we use a band model based on MODTRAN [28] to generate absorption coefficients and optical depths due to atmospheric gases including O_2, H_2O, CO_2, O_3, CH_4, and NO_2. An example of how to use this approach to deal with gaseous absorption in shortwave near infrared bands for an atmosphere overlying a snow surface is provided in [29].

3.4.2. Aerosol IOPs

If we know the size distribution and the refractive index of the aerosol particles, we may use available aerosol models to generate aerosol IOPs. For example, one may use the aerosol models employed in the Sea-viewing Wide Field-of-view Sensor (SeaWiFS) Database Analysis System (SeaDAS), and described by Ahmad et al. [30]. Alternatively, we may use the OPAC models [31]. For atmospheric correction of ocean color imagery, it is customary to assume a lognormal distribution of aerosol sizes [32]. Based on AERONET data [33,34], Ahmad et al. [30] adopted a bimodal lognormal volume size distribution:

$$v(r) = \frac{dV(r)}{dr} = \frac{1}{r}\frac{dV(\ln r)}{d\ln r} = \sum_{i=1}^{2} \frac{V_i}{\sqrt{2\pi}\sigma_i}\frac{1}{r} \exp\left[-\left(\frac{\ln r - \ln r_{\rm vi}}{\sqrt{2}\sigma_i}\right)^2\right] \tag{66}$$

where the subscript i represents the mode, V_i is the total volume of particles with mode i, $r_{\rm vi}$ is the mode radius, also called the volume geometric mean radius, and σ_i is the geometric standard deviation. Please note that since the numerator in the exponential of Equation (66), $\ln(r/r_{\rm vi})$, is dimensionless, so is σ_i. Since

$$\int_0^\infty \frac{dr}{\sqrt{2\pi}\sigma}\frac{1}{r} \exp\left[-\left(\frac{\ln r - \ln r_{\rm v}}{\sqrt{2}\sigma}\right)^2\right] = 1$$

integration over all sizes for both modes, yields:

$$\int_0^\infty v(r)dr = V_1 + V_2 = V.$$

In terms of the number density (concentration), Equation (66) becomes

$$n(r) = \frac{dN(r)}{dr} = \frac{1}{r}\frac{dN(r)}{d(\ln r)} = \sum_{i=1}^{2} \frac{N_i}{\sqrt{2\pi}\sigma_i}\frac{1}{r}\exp\left[-\left(\frac{\ln r - \ln r_{ni}}{\sqrt{2}\sigma_i}\right)^2\right] \tag{67}$$

where the number of particles N_i and the mean geometric (or mode) radius r_{ni} are related to V_i and r_{vi} as follows

$$\ln r_{ni} = \ln r_{vi} - 3\sigma_i^2 \tag{68}$$

$$N_i = \frac{V_i}{\frac{4}{3}\pi r_{ni}^3}\exp(-4.5\sigma_i^2) \tag{69}$$

and integration over all sizes for both modes, yields:

$$\int_0^\infty n(r)dr = N_1 + N_2 = N.$$

If we use the subscript $i = f$ to denote the fine mode, and the subscript $i = c$ to denote the coarse mode, we have $V = V_f + V_c$, and the volume fraction of fine mode particles becomes $f_v = V_f/V$.

Relationship between Effective Radius and Mode Radius

The particle size distribution may also be characterized by an effective radius

$$r_{\text{eff}} = \frac{\int_{r_{\text{min}}}^{r_{\text{max}}} n(r)r^3 dr}{\int_{r_{\text{min}}}^{r_{\text{max}}} n(r)r^2 dr} \tag{70}$$

and an effective variance

$$v_{\text{eff}} = \frac{\int_{r_{\text{min}}}^{r_{\text{max}}} (r - r_{\text{eff}})^2 n(r)r^2 dr}{r_{\text{eff}}^2 \int_{r_{\text{min}}}^{r_{\text{max}}} n(r)r^2 dr} \tag{71}$$

where r_{eff}^2 is included in the denominator of Equation (71) to make v_{eff} dimensionless [35]. The effective radius, r_{eff}, can be used to describe the IOPs in an approximate manner as will be discussed below for cloud as well as snow/ice materials. For a single mode, the lognormal size distribution is given by [see Equation (67)]

$$n(r) = \frac{dN(r)}{dr} = \frac{N}{\sqrt{2\pi}\sigma}\frac{1}{r}\exp\left[-\left(\frac{\ln r - \ln r_n}{\sqrt{2}\sigma}\right)^2\right]$$

where r_n is the mode radius, $n(r)$ is the number density or PSD in units of $[\text{m}^{-3} \cdot \text{m}^{-1}]$ and $N = \int_0^\infty n(r)dr$ $[\text{m}^{-3}]$ is the total number of particles per unit volume since

$$\int_0^\infty \frac{dr}{\sqrt{2\pi}\sigma}\frac{1}{r}\exp\left[-\left(\frac{\ln r - \ln r_n}{\sqrt{2}\sigma}\right)^2\right] = 1. \tag{72}$$

With the change of variable $x = \frac{\ln(r/r_n)}{\sqrt{2}\sigma}$, Equation (72) becomes

$$\frac{1}{\sqrt{\pi}}\int_{-\infty}^{+\infty} \exp(-x^2)dx = 1 \tag{73}$$

and it can be shown that [35]:

$$r_{\text{eff}} = r_n \exp[2.5\sigma^2], \tag{74}$$

and

$$v_{\text{eff}} = \exp\left[\sigma^2\right] - 1 \tag{75}$$

(see [3] for details).

Impact of Relative Humidity

A change in the relative humidity (RH) will affect bot the size and refractive index of a particle. The particle radius can be parameterized as a function of RH from the wet-to-dry mass ratio:

$$r(a_{\text{w}}) = r_0\left[1 + \rho\frac{m_{\text{w}}(a_{\text{w}})}{m_0}\right]^{1/3} \tag{76}$$

where the water activity a_{w} of a soluble aerosol at radius r [μm] can be expressed as

$$a_{\text{w}} = \text{RH}\exp\left[\frac{-2\sigma V_m}{R_{\text{w}}T}\frac{1}{r(a_{\text{w}})}\right]. \tag{77}$$

Here r_0 is the dry particle radius (RH = 0), ρ is the particle density relative to that of water, $m_{\text{w}}(a_{\text{w}})$ is the mass of condensed water, m_0 is the dry particle mass (RH = 0), σ is the surface tension on the wet surface, V_m is the specific volume of water, R_{w} is the gas constant for water vapor, and T is the absolute temperature [K] [36]. Similarly, the change in refractive index with RH can be determined from [36]

$$\tilde{m}_{\text{c}} = \tilde{m}_{\text{c,w}} + (\tilde{m}_{\text{c,0}} - \tilde{m}_{\text{c,w}})\left[\frac{r_0}{r_{\text{RH}}}\right]^3 \tag{78}$$

where $\tilde{m}_{\text{c,w}}$ and $\tilde{m}_{\text{c,0}}$ are the complex refractive indices of water and dry aerosols, respectively, and r_0 and r_{RH} are the radii of the aerosols in the dry state and at the given RH, respectively. From these formulas we note that the magnitude of the particle growth and the change of refractive index with increasing RH depend on the size r_0 of the dry aerosol but also on the type of aerosol through the water uptake [the ratio $m_{\text{w}}(a_{\text{w}})/m_0$ in Equation (76)] [36–38].

A Mie code [39] is needed to compute the IOPs of aerosol particles [$Q'_\alpha(r)$, $Q'_\beta(r)$, and $p_{\text{p}}(\lambda, \Theta, r)$ in Equations (58)–(60)], and numerical integration is required to evaluate the integrals over the lognormal size distributions to obtain $\alpha_{\text{p}}(\lambda)$, $\beta_{\text{p}}(\lambda)$, and $p_{\text{p}}(\lambda, \Theta)$. For polarized radiation all elements of the scattering phase matrix as well as the Greek constants appearing in Equation (48) must be computed. For a bimodal lognormal volume size distribution [Equation (66)] one must specify the fine mode volume fraction $f_v = V_f/V$, where $V = V_f + V_c$, the volume mode radii r_{vf} and r_{vc} as well as the corresponding standard deviations σ_f and σ_c in addition to the refractive index of the particles relative to air.

In analogy to the liquid water content (see Equation (81) below), we may introduce the aerosol mass content (AMC) for each mode defined as

$$\text{AMC} = \rho_{\text{a}}\int_{r_{\min}}^{r_{\max}}\left(\frac{4\pi}{3}\right)r^3 n(r)dr \equiv \rho_{\text{a}} f_{\text{V}} \qquad [\text{kg}\cdot\text{m}^{-3}] \tag{79}$$

where $n(r)$ is the aerosol size distribution [$\text{m}^{-3}\cdot\text{m}^{-1}$], ρ_{a} is the bulk aerosol density [$\text{kg}\cdot\text{m}^{-3}$], and f_{V} is the aerosol volume fraction (not to be confused with the fine mode volume fraction, f_v) given by:

$$f_{\text{V}} \equiv \int_{r_{\min}}^{r_{\max}}\left(\frac{4\pi}{3}\right)r^3 n(r)dr = \text{AMC}/\rho_{\text{a}} \qquad (\text{dimensionless}). \tag{80}$$

Typical values of atmospheric aerosol densities are $\rho_{\text{a}} \approx 1\,\text{g}\cdot\text{cm}^{-3} = 1\times10^6\,\text{g}\cdot\text{m}^{-3}$. Hence, an AMC value of $10^{-6}\,\text{g}\cdot\text{m}^{-3}$ would yield $f_{\text{V}} = 10^{-12}$.

3.4.3. Cloud IOPs

Clouds consist of liquid water droplets or ice (frozen water) particles. While liquid water droplets can be assumed to have spherical shape, ice crystals can have a variety of non-spherical shapes. If we assume for simplicity that all cloud particles consist of spherical water droplets or spherical ice particles, (For ice crystals, a spherical model may be unrealistic. For a good introduction to this topic, see the textbook by Wendisch and Yang [40].) we can use a Mie code to compute their IOPs because their refractive index is known. Hence, we may use Equations (58)–(60) to compute $\alpha_{\text{p}}(\lambda), \beta_{\text{p}}(\lambda)$, and $p_{\text{p}}(\lambda, \Theta)$.

The real part of the refractive index of pure water needed in the Mie computations may be taken from [41], while the imaginary part $\tilde{m}_{i,\text{w}}$ is calculated from the absorption coefficient ($\alpha_{\text{w}}(\lambda) = 4\pi\tilde{m}_{i,\text{w}}/\lambda$) obtained from published data [42–44] for wavelengths between 340 and 700 nm, and from another source [45] for wavelengths between 720 and 900 nm.

It is customary to introduce the liquid water content (LWC) defined as

$$\text{LWC} \equiv \rho_{\text{w}} \int_{r_{\min}}^{r_{\max}} \left(\frac{4\pi}{3}\right) r^3 n(r) dr \equiv \rho_{\text{w}} f_{\text{V}} \qquad [\text{kg} \cdot \text{m}^{-3}] \tag{81}$$

where $n(r)$ is the cloud droplet size distribution $[\text{m}^{-3} \cdot \text{m}^{-1}]$ and ρ_{w} is the liquid water mass density $[\text{kg} \cdot \text{m}^{-3}]$ and f_{V} stands for the dimensionless liquid (cloud) particle volume fraction defined in a similar manner as AMC in Equation (80), i.e., $f_{\text{V}} = \text{LWC}/\rho_{\text{W}}$. For a liquid water cloud, a typical value for LWC is about 0.5 $\text{g} \cdot \text{m}^{-3}$, implying that $f_{\text{V}} = 5 \times 10^{-7}$ for $\rho_{\text{w}} = 10^3$ $\text{kg} \cdot \text{m}^{-3}$. In Equation (70) for the effective radius, the numerator is proportional to the concentration or LWC, while the denominator is related to the scattering coefficient:

$$\beta_c = \int_0^{\infty} dr(\pi r^2) Q_{\beta}(r)\, n(r) dr \qquad [\text{m}^{-1}].$$

If the size of the droplet is large compared to the wavelength λ, then $Q_{\beta}(r) \to 2$. Therefore, in the visible spectral range where $2\pi r/\lambda \gg 1$, we find:

$$\beta_c \approx \frac{3}{2} \frac{1}{\rho_{\text{w}}} \frac{\text{LWC}}{r_{\text{eff}}} = \frac{3}{2} \frac{f_{\text{V}}}{r_{\text{eff}}} \qquad [\text{m}^{-1}]. \tag{82}$$

For ice cloud particles assumed to be spherical in shape a similar expression for the scattering coefficient is obtained with f_{V} being the ice particle volume fraction. For a liquid water cloud with $f_{\text{V}} = 5 \times 10^{-7}$ and $r_{\text{eff}} = 5 \times 10^{-6}$ m, we get $\beta_c = \frac{3}{2} \frac{f_{\text{V}}}{r_{\text{eff}}} = 0.15\,\text{m}^{-1}$, and hence an optical thickness of 15 for a 100 m thick cloud layer.

3.5. Snow and Ice IOPs

3.5.1. General Approach

Assuming that snow grains and sea ice inclusions have spherical shape, we may obtain their IOPs from Mie computations, which require the refractive index and the size distribution of the particles as input. Then, the IOPs, i.e., the absorption and scattering coefficients and the scattering phase function, $\alpha_{\text{p}}(\lambda), \beta_{\text{p}}(\lambda)$, and $p_{\text{p}}(\lambda, \Theta)$, can be obtained from Equations (58)–(60). This approach leads to computed snow albedo values that agree surprisingly well with available observations [46,47]. The following reasons why large errors are not incurred by assuming spherical shape have been advocated by Craig Bohren as quoted elsewhere [48]: *The orientationally averaged extinction cross section of a convex particle that is large compared with the wavelength is one-half its surface area. The absorption cross section of a large, nearly transparent particle is proportional to its volume almost independent of its shape. The closer the real part of the particle's refractive index is to 1, the more irrelevant the particle shape.*

The asymmetry parameter of a large particle is dominated by near-forward scattering, which does not depend greatly on particle shape.

Hence, we may assume that snow grains and ice inclusions (air bubbles and brine pockets) consist of homogeneous spheres with a single-mode lognormal volume size distribution [see Equation (66)], and use the refractive index data base for ice compiled by [49]. Specifying the effective radius r_{eff} and the width of the distribution σ, one obtains the geometrical mean radius r_{n} from Equation (74). Since the complex refractive index is prescribed, r_{n} and σ constitute the only input required for a Mie code (see Section 3.4.2), which can be used to compute absorption and scattering coefficients as well the scattering phase function. One may choose to use only the first moment of the scattering phase function in conjunction with the Henyey-Greenstein scattering phase function because the Mie scattering phase function is unrealistic for non-spherical snow grains and ice inclusions.

3.5.2. Fast, yet Accurate Parameterization of Snow/Ice IOPs

Building on previous work [50–52], Stamnes et al. [53] created a generic tool for computing snow/ice IOPs (τ, ϖ, and g). This tool can be used to generate wavelength-dependent ice/snow IOPs from ice/snow physical parameters: real and imaginary parts of the ice/snow refractive index, brine pocket concentration and effective size (sea ice), air bubble concentration and effective size (sea ice), volume fraction and absorption coefficient of sea ice impurities, asymmetry factors for scattering by snow grains, brine pockets, and air bubbles, and sea ice thickness. We can compute $Q'_\alpha(r)$, $Q'_\beta(r)$, and $p_{\text{p}}(\lambda, \Theta, r)$ using a Mie code, but evaluation of Equations (58)–(60) requires knowledge of the particle size distribution $n(r)$, which is usually unknown. Equations (58)–(60) can be considerably simplified by making the following assumptions [53]:

- The particle distribution is characterized by an effective radius [Equation (70)], which obviates the need for an integration over r.
- The particles are weakly absorbing, so that [51]

$$Q'_\alpha(r) \equiv Q'_\alpha \approx \frac{16\pi\, r_{\text{eff}}\, \tilde{m}_{i,\text{p}}}{3\lambda} \frac{1}{m_{\text{rel}}} [m_{\text{rel}}^3 - (m_{\text{rel}}^2 - 1)^{3/2}] \tag{83}$$

  where $\tilde{m}_{i,\text{p}}$ is the imaginary part of the refractive index of the particle, λ is the wavelength in vacuum, and $m_{\text{rel}} = \tilde{m}_{r,\text{p}}/\tilde{m}_{r,\text{med}}$ is the ratio of the real part of the refractive index of the particle ($\tilde{m}_{r,\text{p}}$) to that of the surrounding medium ($\tilde{m}_{r,\text{med}}$).
- The particles are large compared to the wavelength ($2\pi r/\lambda \gg 1$) which implies

$$Q'_\beta(r) \equiv Q'_\beta = 2. \tag{84}$$

The scattering phase function may be represented by the one-parameter Henyey-Greenstein scattering phase function [see Equation (10)], which depends only on the asymmetry factor defined in Equation (7).

With these assumptions, Equations (58)–(59) become:

$$\alpha_{\text{p}}(\lambda) = \alpha(\lambda) \frac{1}{m_{\text{rel}}} [1 - (m_{\text{rel}}^2 - 1)^{3/2}] f_{\text{V}} \tag{85}$$

$$\beta_{\text{p}}(\lambda) = \frac{3}{2} \frac{f_{\text{V}}}{r_{\text{eff}}}. \tag{86}$$

Here $\alpha(\lambda) = 4\pi\tilde{m}_{i,\text{p}}/\lambda$ is the absorption coefficient of the material of which the particle is composed, and $f_{\text{V}} \equiv \frac{4\pi}{3} \int n(r) r^3 dr \approx \frac{4}{3}\pi r_{\text{eff}}^3 n_{\text{e}}$, where n_{e} = number of particles per unit volume with radius r_{eff}. Since Equation (86) is identical to Equation (82), it is clear that f_{V} represents the volume fraction of the particles as defined in Equation (80).

For wavelengths $\lambda \leq 1.2$ μm, the absorption and scattering efficiency for snow grains, brine inclusions in sea ice, and air bubbles in ice may be parameterized by Equations (83) and (84), and the asymmetry factor g can be held constant with wavelength and set equal to 0.85, 0.89, and 0.997 for air bubbles, snow grains, and brine pockets, respectively. To extend the validity to NIR wavelengths, we may use the following modified parameterizations [53]:

$$Q_\alpha = 0.94[1 - \exp(-Q'_\alpha/0.94)]; \qquad Q_\beta = 2 - Q_\alpha; \qquad g = g_0^{(1-Q_\alpha)^{0.6}} \tag{87}$$

where Q'_α is given by Equation (83). Here g is the asymmetry factor of the scattering phase function, and g_0 is the asymmetry factor for non-absorbing particles. For large particles ($r > \sim 50$ μm) g_0 depends only on the real part of the refractive index. For a medium consisting of several absorbing and scattering constituents the total absorption and scattering efficiencies are just the sum of those due to the separate constituents. The optical thickness τ and single-scattering albedo ϖ for a slab of thickness h become [53]:

$$\tau = \pi r_{\rm eff}^2 N h (Q_\alpha + Q_\beta); \qquad \varpi = \frac{Q_\beta}{Q_\alpha + Q_\beta} \tag{88}$$

where N is the total number of particles per unit volume, and Q_α and Q_β are the total absorption and scattering efficiencies, each equal to the sum of those due to the separate constituents. These modified parameterizations work well for all wavelengths for Q_α, while for Q_β and g they work well for wavelengths shorter than about 2.8 μm, but deviate significantly from predictions by Mie theory for longer wavelengths. Thus, for wavelengths longer than 2.8 μm one should preferably use results from the computationally less efficient Mie theory. Note that for wavelengths shorter than 2.8 μm, where the parameterizations work well, the variations in $\tilde{m}_{r,\rm p}$ and $\tilde{m}_{i,\rm p}$ are large. Thus, one would expect these parameterizations to be valid for most types of large particles [53].

3.5.3. Impurities, Air Bubbles, Brine Pockets, and Snow

If the volume fraction of impurities within a snow grain or brine pocket is not too large, which is the case for typical situations occurring in nature, scattering by impurities can be ignored, so that their effects can be included by simply adding the imaginary part $\tilde{m}_{i,\rm imp}$ of the refractive index for impurities to $\tilde{m}_{i,\rm p}$ in Equation (83). For typical impurities in snow and ice, the wavelength dependence of $\tilde{m}_{i,\rm imp}$ can be parameterized as [53]

$$\tilde{m}_{i,\rm imp}(\lambda) = \tilde{m}_{i,\rm imp}(\lambda_0)\,(\lambda_0/\lambda)^\eta \tag{89}$$

where η would be close to zero for black carbon, but larger for other impurities, and $\tilde{m}_{i,\rm imp}(\lambda_0 = 440\,\rm nm)$ has values that depend on the type of impurity. Equation (89) is based on the observation that the absorption coefficient α of non-algal impurities tend to have a smooth increase towards shorter wavelengths [54–57], and α is connected to the imaginary part of the refractive index through $\alpha = 4\pi\,\tilde{m}_{i,\rm imp}/\lambda$. For snow, the number of snow grain particles per unit volume is $N = \frac{1}{\frac{4}{3}\pi r_{\rm eff}^3}\frac{\rho_{\rm s}}{\rho_{\rm i}}$, where $r_{\rm eff}$ is the effective particle radius, while $\rho_{\rm s}$ and $\rho_{\rm i}$ are the mass densities of snow and pure ice, respectively. The optical thickness and the single-scattering albedo can be calculated from Equations (87) and (88), using the refractive indices of pure ice [49] and impurities [Equation (89)].

We assume that sea ice consists of pure ice with embedded brine pockets, air bubbles, and impurities. To include the effects of the embedded components, we first calculate the absorption coefficient α for sea ice [53]

$$\alpha = \pi r_{\rm br}^2 N_{\rm br} Q_{\alpha,\rm br} + \left[1 - \frac{4}{3}\pi r_{\rm br}^3 N_{\rm br} - \frac{4}{3}\pi r_{\rm bu}^3 N_{\rm bu}\right] \frac{4\pi(\tilde{m}_{i,\rm p} + f_{\rm imp}\tilde{m}_{i,\rm imp})}{\lambda} \tag{90}$$

where $f_{\rm imp}$ is the volume fraction of impurities, $N_{\rm br}$ and $N_{\rm bu}$ are the number concentrations of brine pockets and air bubbles, respectively, $r_{\rm br}$ and $r_{\rm bu}$ are the corresponding effective radii, and $Q_{\alpha,\rm br}$ is the absorption efficiency for brine pockets. The two terms on the right side of Equation (90) represent the absorption coefficients of brine pockets and surrounding ice (including impurities), respectively. In Equation (90), we have used the general relation $\alpha = 4\pi \tilde{m}_{i,p}/\lambda$, where λ is the wavelength in vacuum, and the expression inside the square brackets is the volume fraction of the ice surrounding all brine pockets and bubbles.

The air bubbles were assumed to be non-absorbing ($Q_{\alpha,\text{bu}} = 0$), and the impurities were assumed to be uniformly distributed in the ice with $\tilde{m}_{i,p}$ and $\tilde{m}_{i,\text{imp}}$ being the imaginary parts of the refractive indices for pure ice and impurities, respectively. For brine pockets, which are in the liquid phase, the refractive index of sea water was used. The volume fraction f_{imp} of impurities typically lies in the range between 1×10^{-7} and 1×10^{-5}. The scattering coefficient β of sea ice is given by [53]

$$\beta = \beta_{\text{br}} + \beta_{\text{bu}}; \quad \beta_{\text{br}} = \pi r_{\text{br}}^2 N_{\text{br}} Q_{\beta,\text{br}}; \quad \beta_{\text{bu}} = \pi r_{\text{bu}}^2 N_{\text{bu}} Q_{\beta,\text{bu}} \tag{91}$$

where β_{br} and β_{bu} are the scattering coefficients for brine pockets and air bubbles, respectively, and $Q_{\beta,\text{br}}$ and $Q_{\beta,\text{bu}}$ are the corresponding scattering efficiencies. Here we have ignored the scattering coefficient for pure sea ice because it is very small compared to either β_{br} or β_{bu}. The optical thickness τ, the single-scattering albedo ϖ, and the asymmetry factor g for sea ice now become

$$\tau = (\alpha + \beta)h; \qquad \varpi = \frac{\beta}{\alpha + \beta}; \qquad g = \frac{\beta_{\text{br}} g_{\text{br}} + \beta_{\text{bu}} g_{\text{bu}}}{\beta_{\text{br}} + \beta_{\text{bu}}} \tag{92}$$

where h is the sea ice thickness.

The merit of these IOP parameterizations have been provided by comparisons with field measurements and laboratory data [50–53].

3.6. Ocean IOPs—Bio-Optical Models

In open ocean water, it is customary to assume that the IOPs of particulate matter can be parameterized in terms of the chlorophyll concentration. In coastal water, the IOPs will depend on the presence of "impurities" consisting of inorganic (mineral) particles, organic (algae) particles, and Colored Dissolved Organic Matter (CDOM) in addition to pure water. Due to the complexity of coastal water, we introduce three bio-optical models that have been adopted to represent different types of water. The CoastColour Round Robin (CCRR) model [58] is a useful proxy for turbid coastal water frequently observed in estuary areas, where suspended sediment (i.e., mineral) particles have a strong influence on water IOPs. The Santa Barbara Channel (SBC) and Garver-Siegel-Maritorena (GSM) bio-optical models described below provide useful representations of clean to moderately turbid water, where the IOPs are primarily dominated by algae. By varying the slope parameter S that describe the CDOM spectral absorption (see Equation (109) below), the GSM model may be used to represent CDOM dominated water.

As mentioned in Section 3.4.3, for pure water we may adopt the real part of the refractive index of pure water from [41], and we use the absorption coefficient $\alpha_{\text{w}}(\lambda)$ based on published data [42–44] for wavelengths between 340 and 700 nm, and other data [45] for wavelengths between 720 and 900 nm. Pure water scattering coefficients $\beta_{\text{w}}(\lambda)$ are based on published data [7], and the Rayleigh scattering phase function is given by Equation (9) with depolarization ratio $\rho = 0.039$, and thus $f = (1 - \rho)/(1 + \rho) = 0.925$ (see Section 3.1.1).

3.6.1. The CCRR Water Impurity IOPs

Here we first describe a bio-optical model used in the CoastColour Round Robin (CCRR) effort [58]. The CCRR bio-optical model consists of the three input parameters chlorophyll concentration (CHL), mineral concentration (MIN), and $\alpha_{\text{CDOM}}(443)$, which are allowed to vary. According to this decomposition into three basic components, the "mineral particle" component can include also non-algae particles whose absorption does not covary with that of the algae particles [58].

Mineral Particle IOPs

The absorption coefficient for mineral particles at 443 nm is given by [59] (Note on units: $\alpha_{\rm MIN}(\lambda)/{\rm MIN} = 0.041$ has units [$\rm m^2 \cdot g^{-1}$], so that if MIN has units of [$\rm g \cdot m^{-3}$], then the units of $\alpha_{\rm MIN}(\lambda)$ will be [$\rm m^{-1}$].):

$$\alpha_{\rm MIN}(443) = 0.041 \times 0.75 \times {\rm MIN}$$

and its spectral variation is described by [58,59]:

$$\alpha_{\rm MIN}(\lambda) = \alpha_{\rm MIN}(443)[\exp(-0.0123(\lambda - 443))]. \tag{93}$$

The scattering coefficient at 555 nm is given by [60]

$$\beta_{\rm MIN}(555) = 0.51 \times {\rm MIN}$$

and the spectral variation of the attenuation coefficient is

$$\gamma_{\rm MIN}(\lambda) = \gamma_{\rm MIN}(555) \times (\lambda/\lambda_0)^{-c}; \qquad c = 0.3749, \qquad \lambda_0 = 555\,{\rm nm} \tag{94}$$

where

$$\begin{aligned}\gamma_{\rm MIN}(555) &= \alpha_{\rm MIN}(555) + \beta_{\rm MIN}(555)\\ &= [0.041 \times 0.75 \exp(-0.0123(555 - 443)) + 0.51] \times {\rm MIN}\\ &= 0.52 \times {\rm MIN}.\end{aligned}$$

The spectral variation of the scattering coefficient for mineral particles follows from

$$\beta_{\rm MIN}(\lambda) = \gamma_{\rm MIN}(\lambda) - \alpha_{\rm MIN}(\lambda). \tag{95}$$

The average Petzold phase function with a backscattering ratio of 0.019 [4], may be used to describe the scattering phase function for mineral particles.

Algae Particle IOPs

The absorption coefficient for *pigmented* particles (algae particles or phytoplankton) can be written [61]:

$$\alpha_{\rm pig}(\lambda) = A_\phi(\lambda) \times [{\rm CHL}]^{E_\phi(\lambda)} \tag{96}$$

where $A_\phi(\lambda)$ and $E_\phi(\lambda)$ are given by [61], and where CHL is the chlorophyll concentration, which represents the concentration of pigmented particles (algae particles or phytoplankton).

The attenuation coefficient for pigmented particles at 660 nm is given by [62]:

$$\gamma_{\rm pig}(660) = \gamma_0 \times [{\rm CHL}]^\eta; \qquad \gamma_0 = 0.407; \qquad \eta = 0.795$$

and its spectral variation is taken to be [63]:

$$\gamma_{\rm pig}(\lambda) = \gamma_{\rm pig}(660) \times (\lambda/660)^\nu \tag{97}$$

where

$$\nu = \begin{cases} 0.5 \times [\log_{10}{\rm CHL} - 0.3] & 0.02 < {\rm CHL} < 2.0\\ 0 & {\rm CHL} > 2.0.\end{cases}$$

The spectral variation of the scattering coefficient for pigmented particles follows from the difference:

$$\beta_{\rm pig}(\lambda) = \gamma_{\rm pig}(\lambda) - \alpha_{\rm pig}(\lambda). \tag{98}$$

The scattering phase function for pigmented particles may be described by the Fournier-Forand phase function (see below) with a backscattering ratio equal to 0.006 [63,64].

CDOM IOPs

The absorption by CDOM is given by [60]:

$$\alpha_{\rm CDOM}(\lambda) = \alpha_{\rm CDOM}(443) \times \exp[-S(\lambda - 443)]; \qquad S = 0.0176. \tag{99}$$

The total absorption and scattering coefficients due to water impurities for the CCRR IOP model are given by:

$$\alpha_{\rm tot}(\lambda) = \alpha_{\rm MIN}(\lambda) + \alpha_{\rm pig}(\lambda) + \alpha_{\rm CDOM}(\lambda) \tag{100}$$

$$\beta_{\rm tot}(\lambda) \equiv \beta_{\rm p}(\lambda) = \beta_{\rm MIN}(\lambda) + \beta_{\rm pig}(\lambda). \tag{101}$$

Scattering Phase Function for Particles

Measurements have shown that the particle size distribution (PSD) function in oceanic water can be accurately described by an inverse power law (Junge distribution) $F(r) = C_r/r^{\xi}$, where $F(r)$ is the number of particles per unit volume per unit bin width, and r [μm] is the radius of the assumed spherical particle. C_r [$\text{cm}^{-3} \cdot \mu\text{m}^{\xi-1}$] is the Junge coefficient, and ξ is the PSD slope, which typically varies between 3.0 and 5.0 [65,66]. By assuming an inverse power law (Junge distribution) for the PSD, [67] derived an analytic expression for the scattering phase function of oceanic water (hereafter referred to as the FF scattering phase function), given by [64]

$$\begin{aligned} p_{\rm FF}(\Theta) &= \frac{1}{4\pi(1-\delta)^2\delta^{\nu}}\left\{\nu(1-\delta) - (1-\delta^{\nu}) + \frac{4}{u^2}[\delta(1-\delta^{\nu}) - \nu(1-\delta)]\right\} \\ &\quad + \frac{1-\delta_{180}^{\nu}}{16\pi(\delta_{180}-1)\delta_{180}^{\nu}}[3\cos^2\Theta - 1] \end{aligned} \tag{102}$$

where $\nu = 0.5(3-\xi)$, $u = 2\sin(\Theta/2)$, $\delta \equiv \delta(\Theta) = \frac{u^2}{3(\tilde{m}_{\rm r}-1)^2}$, $\delta_{180} = \delta(\Theta = 180°) = \frac{4}{3(\tilde{m}_{\rm r}-1)^2}$, Θ is the scattering angle, and $\tilde{m}_{\rm r}$ is the real part of the refractive index.

Integrating $p_{\rm FF}(\Theta)$ over the backward hemisphere (setting $x = \cos\Theta$), one obtains the backscattering ratio or backscatter fraction defined in Equation (8) [64]

$$\begin{aligned} b_{\rm FF} &= \frac{1}{2}\int_{\pi/2}^{\pi} p_{\rm FF}(\cos\Theta)\sin\Theta d\Theta = \frac{1}{2}\int_0^1 p_{\rm FF}(-x)dx \\ &= 1 - \frac{1-\delta_{90}^{\nu+1} - 0.5(1-\delta_{90}^{\nu})}{(1-\delta_{90})\delta_{90}^{\nu}} \end{aligned} \tag{103}$$

where $\delta_{90} = \delta(\Theta = 90°) = \frac{4}{3(\tilde{m}_{\rm r}-1)^2}\sin^2(45°) = \frac{2}{3(\tilde{m}_{\rm r}-1)^2}$. Equation (103) can be solved for ν in terms of $b_{\rm FF}$ and δ_{90}, implying that ν and thus ξ can be determined if the real part of the refractive index $\tilde{m}_{\rm r}$ and the backscatter ratio $b_{\rm FF}$ are specified. As a consequence, the FF scattering phase function can be evaluated from a measured value of $b_{\rm FF}$ if the real part of the refractive index $\tilde{m}_{\rm r}$ is known.

As already mentioned, in the CCRR bio-optical model, the Petzold scattering phase function with a backscattering ratio of 0.019 is used to represent mineral (non-algal) particles. These scattering phase functions are shown in Figure 2 together with the Rayleigh scattering phase function, which represents scattering by water molecules.

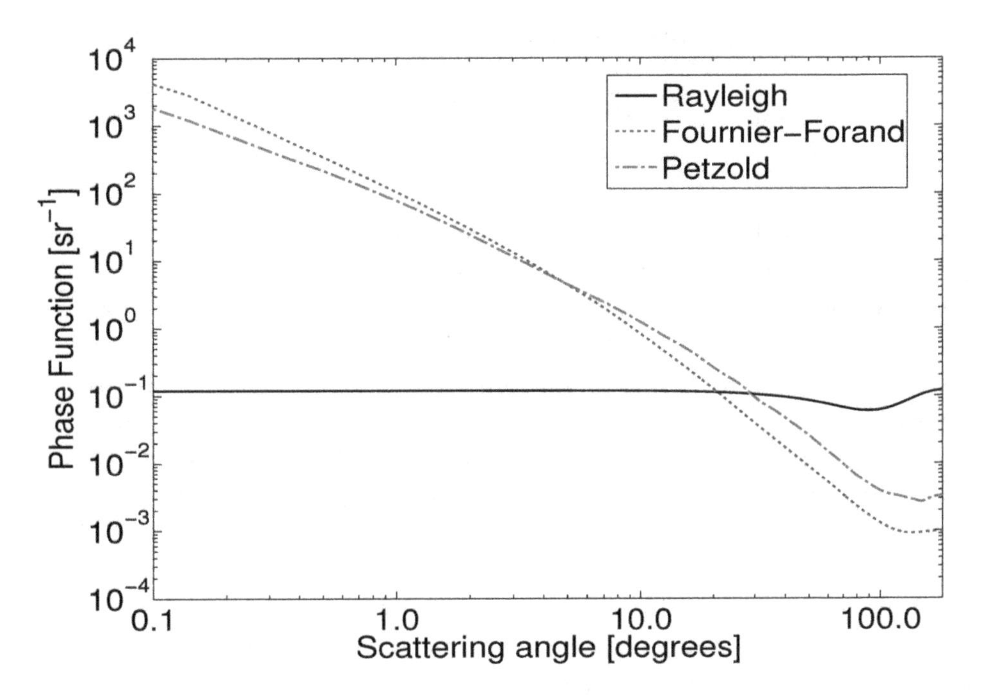

Figure 2. Rayleigh, Fournier-Forand, and Petzold scattering phase functions used to represent scattering by water molecules, pigmented particles, and non-algal particles, respectively, in the CCRR bio-optical model. To generate the FF scattering phase function the values $\xi = 3.38$ and $\tilde{m}_r = 1.068$ were used.

A moment-fitting program [68] may be used to create Legendre expansion coefficients $\chi_{\ell,\mathrm{PET}}$ and $\chi_{\ell,FF}$ for the Petzold and FF scattering phase functions. Hence, the total scattering phase function Legendre expansion coefficients are given by:

$$\chi_\ell = \frac{\beta_{\mathrm{pig}}(\lambda)\chi_{\ell,\mathrm{FF}} + \beta_{\mathrm{MIN}}(\lambda)\chi_{\ell,\mathrm{PET}} + \beta_{\mathrm{w}}(\lambda)\chi_{\ell,\mathrm{water}}}{\beta_{\mathrm{pig}}(\lambda) + \beta_{\mathrm{MIN}}(\lambda) + \beta_{\mathrm{w}}(\lambda)}. \quad (104)$$

Thus, to use the CCRR bio-optical model one must specify the three input parameters CHL, MIN, and $\alpha_{\mathrm{CDOM}}(443)$.

3.6.2. The SBC and GSM Bio-Optical Models

Based on field measurements conducted in the Santa Barbara Channel and compiled in the NOMAD data base [69], Li et al. [70] constructed a local bio-optical model representative for the SBC coastal waters. Another frequently used bio-optical model is the GSM model [71,72], which is included in NASA's SeaDAS software package. GSM is a global model which has the same structure as the SBC model, but with different coefficients.

In the SBC model, the water body, in addition to pure water, is assumed to be described by three parameters that can be varied: (i) the chlorophyll concentration [CHL] (a proxy for the concentration of pigmented particles), the CDOM absorption coefficient at 443 nm [$\alpha_{\mathrm{CDOM}}(443)$], and the total scattering coefficient at 443 nm [$\beta_{\mathrm{tot}}(443)$]. The IOPs of the SBC model are described by:

$$\alpha^{\mathrm{SBC}}_{\mathrm{pig}}(\lambda) = a^{\mathrm{SBC}}_1(\lambda)[\mathrm{CHL}]^{a_2(\lambda)} \quad (105)$$

$$\alpha^{\mathrm{SBC}}_{\mathrm{CDOM}}(\lambda) = \alpha^{\mathrm{SBC}}_{\mathrm{CDOM}}(443)\exp[-S(\lambda - 443)]; \qquad S = 0.012 \quad (106)$$

$$\beta^{\mathrm{SBC}}_{\mathrm{tot}}(\lambda) = \beta^{\mathrm{SBC}}_{\mathrm{tot}}(443)(\lambda/443). \quad (107)$$

Similarly, for the GSM model the IOPs are described by

$$\alpha_{\rm pig}^{\rm GSM}(\lambda) = a_1^{\rm GSM}(\lambda)[{\rm CHL}] \tag{108}$$

$$\alpha_{\rm CDOM}^{\rm GSM}(\lambda) = \alpha_{\rm CDOM}^{\rm GSM}(443)\exp[-S(\lambda - 443)]; \qquad S = 0.0206 \tag{109}$$

$$\beta_{\rm tot}^{\rm GSM}(\lambda) = \beta_{\rm tot}^{\rm GSM}(443)(\lambda/443)^{1.0337}. \tag{110}$$

Please note that the wavelength dependent factors $a_1^{\rm SBC}(\lambda)$ and $a_2(\lambda)$ in Equation (105) as well as $a_1^{\rm GSM}(\lambda)$ in Equation (108) are determined from field measurements compiled in the NOMAD data base. For simplicity we will assume here that $\alpha_{\rm pig}^{\rm SBC}(\lambda) = \alpha_{\rm pig}^{\rm GSM}(\lambda) \equiv \alpha_{\rm pig}^{\rm CCRR}(\lambda)$, so that the difference between the three models lies in the treatment of scattering and CDOM absorption.

For both the SBC and GSM models, CDOM represents a combination of colored dissolved organic matter and mineral particles. Hence, values of $\alpha_{\rm CDOM}^{\rm SBC}(443)$ in Equation (106) and $\alpha_{\rm CDOM}^{\rm GSM}(443)$ in Equation (109) should be compared to the sum $\alpha_{\rm CDOM}^{\rm CCRR}(443) + 0.041 \times 0.75$ MIN. The total suspended particle scattering coefficient is $\beta_{\rm tot}^{\rm SBC}(\lambda)$ and $\beta_{\rm tot}^{\rm GSM}(\lambda)$ in the SBC and GSM model, respectively, each being comparable to $\beta_{\rm tot}(\lambda) = \beta_{\rm MIN}(\lambda) + \beta_{\rm pig}(\lambda)$ in the CCRR model.

The most significant difference between these three models is that the SBC and GSM models do not separately include mineral particles although the total scattering coefficient does include the total suspended particle scattering. Another difference is that the CCRR model is based on three different reference wavelengths, namely 443 nm for $\alpha_{\rm MIN}$, 555 nm for $\gamma_{\rm MIN}$, and 660 nm for $\alpha_{\rm pig}$, while the SBC and the GSM models are based only on 443 nm as a reference wavelength. In the SBC and GSM models one may use the FF scattering phase function for pigments, but other scattering phase functions, such as the Petzold scattering phase function may work better in coastal areas.

4. Radiative Transfer in Coupled Atmosphere-Water (Including Snow/ice) Systems

4.1. *Radiative Transfer Equation—Unpolarized Radiation*

In the AccuRT computational tool, one considers a *coupled* system consisting of two adjacent slabs (atmosphere overlying a water body) separated by a plane, horizontal interface. The refractive index changes abruptly across this interface from a value $\tilde{m}_{c,1}$ in the upper slab (hereafter slab$_1$, the atmosphere) to a value $\tilde{m}_{c,2}$ in the lower slab (hereafter slab$_2$, a water body). If the IOPs in each of the two slabs vary only in the vertical direction denoted by z, where z increases upward, the corresponding vertical optical depth, denoted by $\tau(z)$, is defined by

$$\tau(z) = \int_z^\infty [\alpha(z') + \beta(z')]dz' \tag{111}$$

where the absorption and scattering coefficients α and β are defined in Equations (1). Please note that the vertical optical depth is defined to increase downward from $\tau(z = \infty) = 0$ at the top of the atmosphere. In either of the two slabs, assumed to be in local thermodynamic equilibrium so that they emit radiation according to the local temperature $T(\tau(z))$, the diffuse radiance distribution $I(\tau, u, \phi)$ can be described by the radiative transfer equation (RTE)

$$\mu\frac{dI(\tau, u, \phi)}{d\tau} = I(\tau, u, \phi) - S(\tau, u', \phi') \tag{112}$$

where

$$S(\tau, u', \phi') = S^*(\tau, u', \phi') + [1 - \varpi(\tau)]B(\tau) + \frac{\varpi(\tau)}{4\pi}\int_0^{2\pi} d\phi' \int_{-1}^{1} p(\tau, u', \phi'; u, \phi)I(\tau, u', \phi')du'. \tag{113}$$

Here u is the cosine of the polar angle θ, ϕ is the azimuth angle, $\varpi(\tau) = \beta(\tau)/[\alpha(\tau) + \beta(\tau)]$ is the single-scattering albedo, $p(\tau, u', \phi'; u, \phi)$ is the scattering phase function defined by Equation (5), and $B(\tau)$ is the thermal radiation field given by the Planck function. The differential vertical optical depth is (see Equation (111))

$$d\tau(z) = -[\alpha(\tau) + \beta(\tau)]dz \tag{114}$$

where the minus sign indicates that τ increases in the downward direction, whereas z increases in the upward direction, as noted above. The scattering angle Θ and the polar and azimuth angles are related by (see Equation (25))

$$\hat{\Omega}' \cdot \hat{\Omega} = \cos\Theta = \cos\theta \, \cos\theta' + \sin\theta' \, \sin\theta \, \cos(\phi' - \phi).$$

By definition, $\theta = 180°$ is directed toward nadir (straight down) and $\theta = 0°$ toward zenith (straight up). Thus, $u = \cos\theta$ varies in the range $[-1, 1]$ (from nadir to zenith). For cases of oblique illumination of the system, $\phi = 180°$ is defined to be the azimuth angle of the incident light.

4.1.1. Isolation of Azimuth Dependence

The azimuth dependence in Equation (112) may be isolated by expanding the scattering phase function in *Legendre polynomials*, $P_\ell(\cos\Theta)$, and making use of the addition theorem for spherical harmonics [14]

$$p(\cos\Theta) = p(u', \phi'; u, \phi) = \sum_{m=0}^{2N-1} (2 - \delta_{0,m}) p^m(u', u) \cos m(\phi' - \phi) \tag{115}$$

where $\delta_{0,m}$ is the Kronecker delta, i.e., $\delta_{0,m} = 1$ for $m = 0$ and $\delta_{0,m} = 0$ for $m \neq 0$, and

$$p^m(u', u) = \sum_{\ell=m}^{2N-1} (2l + 1) \chi_\ell \Lambda_\ell^m(u') \Lambda_\ell^m(u). \tag{116}$$

Here

$$\chi_\ell = \frac{1}{2} \int_{-1}^{1} P_\ell(\cos\Theta) p(\cos\Theta) d(\cos\Theta) \tag{117}$$

is an expansion coefficient and $\Lambda_\ell^m(u)$ is given by

$$\Lambda_\ell^m(u) \equiv \sqrt{\frac{(\ell - m)!}{(\ell + m)!}} P_\ell^m(u) \tag{118}$$

where $P_\ell^m(u)$ is an associated Legendre polynomial of order m. Expanding the radiance in a similar way,

$$I(\tau, u, \phi) = \sum_{m=0}^{2N-1} I^m(\tau, u) \cos m(\phi - \phi_0) \tag{119}$$

where ϕ_0 is the azimuth angle of the incident light, one finds that each Fourier component satisfies the following RTE (see [14] for details)

$$\mu \frac{dI^m(\tau, u)}{d\tau} = I^m(\tau, u) - \frac{\varpi(\tau)}{2} \int_{-1}^{1} p^m(\tau, u', u) \, I^m(\tau, u) d\mu - S^{*m}(\tau, u) \tag{120}$$

where $m = 0, 1, 2, \ldots, 2N - 1$ and $p^m(\mu', \mu)$ is given by Equation (116).

4.1.2. The Interface between the Two Slabs—Calm (Flat) Water Surface

When a beam of light is incident upon a plane interface between two slabs of different refractive indices, one fraction of the incident light will be reflected and another fraction will be transmitted or refracted. For unpolarized light incident upon the interface between the two slabs, the Fresnel reflectance ρ_F is given by

$$\rho_{\rm F} = \frac{1}{2}(\rho_\perp + \rho_\parallel) \tag{121}$$

where $\rho_\perp$ is the reflectance for light polarized with the electric field perpendicular to the plane of incidence, and $\rho_\parallel$ is the reflectance for light polarized with the electric field parallel to the plane of incidence [3,14,73,74]. Thus, one finds

$$\rho_{\rm F} = \frac{1}{2}\left[\left|\frac{\mu_1 - m_{\rm rat}\mu_t}{\mu_1 + m_{\rm rat}\mu_2}\right|^2 + \left|\frac{\mu_2 - m_{\rm rat}\mu_1}{\mu_2 + m_{\rm rat}\mu_1}\right|^2\right] \tag{122}$$

where $\mu_1 \equiv \mu_{\rm air} = \cos\theta_1$, θ_1 being the angle of incidence, $\mu_2 \equiv \mu_{\rm ocn} = \cos\theta_2$, θ_2 being the angle of refraction determined by Snell's law ($\tilde{m}_{r,1}\sin\theta_1 = \tilde{m}_{r,2}\sin\theta_2$), and $m_{\rm rat} = \tilde{m}_{c,2}/\tilde{m}_{c,1}$. Similarly, the Fresnel transmittance becomes

$$\mathcal{T}_{\rm F} = 2m_{\rm rel}\mu_i\mu_t\left[\left|\frac{1}{\mu_i + m_{\rm rat}\mu_t}\right|^2 + \left|\frac{1}{\mu_t + m_{\rm rat}\mu_i}\right|^2\right] \tag{123}$$

where $m_{\rm rel} = \tilde{m}_{r,2}/\tilde{m}_{r,1}$.

4.1.3. A Wind-Blown (Rough) Air-Water Interface—Pseudo-Two-Dimensional BRDF Treatment

A calm (flat) atmosphere-water interface occurs only for very low wind speeds. A wind-roughened water surface occurs more frequently and is therefore more realistic.

Consider a Cartesian coordinate system (x, y, z) in which z is the vertical coordinate. To calculate the slope distribution, $p(z_x, z_y)$, we consider a plane wave incident on a rough surface characterized by a Gaussian random height distribution $z = f(x, y) = f(\mathbf{r}_\perp)$ where $f(\mathbf{r}_\perp) = f(x, y)$ with mean height $\langle z\rangle = \langle f(x, y)\rangle \equiv \langle f(\mathbf{r}_\perp)\rangle = 0$. We now focus on a particular tilted surface facet that makes a polar angles θ_n with respect to the vertical direction and a relative azimuth angle α. Let the incident solar radiance I_i be at a zenith angle θ_0, the reflected radiance I_r be at zenith angle θ, and the relative azimuth between I_i and I_r be at angle $\Delta\phi$. Then the slope of the tilted surface facet has components z_x and z_y defined by:

$$z_x = \frac{\partial z}{\partial x} = \frac{\partial f(x, y)}{\partial x} = \sin\alpha\tan\theta_n \qquad z_y = \frac{\partial z}{\partial y} = \frac{\partial f(x, y)}{\partial y} = \cos\alpha\tan\theta_n.$$

For an anisotropic distribution of slope components (dependent on the wind direction), we define new slope components as follows:

$$z'_x = \cos(\chi)z_x + \sin(\chi)z_y \qquad z'_y = -\sin(\chi)z_x + \cos(\chi)z_y$$

where $\chi = \phi_s - \phi_W$ = rotation from the sun-observation system (x, y, z) and ϕ_W = the wind direction. The slope distribution can be written as a Gram-Charlier series [75]:

$$p(z'_x, z'_y) = \frac{1}{2\pi\sigma_x\sigma_y}\exp\left[-\frac{1}{2}\left(\frac{{z'_x}^2}{\sigma_x^2} + \frac{{z'_y}^2}{\sigma_y^2}\right)\right]\left[1 - \Delta(\xi, \eta)\right] \tag{124}$$

where $\xi = \frac{z'_x}{\sigma_x}$, $\eta = \frac{z'_y}{\sigma_y}$, σ_x^2 and σ_y^2 are variances of z'_x and z'_y, and the function $\Delta(\xi, \eta)$ represents the departure of the slope distribution from a strict two-dimensional (2D) Gaussian due to skewness and peakedness.

Please note that in the absence of skewness and peakedness ($\Delta(\xi, \eta) = 0$) the Gram-Charlier series reduces to a 2D Gaussian distribution:

$$p(z'_x, z'_y) = \frac{1}{2\pi\sigma_x\sigma_y}\exp\left[-\frac{1}{2}\left(\frac{{z'_x}^2}{\sigma_x^2} + \frac{{z'_y}^2}{\sigma_y^2}\right)\right]. \tag{125}$$

Furthermore, for an isotropic slope distribution $\chi = 0$ so that $z'_x = z_x$ and $z'_y = z_y$, $\sigma_x = \sigma_y$. Therefore $\sigma^2 = \sigma_x^2 + \sigma_y^2 = 2\sigma_x^2 = 2\sigma_x\sigma_y$, and hence

$$\frac{1}{2}(\frac{z_x'^2}{\sigma_x^2} + \frac{z_y'^2}{\sigma_y^2})] = \frac{z_x^2 + z_y^2}{2\sigma_x\sigma_y} = \frac{z_x^2 + z_y^2}{\sigma^2}; \quad \text{and} \quad z_x^2 + z_y^2 = \tan^2\theta_n(\sin^2\alpha + \cos^2\alpha) = \tan^2\theta_n$$

since $z_x = \sin\alpha\tan\theta_n$ and $z_y = \cos\alpha\tan\theta_n$. Thus, we obtain a 1D Gaussian:

$$p(\mu_n, \sigma) = p(z_x, z_y) = \frac{1}{\pi\sigma^2}\exp(-\frac{z_x^2 + z_y^2}{\sigma^2}) = \frac{1}{\pi\sigma^2}\exp(-\frac{\tan^2\theta_n}{\sigma^2}) = \frac{1}{\pi\sigma^2}\exp(-\frac{1-\mu_n^2}{\sigma^2\mu_n^2}). \quad (126)$$

where

$$\mu_n = \cos\theta_n = \frac{\mu + \mu'}{\sqrt{2(1-\cos\Theta)}} \quad (127)$$

$$\cos\Theta = -\mu\mu' + \sqrt{1-\mu^2}\sqrt{1-\mu'^2}\cos(\Delta\phi). \quad (128)$$

A Pseudo Two-Dimensional (Wind-Direction Dependent) Treatment of the BRDF

At the bottom of the atmosphere ($\tau = \tau_{atm}$), the upward reflected radiance $I^+_{refl}(\tau_{atm}, \mu', \phi')$ is connected to the downward incident diffuse radiance $I^-_{inc}(\tau_{atm}, \mu, \phi)$ and the attenuated direct radiance $F_0 e^{-\tau_{atm}/\mu_0}$ through the sea surface reflection that is described by the BRDF $\rho(\mu, \mu', \Delta\phi)$:

$$I^+_{refl}(\tau_{atm}, \mu', \phi') = \mu_0\, \rho(\mu_0, \mu', \phi')\, F_0\, e^{-\tau_{atm}/\mu_0} + \int_0^{2\pi}\int_0^1 \mu\, \rho(\mu, \mu', \Delta\phi) I^-_{inc}(\tau_{atm}, \mu, \phi) d\mu d\phi \quad (129)$$

where $\Delta\phi = \phi' - \phi$, and the solar azimuth angle was set to $\phi_0 = 0°$ so that $\Delta\phi = \phi' - \phi_0 = \phi'$ for the direct beam reflection $\rho(\mu_0, \mu', \phi')$.

The 1D BRDF $\rho(\mu, \mu', \Delta\phi')$ can be written as:

$$\rho(\mu, \mu', \Delta\phi) = \frac{1}{4\mu'\mu(\mu_n)^4} \cdot p(\mu_n, \sigma) \cdot \rho_F \cdot s(\mu, \mu', \sigma) \quad (130)$$

$$\mu_n = \frac{\mu + \mu'}{\sqrt{2(1-\cos\Theta)}} \quad (131)$$

$$\cos\Theta = -\mu\mu' + \sqrt{1-\mu^2}\sqrt{1-\mu'^2}\cos(\Delta\phi). \quad (132)$$

Here $\mu = \cos\theta$, θ being the view zenith angle for the incident light, $\mu' = \cos\theta'$, θ' being the view zenith angle for the reflected light, $\Delta\phi$ is the relative azimuth angle, and Θ is the scattering angle. In Equation (130), ρ_F is the Fresnel reflectance (see Equation (121)), $s(\mu, \mu', \sigma)$ describes the effect of shadowing, and $p(\mu_n, \sigma)$ is the surface slope distribution (Equation (126)).

In a plane-parallel (1D) geometry, the radiance and the BRDF depend only on the difference $\Delta\phi = \phi' - \phi$ in azimuth between the direction of incidence (θ', ϕ') and observation (θ, ϕ). Hence, in a strict plane-parallel geometry, it is impossible to model a wind-direction dependent (azimuthally-asymmetric) BRDF. As a consequence, most treatments of water surface roughness effects are limited to a 1D treatment [76,77].

In the 1D discrete ordinate method, the radiance $I(\tau, \mu, \phi)$ and the BRDF $\rho(\mu, \mu', \Delta\phi)$ are expanded into a Fourier cosine series to isolate the azimuth dependence [14]. The pseudo two-dimensional treatment of the BRDF employs a 2D BRDF to compute the direct beam reflectance, but a 1D (Fourier expanded) BRDF to compute the reflectance due to diffuse, multiply scattered light. A post-processing step, which corrects the direct beam reflectance (1D $\rightarrow$ 2D), is used for implementation. This post-processing method is similar to the Nakajima-Tanaka (NT) single-scattering correction [78], which retains the multiply scattered radiance, but corrects the singly scattered radiance without considering boundary reflection. In DISORT3, Lin et al. [79] improved the NT procedure by adding a BRDF correction and the same strategy was used to add a 2D BRDF correction [80]. Hence, the correction term for radiance $I_{\text{ss corr}}(\hat{\tau}, \pm\mu, \phi)$ can be written as:

$$I^{\pm}_{\text{ss corr}}(\hat{\tau}, \mu, \phi) = I^{\pm *}_{\text{ss}}(\hat{\tau}, \mu, \phi) - \tilde{I}^{\pm *}_{\text{ss}}(\hat{\tau}, \mu, \phi) + \mu_0 F_0 \Big\{ \rho_{2D}(\mu, \phi; -\mu_0, \phi_0) - \rho_{1D}(\mu, \phi; -\mu_0, \phi_0) \Big\} e^{-\frac{\hat{\tau}_b}{\mu_0} + \frac{\hat{\tau} - \hat{\tau}_b}{\mu}}. \quad (133)$$

Here $\hat{\tau}$ and $\hat{\tau}_b$ are the scaled optical thicknesses at the height of interest and the lower boundary, respectively [78,79]. On the right hand side of Equation (133), $I^{\pm *}_{\text{ss}}(\hat{\tau}, \mu, \phi) - \tilde{I}^{\pm *}_{\text{ss}}(\hat{\tau}, \pm\mu, \phi)$ is the original NT correction [78], $\rho_{2D}(\mu, \phi; -\mu_0, \phi_0)$ is the new 2D BRDF used to compute the 2D single-scattering contribution, $\rho_{1D}(\mu, \phi; -\mu_0, \phi_0)$ is the 1D (Fourier expanded) BRDF used to compute the approximate multiple scattering contribution, and $e^{-\frac{\hat{\tau}_b}{\mu_0} + \frac{\hat{\tau} - \hat{\tau}_b}{\mu}}$ is the beam attenuation coefficient.

For multiply scattered light, we use a 1D Gaussian surface slope distribution given by Equation (126), which is widely used in remote sensing applications to represent the slope statistics of water waves with the numerical value of the slope variance parameterized in terms of the wind speed [75]. The 1D BRDF given by Equations (130)–(132) and (126) is suitable for describing "skyglint", that is, the reflectance of downward diffuse light from a rough water surface, because multiple scattering in the atmosphere has made the radiation field approximately 1D, implying that 2D BRDF effects become relatively unimportant for the reflected diffuse skylight [76,77]. Similarly, the slope distribution for a 2D Gaussian surface is given by Equation (125). It will be shown in Section 5.3 that a 2D BRDF treatment is required in RT models to reproduce detailed measurements of the BRDF [80].

4.2. Radiative Transfer Equation—Polarized Radiation

To generalize Equation (112) to apply to polarized radiation, we note that the multiple scattering term $S^{\text{ms}}(\tau, u, \phi) = \frac{\varpi(\tau)}{4\pi} \int_0^{2\pi} d\phi' \int_{-1}^{1} du' p(\tau, u', \phi'; u, \phi) I(\tau, u', \phi')$ in Equation (113) must be replaced by

$$\mathbf{S}^{\text{ms}}(\tau, u, \phi) = \frac{\varpi(\tau)}{4\pi} \int_0^{2\pi} d\phi' \int_{-1}^{1} du' \mathbf{P}(\tau, u', \phi'; u, \phi) \mathbf{I}(\tau, u', \phi') \quad (134)$$

where $\mathbf{I}(\tau, u', \phi')$ is the Stokes vector, and $\mathbf{P}(\tau, u', \phi'; u, \phi)$ is the scattering phase matrix (see Section 3.2). The first element of the vector $\mathbf{S}^{\text{ms}}(\tau, u, \phi)$ represents the energy per unit solid angle, per unit frequency interval, and per unit time that is scattered by a unit volume in the direction $(u = \cos\theta, \phi)$. Hence, in a plane-parallel (slab) geometry, the integro-differential equation for polarized radiative transfer is expressed in terms of a Stokes vector $\mathbf{I}(\tau, u, \phi)$ as

$$u \frac{d\,\mathbf{I}(\tau, u, \phi)}{d\tau} = \mathbf{I}(\tau, u, \phi) - \mathbf{S}(\tau, u, \phi) \quad (135)$$

where the source vector is

$$\mathbf{S}(\tau, u, \phi) = \frac{\varpi(\tau)}{4\pi} \int_0^{2\pi} d\phi' \int_{-1}^{1} du' \, \mathbf{P}(\tau, u', \phi'; u, \phi) \mathbf{I}(\tau, u', \phi') + \mathbf{Q}(\tau, u, \phi). \quad (136)$$

In the upper slab (slab$_1$, atmosphere), the source term $\mathbf{Q}(\tau, u, \phi)$, due to thermal and beam sources, is given by:

$$\mathbf{Q}_1(\tau, u, \phi) = \frac{\varpi(\tau)}{4\pi} \mathbf{P}(\tau, -\mu_0, \phi_0; u, \phi) \mathbf{S}_b e^{-\tau/\mu_0} + [1 - \varpi(\tau)] \, \mathbf{S}_t(\tau) + \frac{\varpi(\tau)}{4\pi} \mathbf{P}(\tau, \mu_0, \phi_0; u, \phi) \mathbf{R}_F(-\mu_0, m_{\text{rel}}) \mathbf{S}_b \, e^{-\frac{(2\tau_a - \tau)}{\mu_0}}. \quad (137)$$

The first term on the right hand side of Equation (137) describes the incident beam $\mathbf{S}_b$ in direction $(-\mu_0, \phi_0)$, which is attenuated at depth τ by a factor $e^{-\tau/\mu_0}$ and undergoes single scattering into the direction (u, ϕ). For an unpolarized incident beam $\mathbf{S}_b$ has the form

$$\mathbf{S}_b = [I_0/2, I_0/2, 0, 0]^T \text{ or } [I_0, 0, 0, 0]^T \tag{138}$$

where the first or second expression corresponds to the choice of Stokes vector representation, $[I_{\|}, I_{\perp}, U, V]^T$ or $[I, Q, U, V]^T$. The second term on the right hand side of Equation (137) is due to thermal emission, which is unpolarized, and $\mathbf{S}_t(\tau)$ is given by

$$\mathbf{S}_t(\tau) = [B(T(\tau))/2, B(T(\tau))/2, 0, 0]^T \quad \text{or} \quad [B(T(\tau)), 0, 0, 0]^T \tag{139}$$

where B is the Planck function, and where the first or second expression corresponds to the choice of Stokes vector representation. We have set $\mu_0 \equiv |u_0| \equiv |\cos\theta_0|$, where θ_0 is the polar angle of the incident light beam. The third term on the right hand side of Equation (137) describes radiation due to the incident beam $\mathbf{S}_b$ that has been attenuated by the factor $e^{-\tau_a/\mu_0}$ before reaching the air-water interface, undergoing Fresnel reflection given by the reflection matrix $\mathbf{R}_{\mathrm{F}}(-\mu_0, m_{\mathrm{rel}})$, attenuated by the factor $e^{-(\tau_a-\tau)/\mu_0}$ to reach the level τ in the atmosphere, and finally singly scattered from direction (μ_0, ϕ_0) into direction (u, ϕ) described by the factor $\frac{\varpi(\tau)}{4\pi}\mathbf{P}(\tau, \mu_0, \phi_0; u, \phi)$. Thus, the incident beam propagates though the entire atmosphere and a portion of it is reflected upwards by the interface to reach depth τ in the atmosphere, which explains the factor $e^{-(2\tau_a-\tau)/\mu_0}$.

In the lower slab (slab$_2$, water), the source term becomes

$$\begin{aligned}\mathbf{Q}_2(\tau, u, \phi) \quad &= \quad \frac{\varpi(\tau)}{4\pi}\,\mathbf{P}(\tau, -\mu_0^w, \phi_0; u, \phi)\mathbf{S}_b\, e^{-\tau_a/\mu_0} \\ &\times \mathbf{T}_{\mathrm{F}}(-\mu_0, m_{\mathrm{rel}})\frac{\mu_0}{\mu_0^w}e^{-(\tau-\tau_a)/\mu_0^w} \\ &+ [1-\varpi(\tau)]\,\mathbf{S}_t(\tau)\end{aligned} \tag{140}$$

where $\mathbf{T}_{\mathrm{F}}(-\mu_0, m_{\mathrm{rel}})$ is the Fresnel transmission matrix. The first term in Equation (140) is due to the incident beam $\mathbf{S}_b$ that has been attenuated through the atmosphere by the factor $e^{-\tau_a/\mu_0}$, transmitted into the water by the factor $\mathbf{T}_{\mathrm{F}}(-\mu_0, m_{\mathrm{rel}})\frac{\mu_0}{\mu_0^w}$, further attenuated by the factor $e^{-(\tau-\tau_a)/\mu_0^w}$ to reach depth τ in the water, and singly scattered from the direction $(-\mu_0^w, \phi_0)$ into the direction (u, ϕ) which explains the factor $\frac{\varpi(\tau)}{4\pi}\mathbf{P}(\tau, -\mu_0^w, \phi_0; u, \phi)$. The second term in Equation (140) is due to thermal emission in the water.

Isolation of Azimuth Dependence

We start by expanding the scattering phase matrix in a Fourier series:

$$\mathbf{P}(u', u; \phi' - \phi) = \sum_{m=0}^{M} \{\mathbf{P}_{\mathrm{c}}^m(u', u)\cos m(\phi' - \phi) + \mathbf{P}_{\mathrm{s}}^m(u', u)\sin m(\phi' - \phi)\}. \tag{141}$$

To isolate the azimuth dependence of the radiation field we expand the Stokes vector $\mathbf{I}(\tau, u, \phi)$ in Equation (135) and the source term $\mathbf{Q}_1(\tau, u, \phi)$ in Equation (137) or $\mathbf{Q}_2(\tau, u, \phi)$ in Equation (140) in a Fourier series in a manner similar to the expansion of the scattering phase matrix in Equation (141):

$$\mathbf{I}(\tau, u, \phi) = \sum_{m=0}^{M} \left\{\mathbf{I}_{\mathrm{c}}^m(\tau, u)\cos m(\phi_0 - \phi) + \mathbf{I}_{\mathrm{s}}^m(\tau, u)\sin m(\phi_0 - \phi)\right\} \tag{142}$$

$$\mathbf{Q}_p(\tau, u, \phi) = \sum_{m=0}^{M} \left\{ \mathbf{Q}_{\mathrm{cp}}^m(\tau, u) \cos m(\phi_0 - \phi) + \mathbf{Q}_{\mathrm{sp}}^m(\tau, u) \sin m(\phi_0 - \phi) \right\} \tag{143}$$

where the subscript s or c denotes sine or cosine mode and the subscript p indicates the slab, p $=1$ for slab$_1$, and p $=2$ for slab$_2$. Using these expansions it can be shown that we obtain the following equations for the Fourier components (see [3] for details)

$$\begin{aligned} u\frac{d\mathbf{I}_{\mathrm{c}}^m(\tau,u)}{d\tau} &= \mathbf{I}_{\mathrm{c}}^m(\tau,u) - \frac{\varpi(\tau)}{4}\int_{-1}^{1} du' \left\{ \mathbf{P}_{\mathrm{c}}^m(\tau,u',u)\,\mathbf{I}_{\mathrm{c}}^m(\tau,u')\,(1+\delta_{0m}) \right. \\ &\quad \left. - \mathbf{P}_{\mathrm{s}}^m(\tau,u',u)\,\mathbf{I}_{\mathrm{s}}^m(\tau,u') \right\} - \mathbf{Q}_{\mathrm{c}}^m(\tau,u) \end{aligned} \tag{144}$$

$$\begin{aligned} u\frac{d\mathbf{I}_{\mathrm{s}}^m(\tau,u)}{d\tau} &= \mathbf{I}_{\mathrm{s}}^m(\tau,u) - \frac{\varpi(\tau)}{4}\int_{-1}^{1} du' \left\{ \mathbf{P}_{\mathrm{c}}^m(\tau,u',u)\,\mathbf{I}_{\mathrm{s}}^m(\tau,u') \right. \\ &\quad \left. + \mathbf{P}_{\mathrm{s}}^m(\tau,u',u)\,\mathbf{I}_{\mathrm{c}}^m(\tau,u') \right\} - \mathbf{Q}_{\mathrm{s}}^m(\tau,u). \end{aligned} \tag{145}$$

The discrete ordinate method consists of replacing the integration over u' by a discrete sum using Gaussian quadrature points u_j (the discrete ordinates) and corresponding weights w_j. One obtains for each Fourier component:

$$\begin{aligned} u_i\frac{d\mathbf{I}_{\mathrm{c}}^m(\tau,u_i)}{d\tau} &= \mathbf{I}_{\mathrm{c}}^m(\tau,u_i) \\ &- \frac{\varpi(\tau)}{4}\sum_{\substack{j=-N\\ j\neq 0}}^{N} w_j \left\{ (1+\delta_{0m})\,\mathbf{P}_{\mathrm{c}}^m(\tau,u_j,u_i)\,\mathbf{I}_{\mathrm{c}}^m(\tau,u_j) \right. \\ &\left. - \mathbf{P}_{\mathrm{s}}^m(\tau,u_j,u_i)\,\mathbf{I}_{\mathrm{s}}^m(\tau,u_j) \right\} - \mathbf{Q}_{\mathrm{c}}^m(\tau,u_i), \quad i=\pm1,\ldots,\pm N \end{aligned} \tag{146}$$

$$\begin{aligned} u_i\frac{d\mathbf{I}_{\mathrm{s}}^m(\tau,u_i)}{d\tau} &= \mathbf{I}_{\mathrm{s}}^m(\tau,u_i) \\ &- \frac{\varpi(\tau)}{4}\sum_{\substack{j=-N\\ j\neq 0}}^{N} w_j \left\{ \mathbf{P}_{\mathrm{c}}^m(\tau,u_j,u_i)\,\mathbf{I}_{\mathrm{s}}^m(\tau,u_j) \right. \\ &\left. + \mathbf{P}_{\mathrm{s}}^m(\tau,u_j,u_i)\,\mathbf{I}_{\mathrm{c}}^m(\tau,u_j) \right\} - \mathbf{Q}_{\mathrm{s}}^m(\tau,u_i), \quad i=\pm1,\ldots,\pm N. \end{aligned} \tag{147}$$

The convention for the indices of the quadrature points is such that $u_j < 0$ for $j < 0$, and $u_j > 0$ for $j > 0$. These points are distributed symmetrically about zero, i.e., $u_{-j} = -u_j$. The corresponding weights are equal, i.e., $w_{-j} = w_j$.

Each of the two slabs (atmosphere and water) is divided into several adjacent layers, large enough to resolve vertical changes in the IOPs of each slab. Equations (147) and (148) apply in each layer in the atmosphere or water. As described in some detail elsewhere [3] the solution involves the following steps:

1. the homogeneous version of Equations (147) and (148) with $\mathbf{Q}_{\mathrm{c}}^m = \mathbf{Q}_{\mathrm{s}}^m = 0$ yields a linear combination of exponential solutions (with unknown coefficients) obtained by solving an algebraic eigenvalue problem;
2. analytic particular solutions are found by solving a system of linear algebraic equations;
3. the general solution is obtained by adding the homogeneous and particular solutions;
4. the solution is completed by imposing boundary conditions at the top of the atmosphere and the bottom of the water;

5. the solutions are required to satisfy continuity conditions across layer interfaces in the atmosphere and the water, and last but not least to satisfy Fresnel's equations and Snell's law at the atmosphere-water interface, where there is an abrupt change in the refractive index;
6. the application of boundary, layer interface, and atmosphere-water interface conditions leads to a sparse system of linear algebraic equations, and the numerical solution of this system of equations yields the unknown coefficients in the homogenous solutions.

4.3. Summary of AccuRT

We have described a computational tool, AccuRT, for radiative transfer simulations in a coupled system consisting of two adjacent horizontal slabs with different refractive indices. The computer code accounts for reflection and transmission at the interface between the two slabs, and allows for each slab to be divided into a sufficiently large number of layers to resolve the variation in the IOPs, described in Section 3, with depth in each slab.

The user interface of AccuRT is designed to make it easy to specify the required input including wavelength range, solar forcing, and layer-by-layer IOPs in each of the two slabs as well as the two types of desired output:

- irradiances and mean radiances (scalar irradiances) at desired vertical positions in the coupled system;
- total radiances and polarized radiances (including degree of polarization) in desired directions and vertical positions in the coupled system.

5. Examples of Forward-Inverse Modeling

5.1. Introduction

A primary goal in *remote sensing* of the Earth from space is to retrieve information about atmospheric and surface properties from measurements of the radiation emerging at the top-of-the-atmosphere (TOA) at several wavelengths [24,81]. These *retrieval parameters* (RPs), including cloud phase and optical depth, aerosol type and loading, and concentrations of aquatic constituents in an open ocean or coastal water area, depend on the *inherent optical properties* (IOPs) of the atmosphere and the water. If there is a model providing a link between the RPs and the IOPs, a forward radiative transfer (RT) model can be used to compute how the measured TOA radiation field should respond to changes in the RPs, and an inverse RT problem can be formulated and solved to derive information about the RPs [3,82]. A *forward RT model*, employing IOPs that describe how atmospheric and aquatic constituents absorb and scatter light, can be used to compute the *multiply scattered light field* in any particular direction (with specified polar and azimuth angles) at any particular depth level (including the TOA) in a vertically *stratified medium*, such as a coupled atmosphere-water system [83]. In order to solve the *inverse RT problem* it is important to have an accurate and efficient forward RT model. Accuracy is important in order to obtain reliable and robust retrievals, and efficiency is an issue because standard iterative solutions of the *nonlinear inverse RT problem* require executing the forward RT model repeatedly to compute the radiation field and its partial derivatives with respect to the RPs (the *Jacobians*) [82].

In addition to scalar forward RT models, vector RT models that consider polarization are important (see Section 4). Numerous RT models that include polarization effects are available (see Zhai et al. [84] and references therein for a list of papers), and the interest in applications based on vector RT models that apply to coupled atmosphere-water systems is growing. Examples of vector RT modeling pertinent to a coupled atmosphere-water system include applications based on the *doubling-adding method* [85–87], the *successive order of scattering method* [84,88,89], the *matrix operator method* [90,91], and *Monte Carlo methods* [92,93].

The purpose of this section is not provide a comprehensive review of forward-inverse methodology, but rather to provide a few examples of how RT modeling involving coupled atmosphere-water systems described in the previous sections can be used to solve the inverse problem with an emphasis of how machine learning techniques (neural networks) can be used to our advantage.

5.2. Bidirectional Reflectance of Water—Why Is It Important?

The **B**idirectional **R**eflectance **D**istribution **F**unction (BRDF) is defined as the ratio of the reflected radiance to the incident power per unit surface area:

$$\rho(\mu,\phi;-\mu',\phi') = \frac{dI_{\text{refl}}(\tau^*,\mu,\phi)}{I(\tau^*,-\mu',\phi')\;\mu'\;d\mu'd\phi'}. \tag{148}$$

Here $dI_{\text{refl}}(\tau^*,\mu,\phi)$ is the reflected radiance in direction (μ,ϕ), while $I(\tau^*,-\mu',\phi')$ is the incident radiance in direction $(-\mu',\phi')$. Understanding bidirectional effects including sunglint is important for several reasons [63,94]:

1. correct interpretation of ocean color data;
2. comparing consistency of spectral radiance data derived from space observations with a single instrument for a variety of illumination and viewing conditions;
3. merging data collected by different instruments operating simultaneously.

The BRDF defined in Equation (148) has unit per steradian [sr^{-1}]. The remote sensing reflectance defined as $R_{\text{rs}} = I(0^+,\mu,\phi)/F^-(0^+)$, where 0^+ refers the level just above the air-water interface, and $F^-(0^+)$ is the downward irradiance, also has unit [sr^{-1}]. It should be noted that the frequently used bidirectional reflectance factor defined as BRF = πR_{rs} is dimensionless, because π has unit [sr].

A BRDF correction algorithm [63] (denoted as MAG02) was developed for application to open ocean water based on the following expression for the normalized water-leaving radiance nL_w

$$nL_w = L_w \times \frac{\mathfrak{R}_o}{\mathfrak{R}} \times \frac{f_0(\tau_a,W,\text{IOP})}{Q_0(\tau_a,W,\text{IOP})} \times \left[\frac{f(\theta_0,\theta,\Delta\phi,\tau_a,W,\text{IOP})}{Q(\theta_0,\theta,\Delta\phi,\tau_a,W,\text{IOP})}\right]^{-1}$$

where W is the wind speed, and the function $\mathfrak{R}$ accounts for refraction and reflection effects when radiances propagate through the air-water interface. The function f relates the irradiance reflectance $(R = F^+/F^-)$ to the IOPs, and the function Q is a bidirectional function, defined as $Q(\theta_0,\theta',\Delta\phi) = F^+(0^-)/I^+(0^-,\theta_0,\theta',\Delta\phi)$. The subscripts "0" on $\mathfrak{R}_0$, f_0 and Q_0 are the values of the three functions evaluated in the nadir direction.

However, the MAG02 algorithm requires knowledge of CHL to derive the f/Q correction factor, and it does not work well in turbid (coastal) water. To remedy these shortcomings, Fan et al. [95] developed a neural network method to correct for bidirectional effects in water-leaving radiances for both clear (open ocean) and turbid (coastal) water. This neural network algorithm directly derives the entire spectral nadir remote sensing reflectances $R_{\text{rs}}(\lambda_i,\theta_0)$ from the angular values $R_{\text{rs}}(\lambda_i,\theta_0,\theta,\Delta\phi)$, without any prior knowledge of the water IOPs. Based on AccuRT simulations, Fan et al. [95] showed that differences in spectral R_{rs} values are significant between clear (open ocean) and turbid (coastal) water, but relatively small between nadir- and slant-viewing directions for a given water type. Consequently, a trained Radial Basis Function Neural Network (RBF-NN) can be used to convert the spectral R_{rs} values from the slant- to the nadir-viewing direction.

To this end, AccuRT was used to simulate R_{rs} values at both nadir- and slant-viewing directions for a 13-layer atmosphere with aerosols added in the bottom 0–2 km layer, by randomly selecting aerosol models based on fraction of small-mode aerosol particles (f_a) and relative humidity (RH). The CCRR bio-optical model, parameterized in terms of CHL, CDOM, and MIN as described above, was used to represent the water IOPs. To obtain the water-leaving radiance, the upward radiance was computed just above the ocean surface twice using AccuRT. Assuming the ocean to be black, i.e., totally absorbing (no scattering), the upward radiance was first computed just above the ocean surface, $I^+_{\text{black}}(0^+,\lambda,\theta_0,\theta,\Delta\phi)$, which includes the radiance due to Fresnel reflection of direct attenuated sunlight and skylight by the air-water interface, but no radiance from the water. The second time the ocean with water and its embedded constituents, was included and the radiance $I^+(0^+, \lambda, \theta_0, \theta, \Delta\phi)$ was computed. Hence, $I^+(0^+, \lambda, \theta_0, \theta, \Delta\phi)$ included the water-leaving radiance as well as the Fresnel reflected direct attenuated sunlight and skylight. Then the water-leaving radiance was obtained from the difference

$$L_{\rm w}(0^+, \lambda, \theta_0, \theta, \Delta\phi) = I^+(0^+, \lambda, \theta_0, \theta, \Delta\phi) - I^+_{\rm black}(0^+, \lambda, \theta_0, \theta, \Delta\phi). \tag{149}$$

To work satisfactorily, a neural network must be properly trained. For this purpose AccuRT was used to generate a training dataset containing 30,000 data points of $I^+(0^+, \lambda, \theta_0, \theta, \Delta\phi)$ and $I^+_{\rm black}(0^+, \lambda, \theta_0, \theta, \Delta\phi)$ at seven wavelengths, 412, 443, 490, 510, 560, 620, and 665 nm, which are similar to the wavelengths used in the MAG02 algorithm. The synthetic dataset was generated by randomly selecting 5000 combinations of the aerosol optical depths at 865 nm ($\tau_a(865)$), the fraction of small aerosol particles (f_a), the relative humidity (RH), and the three ocean parameters: CHL, MIN, and CDOM. To cover a wide range of water and atmospheric IOPs, these six parameters were randomly sampled logarithmically from the following ranges: (i) $\tau_a(865)$: 0.001–0.5, (ii) f_a: 1–95 [%], (iii) RH: 30–95 [%], (iv) CHL: 0.01–100 [mg·m^{-3}], (v) MIN: 0.01–100 [g·m^{-3}], (vi) CDOM: 0.001–10 [m^{-1}]. Then for each case in the 5000 combinations, six combinations of the Sun-sensor geometry were randomly selected in the following ranges: (i) θ_0: 0–80 [°], (ii) θ: 0–70 [°], (iii) $\Delta\phi$: 0–180 [°]. The downward irradiance just above the ocean surface ($F^-(\lambda, 0^+)$) was also computed for each case as well as the remote sensing reflectance:

$$R_{\rm rs}(\lambda, \theta_0, \theta, \Delta\phi) = \frac{L_{\rm w}(0^+, \lambda, \theta_0, \theta, \Delta\phi)}{F^-(\lambda, 0^+)} \tag{150}$$

where $L_{\rm w}(0^+, \lambda, \theta_0, \theta, \Delta\phi)$ is given by Equation (149).

The remote sensing reflectances were arranged into two groups: (i) one consisting of all the angle-dependent remote sensing reflectances $R_{\rm rs}(\lambda_i, \theta_0, \theta, \Delta\phi)$, the other consisting of the the corresponding nadir remote sensing reflectances, $R_{\rm rs}(\lambda_i, \theta_0)$. Then a neural network with two hidden layers was created. The first layer used RBFs as neurons, while the second layer used a linear function as neurons.

The input to the neural network training was the three geometry angles $\theta_0, \theta, \Delta\phi$ plus the angle-dependent remote sensing reflectances $R_{\rm rs}(\lambda_i, \theta_0, \theta, \Delta\phi)$, while the output consisted of the corresponding nadir remote sensing reflectances, $R_{\rm rs}(\lambda_i, \theta_0)$. After the training, the nadir remote sensing reflectances can be derived from a single equation:

$$R_{\rm rs}(\lambda_i, \theta_0) = \sum_{j=1}^{N} a_{ij} \exp[-b^2 \sum_{k=1}^{N_{\rm in}} (p_k - c_{jk})^2] + d_i \tag{151}$$

where N is the number of neurons, b and c_{jk} are the bias and weight in the first layer, a_{ij} and d_i are the weight and bias in the second layer. These weights and biases are optimized from the training procedure to minimize the error between the neural network derived $R_{\rm rs}(\lambda_i, \theta_0)$ and the actual $R_{\rm rs}(\lambda_i, \theta_0)$ values in the training dataset. $N_{\rm in}$ is the number of input parameters, which in our neural network equals 10: three geometry angles plus seven wavelengths. The input parameters are denoted by p_k, which in this case are the three geometry angles and the angle-dependent remote sensing reflectances, $R_{\rm rs}(\lambda, \theta_0, \theta, \Delta\phi)$, at each of the seven wavelengths.

As discussed by Fan et al. [95], this neural network approach to convert remote sensing reflectances from actual slant-viewing to nadir-viewing directions was tested using synthetic data as well as field measurements. The results can be summarized as follows (see [95] for details):

- The generally anisotropic remote sensing reflectance of oceanic water must be corrected in remote sensing applications that make use of the nadir water-leaving radiance (or remote sensing reflectance) to derive ocean color products.

- The standard MAG02 correction method [63], based on the open ocean assumption, is unsuitable for turbid waters, such as rivers, lakes, and coastal water. The MAG02 method requires the chlorophyll concentration as an input, which is a drawback in remote sensing applications, because the chlorophyll concentration is generally produced from the corrected remote sensing reflectance.
- To meet the need for a correction method that works for water that may be dominated by turbidity or CDOM, Fan et al. [95] developed a neural network method that directly converts the remote sensing reflectance from the slant-viewing to the nadir-viewing direction.
- The neural network was trained using remote sensing reflectances at slant and nadir directions generated by a radiative transfer model (AccuRT), in which scattering phase functions for algal and non-algal particles were adopted. Therefore, the remote sensing reflectance implicitly contains information about the shape of the scattering phase function which affects the BRDF.
- This method uses spectral remote sensing reflectances as input. Hence, it does not require any prior knowledge of the water constituents or their optical properties.
- Tests based on synthetic data show that this method is sound and accurate. Validation using field measurements [96] shows that this neural network method works equally well compared to the standard method [63] for open ocean or chlorophyll-dominated water. For turbid coastal water a significant improvement over the standard method was found, especially for water dominated by sediment particles.

5.3. Sunglint: A Nuisance or Can Can It Be Used to Our Advantage?

For clarity, we should note that "glint" here refers only to Fresnel reflectance from the (calm or "wind-roughened") water surface. In the presence of glint, satellite remote sensing remains a challenging problem [97]. The contribution from glint to the TOA radiance is large enough to dominate the signals received by sensors deployed in space. Algorithms developed for current satellite sensors such as the Sea-viewing Wide Field of view Sensor (SeaWiFS), the MODerate-resolution Imaging Spectroradiometer (MODIS), the MEdium Resolution Imaging Spectrometer (MERIS), the Polarization and Directionality of Earth Reflectances (POLDER) instrument, and the Global Imager (GLI), use different correction algorithms [77,98,99] based on the same principle: estimate the glint contribution inferred from a statistical glint model and a direct beam reflectance and then remove its contribution from the signal received by the sensor.

To analyze remotely sensed radiances obtained by instruments such as SeaWiFS, MODIS, and MERIS, NASA has developed a comprehensive data analysis software package (SeaWiFS Data Analysis System, SeaDAS), which performs several tasks, including cloud screening and calibration, required to convert the raw satellite signals into calibrated TOA radiances. The SeaDAS software package also has tools aimed at quantifying and removing the atmospheric contribution to the TOA radiance (atmospheric correction) as well as contributions from whitecaps and sunglint due to reflections from the ocean surface [81].

In the SeaDAS algorithm (and similar algorithms) a sunglint flag is activated to mask out pixels for which the reflectance or BRDF, exceeds a certain threshold. If the reflectance for a given pixel is above the threshold, the signal is not processed. If the reflectance is below the threshold, a directly transmitted radiance (DTR) approach is used to calculate the TOA sunglint radiance in the SeaDAS algorithm. Thus, it is computed assuming that the direct beam and its reflected portion only experience exponential attenuation through the atmosphere [77], that is

$$I_{\text{glint}}^{\text{TOA}}(\mu_0, \mu, \Delta\phi) = F_0(\lambda) T_0(\lambda) T(\lambda) I_{GN} \tag{152}$$

$$T_0(\lambda)T(\lambda) = \exp\Big\{ - [\tau_M(\lambda) + \tau_A(\lambda)]\Big(\frac{1}{\mu_0} + \frac{1}{\mu}\Big)\Big\} \quad (153)$$

where μ_0 and μ are cosines of the solar zenith angle and polar viewing angle, respectively, and the normalized sunglint radiance I_{GN} is the radiance that would result for a transparent atmosphere if the incident solar irradiance were $F_0(\lambda) = 1$. The Rayleigh (molecular) and aerosol optical thicknesses are denoted $\tau_M(\lambda)$ and $\tau_A(\lambda)$, respectively. The downward diffuse incident light (sunlight being multiply scattered by atmospheric molecules and aerosols before hitting the rough sea surface) also contributes to the upward reflectance. In the SeaDAS algorithm, such diffuse light reflectance that accounts for the effect of ocean surface roughness has been included only in the Rayleigh lookup tables [100].

Radiative transfer (RT) simulations may provide a more complete look at the glint problem [87]. An RT model can be used for accurate quantification of contributions not only from direct sunglint, but also from skyglint due to multiply scattered light [101]. Hence, RT simulations can be used to test current correction methods and explore the potential for extending remote sensing into strong glint situations masked out by the current SeaDAS algorithm.

As alluded to in Section 4.1.3, plane parallel RT models assume that the BRDF depends only on the difference in azimuth between the sun-sensor directions. Therefore, they are intrinsically one-dimensional, and cannot be used to simulate the directional dependence of realistic slope distributions that require a wind-direction dependent (hereafter referred to as 2D BRDF) treatment. Also, few studies have focussed on validation of realistic 2D BRDF implementations due to the general lack of a complete set of reflectance measurements that would be suitable for testing purposes.

To enable more realistic simulations of ocean glint reflectance, Lin et al. [80] developed a RT model with a 2D BRDF to mimic the nature of actual sea surface slope distributions (see Section 4.1.3 for details). Reflectance measurements obtained by an instrument deployed on a National Aeronautics and Space Administration (NASA) aircraft were used for validation. The goal was to match simulated reflectances with those measured by the instrument deployed on the NASA airplane, and to use RT simulations as a forward model to invert the measured reflectance in order to retrieve wind direction, sea surface slopes in the crosswind and upwind directions, and aerosol optical thickness. These parameters are important for atmospheric correction, which is the largest source of error and uncertainty in determining water-leaving radiance from space.

5.3.1. BRDF Measurements

NASA CAR Instrument

The measurements used by Lin et al. [80] were obtained under clear sky conditions from the NASA Cloud Absorption Radiometer (CAR) deployed aboard the University of Washington Convair 580 (CV-580) research aircraft [102]. The CAR is an airborne multi-wavelength scanning radiometer that measures scattered light in 14 spectral bands between 0.34 and 2.30 μm. To measure BRDFs, the airplane flew in a circle about 3 km in diameter, taking roughly 2–3 min to complete an orbit about 200 m above the surface. A servo control system is installed to allow the instrument to point at any angle from zenith to nadir, and to compensate for variations in airplane roll angle down to a fraction of a degree. Multiple circular orbits were acquired over a selected surface so that average BRDFs would be smooth. Radiometric calibration was performed at Goddard Space Flight Center prior to and just after the field experiment and a linear change between them is assumed. For more details about the BRDF measurements, see [102].

CAR measurements provide accurate BRDFs for all geometry angles including zenith viewing angles from the nadir ($\theta = 0°$) to the horizon direction ($\theta = 90°$) over all relative azimuth angles (0–360°). The resolution for both polar and azimuth angles is 1°. The BRDF measurements are also accompanied by concurrent measurements of atmospheric aerosol optical thickness above the airplane and wind speeds from the NOAA Marine Environmental Buoy Database.

5.3.2. Radiative Transfer Simulations

The DISORT code [103] implemented in AccuRT provides accurate computations of singly and multiply scattered radiances in a turbid medium. The DISORT code has been used in a great variety of studies including remote sensing applications. Lin et al. [79] developed an upgraded version called DISORT3 with improved BRDF capabilities. The DISORT3 code can be obtained from the following web site: http://lllab.phy.stevens.edu/disort/. DISORT3 was further modified and optimized [80] to simulate the 2D (wind-direction dependent) nature of the surface reflectance.

Atmospheric Input & Output

To minimize the influence of light backscattered from the water, we used a near infrared wavelength at 1036 nm, with significant water absorption, and adopted two atmospheric layers: one Rayleigh (molecular) layer (2–10 km) and one layer with aerosols and molecules homogeneously mixed (0–2 km). Based on the US standard atmosphere [104], the single-scattering albedo at 1036 nm is $\varpi_{\text{mol}} = 0.9610$ in the upper layer (molecular scattering and water vapor absorption, no aerosols) and the upper layer optical thickness is $\tau_{\text{mol}} = 0.00645$. In the lower layer, we adopted an aerosol model implemented in SeaDAS [30]. The inherent optical properties (IOPs) of aerosols and molecules were then combined to give a "mixed" single-scattering albedo $\varpi_{\text{mix}} = (\beta_{\text{mol}} + \beta_{\text{aer}})/(\gamma_{\text{mol}} + \gamma_{\text{aer}}) = 0.9772$, where β_{mol} and β_{aer} are scattering coefficients, and γ_{mol} and γ_{aer} are extinction coefficients for molecules and aerosols, respectively.

The simulated atmospheric output is the radiance $I(\tau, \mu, \phi)$ in arbitrary directions (μ, ϕ). However, for comparison with measurements, we used the bidirectional reflectance factor defined as $\text{BRF}(\tau, \mu, \phi) = \pi I(\tau, \mu, \phi)/\mu_0 F_0$, where $\mu = \cos\theta$, θ is the zenith view angle, ϕ is the azimuth angle, τ is the optical thickness (at aircraft altitude), $\mu_0 = \cos\theta_0$, θ_0 being the solar zenith angle, and F_0 is the extraterrestrial solar irradiance. Please note that this BRF is defined such that it would represent the reflected irradiance normalized by the TOA incident irradiance $\mu_0 F_0$ if the radiance were to be isotropic.

5.3.3. Comparison between Measured and Simulated Reflectances

Retrieval Surface Roughness and Aerosol Parameters

Although the variances in the crosswind and upwind directions, σ_{c}^2 and σ_{u}^2, can be parameterized in terms of wind speed and direction [75], they may instead be considered to be model input parameters describing the 2D surface slope distribution. For a 1D surface, the slope variance is then automatically given as $\sigma^2 = \sigma_{\text{u}}^2 + \sigma_{\text{c}}^2$. This approach has the advantage that the slope variances do not depend on parameterizations in terms of wind speed, which provides more freedom to reproduce the measured 2D glint pattern by varying σ_{c} and σ_{u}.

In addition to reflection from Gaussian surface slope facets, multiple surface reflections, shadowing [see Equation (130)], and polarization effects also influence the glint signal [105,106]. Since multiple reflections and shadowing become important only for very low solar elevations and polarization effects are relatively unimportant at 1035 nm due to weak molecular (and aerosol) scattering, these effects can be ignored [80].

To determine the optimum match between model-simulated and CAR-measured reflectances, we need to invert the measurements to find the best estimate of the model parameters described by the state vector $\mathbf{x} = [\sigma_{\text{c}}^2, \sigma_{\text{u}}^2, \phi_{\text{wind}}, \tau_{\text{mix}}]^T$ including the four retrieval parameters: (i) the slope variance in the crosswind direction, σ_{c}^2, (ii) the slope variance in the upwind direction, σ_{u}^2, (iii) the wind direction, ϕ_{wind}, and (iv) the optical thickness τ_{mix} of the layer with a mixed population of aerosols and molecules. To this end, [80] used a Gauss-Newton/Levenberg-Marquardt non-linear inversion algorithm, in which the residual between reflectances produced by the RT forward model $\mathbf{F}(\mathbf{x})$ and the CAR measurements, stored in the vector $\mathbf{y}$, is minimized. The kth iteration of $\mathbf{x}$ yields:

$$\mathbf{x}_{k+1} = \mathbf{x}_k + \left[\mathbf{J}_k^T \mathbf{J}_k + \gamma_k \mathbf{I}\right] \mathbf{J}_k^T (\mathbf{F}(\mathbf{x}_k) - \mathbf{y}_k) \tag{154}$$

where the vector $\mathbf{J}$ contains the Jacobians of the forward model, $\mathbf{I}$ is identity matrix, and the parameter γ_k $(0 \leq \gamma_k \leq \infty)$ is chosen at each step of the iteration to minimize the residual. If $\gamma_k = 0$ we have a classic Gauss-Newton method, while if γ_k is large we have a steepest descent method.

Figure 3 shows a comparison of model-simulated and measured reflectances at 1036 nm on 20 July 2001. Clearly, model-simulated results agree very well with the measurements, reproducing the main characteristics of the glint pattern. Both the shape of the simulated glint ellipse and its tilt are generally well matched with the measurements.

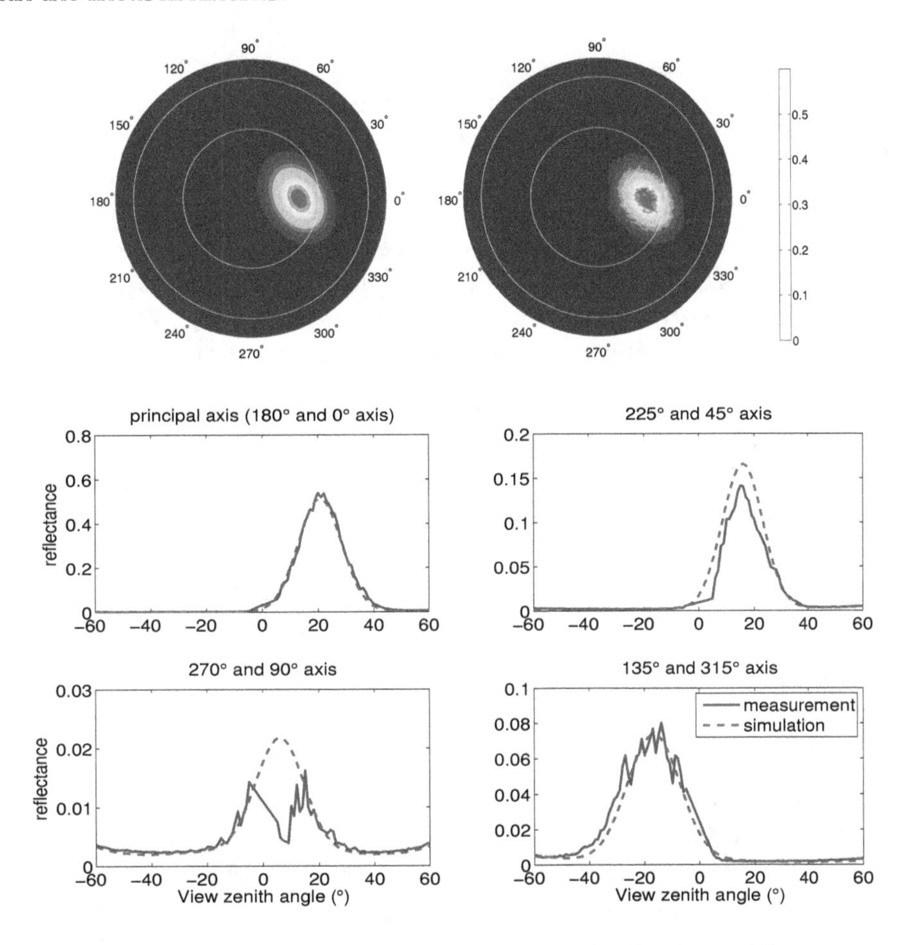

Figure 3. Comparison between model-simulated and measured reflectances. The measurements are in solid blue and the simulations in dashed red lines. The dip in the measured reflectance in the lower left panel is due to aircraft shadowing. [Reproduced from Figures 1a and 2a of [80].]

The shape of the elliptical glint pattern in Figure 3 is determined by the crosswind and upwind slope variances (σ_{c}^2 and σ_{u}^2), while the wind direction is determined from the tilt angle. Such a titled ellipse indicates that use of the 2D asymmetric Gaussian BRDF is needed to fit the angular distribution of the reflectance measurements, and that a 1D Gaussian is insufficient because it averages those slopes ($\sigma^2 = \sigma_{\mathrm{c}}^2 + \sigma_{\mathrm{u}}^2$) to give only a circular glint pattern. Figure 4 shows a comparison of 1D and 2D results for the case shown in Figure 3. It is clear that model simulations based on a 1D Gaussian BRDF are unable to match the measured tilted elliptical glint pattern. Hence, employing a 2D BRDF in the data analysis is essential.

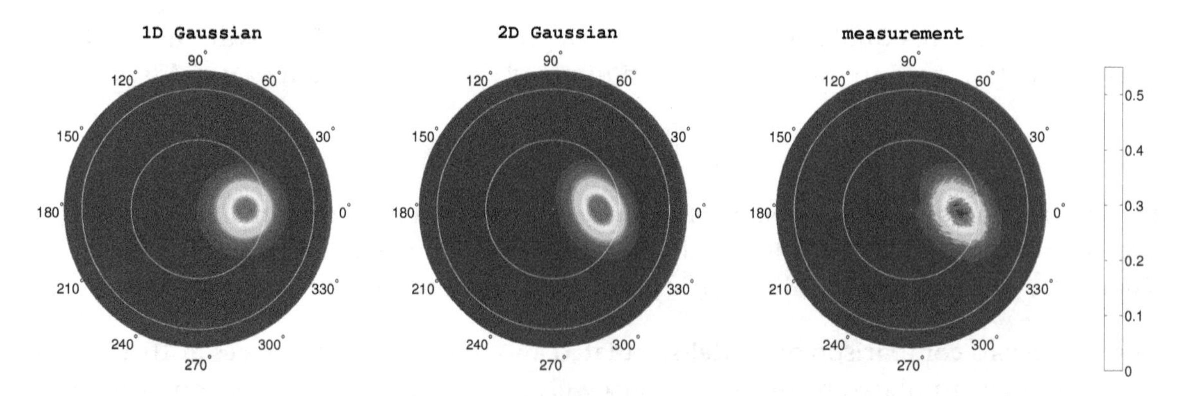

Figure 4. Comparison between model-simulated reflectances assuming a 1D Gaussian BRDF (**left**), a 2D Gaussian BRDF (**middle**), and measurements (**right**) obtained on 10 July 2001.

Since the surface roughness parameters (σ_c^2, σ_u^2, σ^2, and wind direction) inferred from the channel at 1036 nm are independent of wavelength, this information can be applied at shorter wavelengths (472, 672, and 870 nm) to improve the ocean color retrieval in the sun-glint area. The aerosol and ocean parameters can be retrieved from radiances outside the glint region (viewing angles 20–60 degrees and relative azimuth angles 120–240 degrees) using multi-angle reflectances in three CAR channels at 472, 682, and 870 nm. For this purpose, Lin et al. [80] used AccuRT, a RTM for the coupled atmosphere-ocean system (described in Sections 2–4) combined with an optimization technique [see Equation (154)] to retrieve aerosol and water constituent parameters simultaneously [70,107]. The Ahmad et al. aerosol model [30] and the CCRR bio-optical model [58] were used in AccuRT for this retrieval.

In this case, the retrieved aerosol optical depth at 870 nm was 0.086, the Ångstrøm coefficient (472 to 870 nm) was 1.463, the chlorophyll concentration was 0.67 mg·m^{-3}, the colored dissolved organic matter (CDOM) absorption coefficient at 443 nm was 0.07 m^{-1}, and the mineral particle concentration (MIN) was 0.009 g·m^{-3}. When these retrieval results were applied to the BRDF simulation, a very good match to the CAR measurements were obtained, especially in the glint area [80]. Hence, the glint information retrieved from the channel at 1.036 nm can be used in ocean color remote sensing to estimate the glint contribution at visible and NIR ocean color channels.

5.3.4. Summary of Glint Issues

The results discussed above may be summarized as follows:

- A wind-direction dependent Gaussian surface BRDF that uses (1) a 2D slope distribution for singly scattered light, and (2) a 1D slope distribution for multiply scattered light, can be used to successfully simulate BRDF measurements obtained by NASA's Cloud Absorption Radiometer (CAR) at the 1036 nm wavelength.
- Upwind and crosswind slope variances, wind direction, and aerosol optical depth, can be accurately retrieved through forward-inverse modeling.
- The glint parameters (slope variances and wind direction) can be applied to estimate the glint contribution at visible and NIR wavelengths, resulting in a very good match between model-simulated and measured reflectances.
- An advantage of RT simulations of glint reflectance is its inclusion of contributions from the diffuse or multiply scattered light due to scattering by atmospheric molecules and aerosols. The diffuse light reflectance ("skyglint") gives an additional glint signal in addition to "sunglint" resulting from the direct beam reflectance.
- Simulations show that the diffuse glint may contribute more than 4% at 472 nm for a wind speed of 2 m/s, and more than 8% when the wind speed increases to 8 m/s. Hence, the diffuse light reflectance should be considered in the visible bands, especially for large wind speeds.

- A simplified version of the pseudo 2D BRDF and glint reflectance method described above has been implemented in the AccuRT model for the coupled atmosphere-ocean system.

5.4. Retrievals of Atmosphere-Water Parameters from Geostationary Platforms: Challenges and Opportunities

Simultaneous retrieval of aerosol and surface properties by means of inverse techniques based on a coupled atmosphere-surface radiative transfer model and optimal estimation can yield improved retrieval accuracy in complex aquatic environments compared with traditional methods. At high latitudes low solar elevations is a problem, and if one desires to do satellite remote sensing from a geostationary platform in order to study diurnal variations, then large solar zenith and viewing angles become important issues to be resolved.

Satellite remote sensing for such complex situations/environments represents specific challenges due to:

- (i) the complexity of the atmosphere and water inherent optical properties,
- (ii) the unique bidirectional dependence of the water-leaving radiance, and
- (iii) the desire to do retrievals for large solar zenith and viewing angles.

Hence, one needs to consider challenges related to how

1. atmospheric gaseous absorption, absorbing aerosols, and turbid waters can be addressed by using a *coupled* forward model in the retrieval process,
2. corrections for *bidirectional effects* will be accomplished,
3. the *curvature* of the atmosphere will be taken into account, and
4. *uncertainty assessments* and error budgets will be dealt with.

The generic problem is illustrated in Figures 5 and 6, which show that

- there is *a significant change* in sub-surface color with increasing chlorophyll concentration, while at the same time
- there is *only a slight change* in color at the TOA, where *the spectra are dominated by light from atmospheric scattering.*

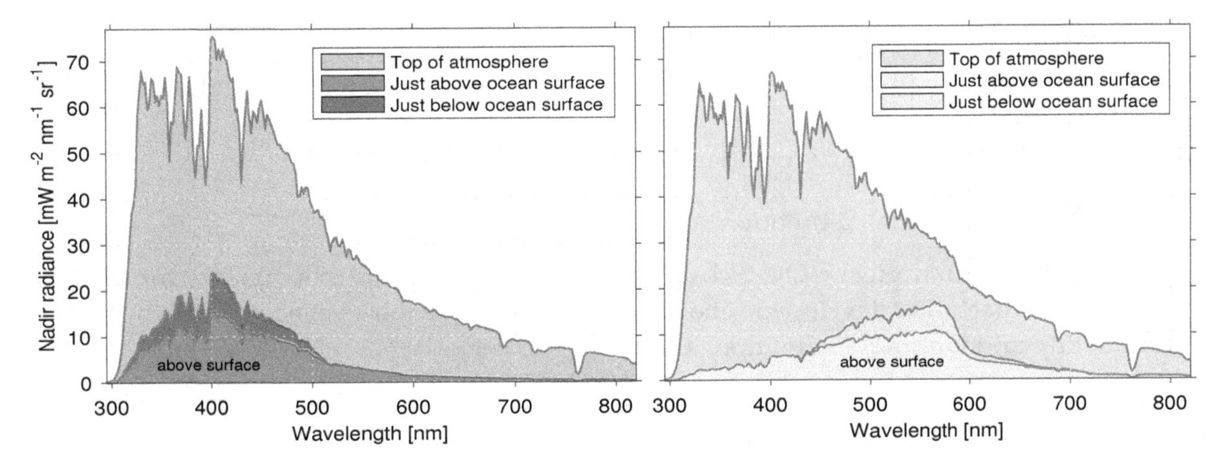

Figure 5. Simulated upward radiance in the nadir direction at the top of the atmosphere and close to the ocean surface. Solar zenith angle = 45°, US Standard atmosphere with aerosol optical depth = 0.23 at 500 nm. (**Left**) Clear water with chlorophyll concentration = 0.1 mg·m^{-3}, MIN = 0.003 g·m^{-3}, CDOM443 = 0.003 m^{-1} (CCRR bio-optical model). (**Right**) Turbid water with chlorophyll concentration = 10 mg·m^{-3}, MIN = 0.1 g·m^{-3}, CDOM443 = 0.1 m^{-1}.

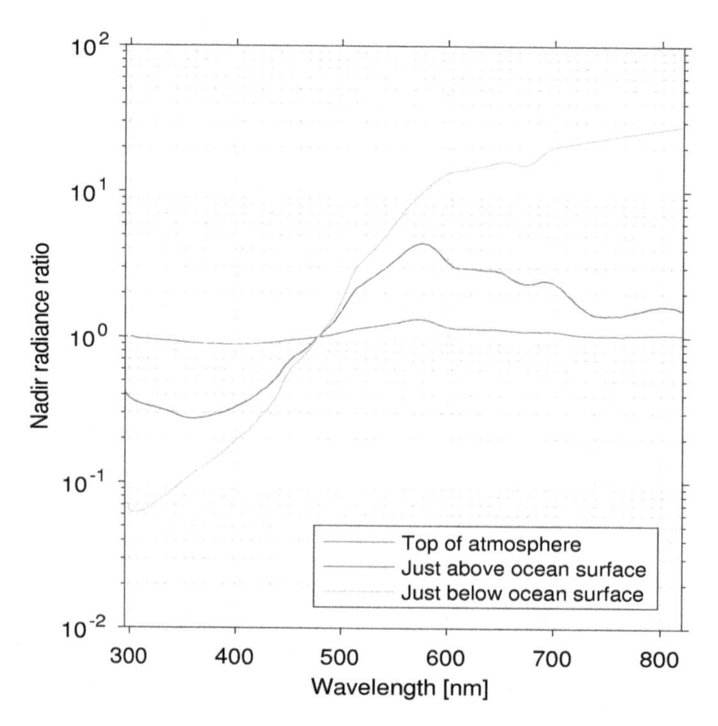

Figure 6. The ratio of the values for turbid water to those for clear water in Figure 5.

Most ocean color algorithms consist of two steps:

1. First, one does an "atmospheric correction" (assuming the water to be black at NIR wavelengths) to determined the water-leaving radiance.
2. Second, one retrieves the desired aquatic parameters from the water-leaving radiance.

In the visible, up to 90% of the radiance measured by a satellite sensor typically comes from the atmosphere implying that:

- Atmospheric correction becomes a very challenging task unless the near-infrared (NIR) *black-pixel approximation (BPA)* is valid.
- Estimation of diffuse transmittance is also important, but difficult because it depends on the angular distribution of the radiance just beneath the water surface.

Also, accurate characterization of the atmosphere is important because:

- a small uncertainty in the atmospheric correction may lead to a big error in the inferred aquatic parameters, and
- aerosol optical properties vary considerably in space and time.

5.4.1. The OC-SMART Optimal Estimation Approach

To address this situation, the OC-SMART (**O**cean **C**olor—**S**imultaneous **M**arine and **A**erosol **R**etrieval **T**ool) approach was developed. The goal was to improve retrieval accuracy by use of AccuRT forward modeling and Optimal Estimation/Levenberg-Marquardt (OE/LM) inversion [108]:

- AccuRT: accurate discrete-ordinates radiative transfer model for the *coupled* atmosphere-ocean system; delivers a complete set of simulated radiances and Jacobians (weighting functions).
- OE/LM inversion: is an iterative, nonlinear least squares cost function minimization with *a priori* and Levenberg-Marquardt regularization.
- For retrievals of aerosol and aquatic parameters from Ocean Color data, we define a 5-element state vector:

$$\mathbf{x} = \{\tau_{865}, f_a, \mathbf{CHL}, \mathbf{CDM}, \mathbf{BBP}\}$$

consisting of

- *2 aerosol parameters* (τ_{865} = optical depth at 865 nm and f_a = bimodal fraction of particles),
- *3 marine parameters* (chlorophyll concentration, **CHL**, combined absorption by detrital and dissolved material at 443 nm, **CDM**, and backscattering coefficient at 443 nm, **BBP**).

At each iteration step, the next estimate of the state vector is given by the OE/LM inversion [70]

$$\mathbf{x}_{n+1} = \mathbf{x}_n + [(1+\gamma_n)\mathbf{S}_a^{-1} + \mathbf{J}_n^T\mathbf{S}_m^{-1}\mathbf{J}_n]^{-1}\{\mathbf{J}_n^T\mathbf{S}_m^{-1}(\mathbf{y_m} - \mathbf{y}_n) - \mathbf{S}_a^{-1}(\mathbf{x}_n - \mathbf{x}_a)\}. \tag{155}$$

In Equation (155) $\mathbf{y}_m$ is the vector of measured TOA radiances; $\mathbf{y}_n = \mathrm{F}(\mathbf{x}_n, \mathbf{b})$ is the vector of simulated TOA radiances generated by the AccuRT forward model; $\mathbf{y}_n$ is a (non-linear) function of the state vector $\mathbf{x}_n$ of retrieval elements, and $\mathbf{b}$ represents model parameters; $\mathbf{J}_n$ is a matrix of simulated radiance partial derivatives with respect to the state vector elements $\mathbf{x}_n$ (the Jacobians); $\mathbf{x}_a$ and $\mathbf{S}_a$ are the *a priori* state vector and covariance matrix, respectively, and $\mathbf{S}_m$ is the measurement error covariance matrix. γ_n is the Levenberg-Marquardt (LM) regularization parameter, When $\gamma_n \to 0$ Equation (155) becomes the standard Gauss-Newton Optimal Estimation (OE), and when $\gamma_n \to \infty$ it tends to the steepest descent method. AccuRT returns simulated radiances ($\mathbf{y}_n$) and Jacobians ($\mathbf{J}_n$) required to update the state vector estimate ($\mathbf{x}_n$) according to Equation (155) above.

One issue with the OC-SMART approach is that it is relatively slow due to the need to call the forward AccuRT model repeatedly in the iterative inversion to compute radiances and Jacobians. To deal with that problem, one could use AccuRT instead to create a training ensemble in order to construct a Radial Basis Function Neural Network. This approach typically leads to an increase in computational speed by a factor of about 1000. In this way one replaces the AccuRT forward model (thousands of lines of code) with the following single equation (similar to Equation (151)):

$$I_i = \sum_{j=1}^{N} a_{ij} \exp[-b \sum_{k=1}^{K} (P_k - c_{jk})^2] + d_i$$

where I_i it the TOA radiance in channel $i = 1, \ldots, 8$, K = # of input parameters, and $a_{i,j}, b, c_{j,k}, d_i$ are the coefficients to be optimized in the network training.

The Jacobians (partial derivatives) $\mathbf{J}$ are also required for the non-linear optimal estimation using Equation (155). These Jacobians can be calculated by taking partial derivatives of I_i respect to each input parameter:

$$J_{ik} = \frac{\partial I_i}{\partial P_k} = -2 \sum_{j=1}^{N} a_{ij} b^2 (P_k - c_{jk}) \exp\{\sum_{k=1}^{K} (P_k - c_{jk})^2\}. \tag{156}$$

Fan et al. [107] applied the OC-SMART algorithm to a MODIS image obtained on 18 April 2014 over a Norwegian coastal area and compared retrieval results with the standard SeaDAS retrievals as shown in Figure 7 with OC-SMART results at the top and SeaDAS results at the bottom. From left to right, the columns are τ_{869}, f_a, CHL, CDOM and b_{bp}, respectively. The first two columns are retrieved aerosol parameters. Our results are very similar to the SeaDAS retrievals, which makes sense because the two algorithms share the same aerosol model. However, the marine parameters show some differences and we should point out that in OC-SMART, we used the GSM bio-optical model when we retrieved the ocean parameters. The SeaDAS CHL retrieval shown in Figure 7 is from the OC4v6 algorithm. SeaDAS failed when using the GSM model due to negative water-leaving radiances resulting from an incorrect atmospheric correction, which is a common issue in traditional atmospheric correction algorithms. The aerosol optical depths was determined at a near infrared (NIR) wavelength (τ_{869}) and then extrapolated to shorter wavelengths. Since the water-leaving radiance is only a small fraction of the TOA radiance measured by the satellite instrument, a small error in the aerosol retrieval may cause the water-leaving radiance to become negative. We also found the OC-SMART results to be close to the field measurements. The CHL value was 0.56 [mg/m^3] while the field measurements showed an average of 0.86 ± 0.34 [mg/m^3] in the spring season. The CDOM absorption coefficient was 0.15 [m^{-1}] and the field measurements show an average of 0.14 ± 0.06 [m^{-1}].

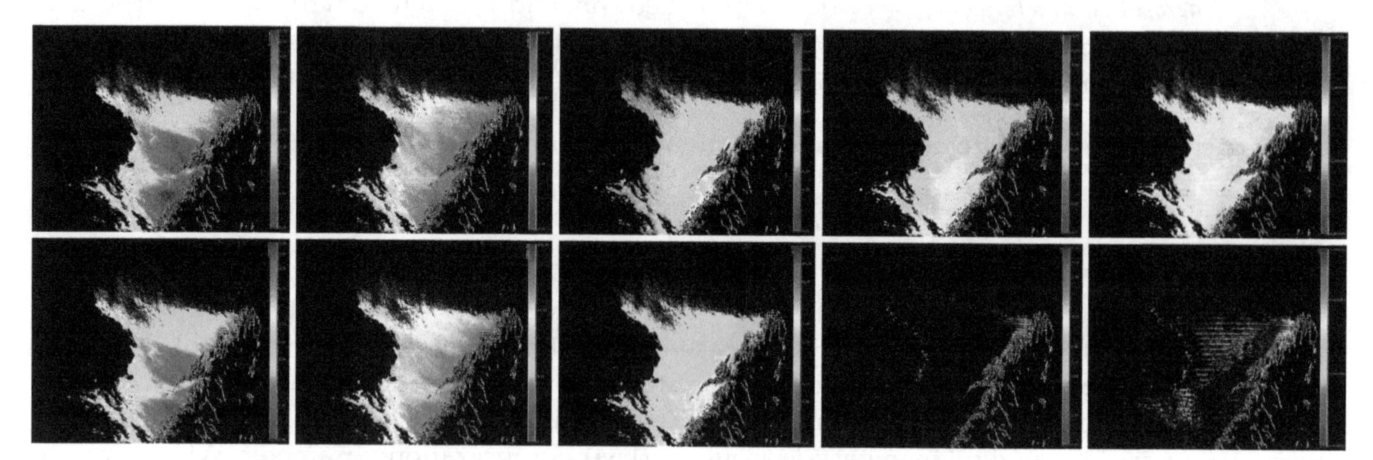

Figure 7. MODIS image comparison between OC-SMART (**top**) and standard SeaDAS (**bottom**) retrievals on 18 April 2014 over a Norwegian coastal area [107]. From left to right: τ_{869}, f, CHL, CDOM and b_{bp}, respectively.

5.4.2. The OC-SMART Multilayer Neural Network Approach

In open ocean areas where the water IOPs are correlated with pigmented particles, standard atmospheric correction (AC) algorithms seem to work reasonably well. However, in turbid coastal water areas the IOPs of suspended inorganic particles, and colored dissolved organic matter (CDOM) may vary independently of pigmented particles. Therefore, in turbid coastal waters standard AC algorithms often exhibit large inaccuracies that may lead to highly uncertain and frequently negative water-leaving radiances (L_w) or remote sensing reflectances (R_{rs} values). To address this problem, Fan et al. [109] introduced a new atmospheric correction algorithm for coastal water areas based on a multilayer neural network (MLNN) method. A coupled atmosphere-ocean radiative transfer model (AccuRT) was used to simulate the Rayleigh-corrected radiance (L_{rc}) at the TOA and the R_{rs} just above the surface simultaneously, and to train a MLNN to derive the aerosol optical depth (AOD) and R_{rs} values directly from the TOA L_{rc}. The method was validated using both a synthetic dataset and Aerosol Robotic Network–Ocean Color (AERONET–OC) measurements.

Extensive testing has shown that this MLNN approach has several advantages [109]:

1. It significantly improved the quality of retrieved remote sensing reflectances (compared to the SeaDAS NIR algorithm) by reducing the average percentage difference (APD) between MODIS retrievals and ground-truth (AERONET-OC) validation data.
2. In highly absorbing coastal water, such as the Baltic Sea, it provides reduction of the APD by more than 60%, and in highly scattering water, such as the Black Sea, it provides reduction of the APD by more than 25%.
3. It is robust and resilient to contamination due to sunglint and adjacency effects of land and cloud edges.
4. It is applicable in extreme conditions such as those encountered for heavily polluted continental aerosols, extreme turbid water, and dust storms.
5. It does not require shortwave infrared (SWIR) bands, and is therefore suitable for all ocean color sensors.
6. It is very fast and suitable for operational use.

Recent addition of training data to make it representative for a variety of water types has shown that this approach can produce a seamless (smooth) transition between turbid coastal water areas and clean open ocean areas.

5.4.3. Issues Specific to Geostationary Platforms

Low Solar Elevations

With geostationary platforms one has the opportunity to investigate diurnal variations in water properties. Hence, there is a desire to obtain useful observations throughout the day from sunrise to sunset. However, for solar zenith angles larger than about 75° (and large viewing angles) the *plane-parallel approximation* (PPA) becomes invalid, and we need to take Earth curvature into account, as discussed in a recent paper [110]. An approximate way to deal with this problem is the so-called *pseudo-spherical approximation* (PSA), in which the direct beam single scattering (solar pseudo-source) term is treated in spherical geometry: $e^{-\tau/\mu_0} \rightarrow e^{-\tau\, Ch(\mu_0)}$, where the Chapman function $Ch(\mu_0)$ takes curvature into account [14,110], while the multiple scattering term is treated using the PPA. Hence, in the PSA the RTE becomes:

$$
\begin{aligned}
u\frac{dI(\tau,u,\phi)}{d\tau} &= I(\tau,u,\phi) - \overbrace{\frac{\varpi(\tau)}{4\pi}\int_0^{2\pi} d\phi' \int_{-1}^{1} du' p(\tau,u',\phi';u,\phi) I(\tau,u',\phi')}^{\text{multiple scattering}} \\
&\quad - \overbrace{\frac{\varpi(\tau)}{4\pi} p(\tau,-\mu_0,\phi_0;u,\phi) F_0 e^{-\tau\, Ch(\mu_0)}}^{\text{single scattering}} .
\end{aligned}
\tag{157}
$$

It should be pointed out that low solar elevations is generally a problem at high latitudes also for polar-orbiting platforms.

Surface Roughness Considerations

As alluded to above, it would be worthwhile considering the advantage gained by using a 2D rather than a 1D slope distribution to deal with a wind-roughened surface (see Figure 4). Hence, one should consider using

1. a 2D Gaussian surface slope distribution for singly scattered light, and
2. a 1D Gaussian surface slope distribution for multiply scattered light.

Such an approach may be quite successful, as demonstrated in Figure 3, because a 2D BRDF simulates sunglint very well, while a 1D BRDF is sufficient to simulate the smoother (more directionally uniform) skylight.

Use of Vector (Polarized) RT Simulations

Analysis of polarization measurements require vector RT modeling [111], and retrieval results from radiance-only measurements are likely to be improved by employing a vector (polarized) forward RT model to compute the radiances and Jacobians used in the inversion step, particularly for modeling the radiance due to scattering by small aerosol particles at short wavelengths. Also, the increased information from polarization measurements can lead to significant improvements in atmospheric correction due to aerosols [112]. Hence, for ocean color retrievals from geostationary platforms, we should explore the advantage of using the pseudo-spherical approximation combined with

- polarized (vector) radiative transfer simulations,
- a 2D Gaussian distribution of surface slopes,

and neural networks and optimal estimation for

- simultaneous retrieval of atmospheric and marine parameters from multi-spectral as well as hyperspectral measurements of total and polarized (if available) radiances, and
- assessments of retrieval accuracy and error budgets.

6. Remaining Problems

6.1. 3-D Radiative Transfer: The LiDAR Problem

Although we have mostly considered *plane-parallel* systems so far with an emphasis on the coupling between the atmosphere and the underlying surface consisting of a water body or a snow/ice surface, there are many applications that require a three-dimensional (3-D) RT treatment. For a clear (cloud- and aerosol-free) atmosphere, 3-D effects are related to the impact of the Earth's curvature on the radiation field as discussed briefly in Section 5.4.3. To include Earth curvature effects it may be sufficient to employ a *"pseudo-spherical"* treatment, in which the direct solar beam illumination is treated in spherical geometry [110,113], whereas *multiple scattering* is treated in plane-parallel geometry. In fact, this pseudo-spherical approach has been implemented in many RT codes [114,115]. Also, there is a large body of literature on 3-D RT modeling with applications to broken clouds, and readers interested in this topic may want to consult [116] or visit the web-site http://i3rc.gsfc.nasa.gov/.

Finally, 3-D RT modeling may also be important for analysis and interpretation of LiDAR data. The classical "searchlight problem" [117], which considers the propagation of a laser beam through a turbid medium, is relevant in this context. Long-range propagation of a LiDAR beam has been studied both theoretically and experimentally [118]. Monte Carlo simulations are well suited for such studies [119], and use of deterministic models such the *discrete-ordinate method*, briefly discussed in Section 4, have also been pursued to investigate this problem [120,121].

6.2. Time-Dependent Radiative Transfer

Most studies of radiative transfer in the ocean have been concerned with understanding the propagation of sunlight through natural water bodies [3,4,14,122]. For such applications, the transient or *time-dependent* term in the RTE can be ignored, because changes in the incident illumination are much slower than the changes imposed by light propagation through the coupled atmosphere-water system. While this assumption is satisfied for solar illumination, LiDAR systems can use pulses that are shorter than the attenuation distance of seawater divided by the speed of light in water. Also, it has been pointed out [123] that due to multiple light scattering, understanding the LiDAR signal requires a solution of the *time-dependent* RTE. The transient RT problem can be reduced to solving a series of *time-independent* RT problems [124].

6.3. Other Issues

Although inelastic scattering processes (Raman and Brillouin) certainly can be very important and indeed essential in some atmospheric [125–127] and aquatic [128,129] applications, we have limited our discussion to elastic scattering. In fact, inelastic scattering effects can become important for very clear water and sky conditions.

Forward radiative transfer models need accurate IOPs as input. Therefore, in situ measurements are needed of the scattering phase function of hydrosols in clear as well as turbid (coastal) water, including measurements at a scattering angle close to 180°. The 180°-backscatter angle is of particular interest for LiDAR applications. In view of the growing interest in applications based on RT models for coupled systems [84,86,87,90–93], there is also a need for systematic and sustained measurements of the scattering phase (or Muller) matrix, since not much work has been done after the 1984 Voss and Fry publication [130].

To process hyperspectral data one needs algorithms that can handle large data volumes and deal with measurement uncertainties in a proper manner. In general, better bio-optical models are also needed. For example, the approximation that the spectral slope $\nu = 0$ for CHL > 2.0 in Equation (97) is problematic. Hyperspectral (and potentially polarimetric) measurements are needed to construct more generally applicable bio-optical models.

In Section 5.4.2 we discussed how a forward-inverse modeling approach, based on AccuRT simulations for the coupled atmosphere-water system combined with a multilayer neural network

(MLNN), can be used to infer accurate remote sensing reflectances (R_{rs} values) from clear as well as turbid water [109]. The logical next step is to use these R_{rs} values to infer water IOPs. In fact, several semi-analytic algorithms have been developed for this purpose including the generalized IOP (GIOP) algorithm described by Werdell and colleagues [131].

This GIOP approach is currently the default algorithm used by NASA for processing of MODIS ocean color data. First, the R_{rs} values are obtained after an atmospheric correction step using NIR channels. Ocean IOP parameters are then retrieved from the approximate R_{rs} values (based on many assumptions) by performing a non-linear inversion. The QAA algorithm [132] is another semi(quasi)-analytic algorithm, similar to the GIOP, that performs better than the GIOP in some situations. The GIOP and QAA algorithms generally perform well over the open ocean, but can have significant issues in coastal water because the atmospheric correction method produces inaccurate R_{rs} values that lead to large errors in derived IOPs. Furthermore, the performance of the atmospheric correction is worse the farther the extrapolation is taken from the NIR to shorter wavelengths, which is a problem for ocean color sensors with UV channels. Also, particle size distribution and scattering phase function constitute *a priori* information in the semi-analytic algorithms that cannot be retrieved. These limitations make them unsuitable for application to new sensors with increased information content, such as hyperspectral ocean color sensors. Instruments that can measure the polarization of light will also be extremely useful for improving aerosol retrievals and thus atmospheric correction capabilities. The next generation of sensors composed of hyperspectral polarimeters and ocean-capable LiDAR instruments will enable the use of more advanced retrieval algorithms to more accurately quantify and monitor the atmosphere-ocean system, particularly in coastal areas. Nevertheless, use of accurate R_{rs} values, obtained from new and improved AC methods, such as the MLNN algorithm [109], in existing semi-analytic algorithms (like GIOP and QAA) is expected to significantly enhance their performance when applied to current ocean color sensor data obtained over turbid water.

7. Summary

A review has been provided of forward and inverse radiative transfer modeling in coupled systems consisting of two adjacent, horizontal slabs with different indices of refraction and a rough interface characterized by a Gaussian distribution of surface slopes in one or two dimensions. Such a configuration can be used to simulate radiative transfer in coupled atmosphere-snow/ice-ocean systems. Input and output parameters including boundary conditions for the forward radiative transfer problem were introduced in Section 2, while in Section 3 a review was provided of inherent optical properties (IOPs) of the atmosphere, snow/ice and ocean water. These IOPs are required inputs to the radiative transfer equation introduced in Section 4 both for unpolarized (scalar) and polarized (vector) radiative transfer. Examples of how to solve inverse problems occurring in remote sensing of the environment employing optimization techniques as well as the unique power of machine learning (neural networks) for convenience and efficiency were provided in Section 5. Finally, in Section 6 some remaining radiative transfer problems were discussed.

Author Contributions: K.S. wrote the first draft of the paper. The other authors (B.H., S.S., N.C., Y.F., W.L., Z.L., and J.S.) contributed by reviewing, editing, and making additions to various parts of the paper.

Acknowledgments: We thank two anonymous reviewers for constructive comments that led to a significant improvement of this paper.

References

1. Ricchiazzi, P.; Yang, S.; Gautier, C.; Sowle, D. SBDART: A research and teaching software tool for plane-parallel radiative transfer in the Earth's atmosphere. *Bull. Am. Meteorol. Soc.* **1998**, *79*, 2101–2114. [CrossRef]

2. Key, J.R.; Schweiger, A.J. Tools for atmospheric radiative transfer: Streamer and FluxNet. *Comput. Geosci.* **1998**, *24*, 443–451. [CrossRef]
3. Stamnes, K.; Stamnes, J.J. *Radiative Transfer in Coupled Environmental Systems*; Wiley-VCH: Weinheim, Germany, 2015.
4. Mobley, C.D. *Light and Water*; Academic Press: Cambridge, MA, USA, 1994.
5. Hovenier, J.W.; der Mee, C.D.V.; Domke, H. *Transfer of Polarized Light in Planetary Atmospheres*; Kluwer Academic Publsihers: Dordrecht, The Netherlands, 2004.
6. Rayleigh, L. A re-examination of the light scattered by gases in respect of polarization. II. Experiments on helium and argon. *Proc. R. Soc.* **1920**, *98*, 57–64. [CrossRef]
7. Morel, A. Optical properties of pure water and pure seawater. In *Optical Aspects of Oceanography*; Jerlov, N.G., Nielsen, E.S., Eds.; Academic Press: Cambridge, MA, USA, 1974; pp. 1–24.
8. Morel, A.; Gentili, B. Diffuse reflectance of oceanic waters: its dependence on sun angle as influenced by the molecular scattering contribution. *Appl. Opt.* **1991**, *30*, 4427–4437. [CrossRef] [PubMed]
9. Rayleigh, L. On the light from the sky, its polarization and colour. *Philos. Mag.* **1871**, *41*, 107–120, 274–279, 447–454.
10. Bodhaine, B.; Wood, N.; Dutton, E.; Slusser, J. On Rayleigh optical depth calculations. *J. Atmos. Ocean. Technol.* **1999**, *16*, 1854–1861. [CrossRef]
11. Farinato, R.S.; Rowell, R.L. New values of the light scattering depolarization and anisotropy of water. *J. Chem. Phys.* **1976**, *65*, 593–595. [CrossRef]
12. Zhang, X.; Hu, L. Estimating scattering of pure water from density fluctuation of the refractive index. *Opt. Express* **2009**, *17*, 1671–1678. [CrossRef] [PubMed]
13. Henyey, L.C.; Greenstein, J.L. Diffuse radiation in the galaxy. *Astrophys. J.* **1941**, *93*, 70–83. [CrossRef]
14. Stamnes, K.; Thomas, G.E.; Stamnes, J.J. *Radiative Transfer in the Atmosphere and Ocean*, 2 ed.; Cambridge University Press: Cambridge, UK, 2017.
15. Hovenier, J.W.; van der Mee, C.V.M. Fundamental relationships relevant to the transfer of polarized light in a scattering atmosphere. *Astron. Astrophys.* **1983**, *128*, 1–16.
16. Chandrasekhar, S. *Radiative Transfer*; Dover Publications: Mineola, NY, USA, 1960.
17. De Haan, J.; Bosma, P.; Hovenier, J. The adding method for multiple scattering calculations of polarized light. *Astron. Astrophys.* **1987**, *183*, 371–391.
18. Siewert, C. A discrete-ordinates solution for radiative-transfer models that include polarization effects. *J. Quant. Spectrosc. Radiat. Transf.* **2000**, *64*, 227–254. [CrossRef]
19. Cohen, D.; Stamnes, S.; Tanikawa, T.; Sommersten, E.R.; Stamnes, J.J.; Lotsberg, J.K.; Stamnes, K. Comparison of Discrete Ordinate and Monte Carlo Simulations of Polarized Radiative Transfer in two Coupled Slabs with Different Refractive Indices. *Opt. Express* **2013**, *21*, 9592–9614. [CrossRef] [PubMed]
20. Siewert, C.E. On the equation of transfer relevant to the scattering of polarized light. *Astrophys. J.* **1981**, *245*, 1080–1086. [CrossRef]
21. Siewert, C.E. On the phase matrix basic to the scattering of polarized light. *Astron. Astrophys.* **1982**, *109*, 195–200.
22. Sommersten, E.R.; Lotsberg, J.K.; Stamnes, K.; Stamnes, J.J. Discrete ordinate and Monte Carlo simulations for polarized radiative transfer in a coupled system consisting of two media with different refractive indices. *J. Quant. Spectrosc. Radiat. Transf.* **2010**, *111*, 616–633. [CrossRef]
23. Mishchenko, M.I. Light scattering by randomly oriented rotationally symmetric particles. *J. Opt. Soc. Am. A* **1991**, *8*, 871–882. [CrossRef]
24. Mishchenko, M.I.; Travis, L.D. Satellite retrieval of aerosol properties over the ocean using polarization as well as intensity of reflected sunlight. *J. Geophys. Res.* **1997**, *102*, 16989–17013. [CrossRef]
25. Anderson, G.P.; Clough, S.A.; Kneizys, F.X.; Chetwynd, J.H.; Shettle., E.P. *AFGL Atmospheric Constituent Profiles (0–120 km), AFGL-TR-86-0110 (OPI)*; Optical Physics Division, Air Force Geophysics Laboratory Hanscom AFB: Bedford, MA, USA, 1986.
26. Berk, A.; Anderson, G.P.; Acharya, P.K.; Bernstein, L.S.; Muratov, L.; Lee, J.; Fox, M.; Adler-Golden, S.M.; Chetwynd, J.H.; Hoke, M.L.; et al. *MODTRAN 5: A Reformulated Atmospheric Band Model with Auxiliary Species and Practical Multiple Scattering Options: Update*; Defense and Security, International Society for Optics and Photonics: Orlando, FL, USA, 2005; pp. 662–667.

27. Berk, A.; Conforti, P.; Kennett, R.; Perkins, T.; Hawes, F.; van den Bosch, J. *MODTRAN6: A Major Upgrade of the MODTRAN Radiative Transfer Code*; SPIE Defense+ Security, International Society for Optics and Photonics: Orlando, FL, USA, 2014; p. 90880H.
28. Kneizys, F.X.; Abreu, L.W.; Anderson, G.; Chetwynd, J.; Shettle, E.; Berk, A.; Bernstein, L.; Roberson, D.; Acharya, P.; Rothman, L.; et al. *MODTRAN2/3 Report and LOWTRAN 7 Model*; Technical Report; Phillips Laboratory, Hanscom AFB: Bedford, MA, USA, 1996.
29. Chen, N.; Li, W.; Tanikawa, T.; Hori, M.; Shimada, R.; Aoki, T.; Stamnes, K. Fast yet accurate computation of radiances in shortwave infrared satellite remote sensing channels. *Opt. Express* **2017**, *25*, A649–A664. [CrossRef] [PubMed]
30. Ahmad, Z.; Franz, B.A.; McClain, C.R.; Kwiatkowska, E.J.; Werdell, J.; Shettle, E.P.; Holben, B.N. New aerosol models for the retrieval of aerosol optical thickness and normalized water-leaving radiances from the SeaWiFS and MODIS sensors over coastal regions and open oceans. *Appl. Opt.* **2010**, *49*, 5545–5560. [CrossRef] [PubMed]
31. Hess, M.; Koepke, P.; Schult, I. Optical properties of aerosols and clouds: The software package OPAC. *Bull. Am. Met. Soc.* **1998**, *79*, 831–844. [CrossRef]
32. Davies, C.N. Size distribution of atmospheric particles. *J. Aerosol Sci.* **1974**, *5*, 293–300. [CrossRef]
33. Holben, B.N.; Eck, T.F.; Slutsker, I.; Tanre, D.; Buis, J.P.; Setzer, A.; Vermote, E.; Reagan, J.A.; Kaufman, Y.; Nakajima, T.; et al. AERONET—A federated instrument network and data archive for aerosol characterization. *Remote Sens. Environ.* **1998**, *66*, 1–16. [CrossRef]
34. Holben, B.N.; Tanre, D.; Smirnov, A.; Eck, T.F.; Slutsker, I.; Abuhassan, N.; Newcomb, W.W.; Schafer, J.; Chatenet, B.; Lavenue, F.; et al. An emerging ground-based aerosol climatology: aerosol optical depth from AERONET. *J. Geophys. Res.* **2001**, *106*, 12067–12097. [CrossRef]
35. Hansen, J.E.; Travis, L.D. Light scattering in planetary atmospheres. *Space Sci. Rev.* **1974**, *16*, 527–610. [CrossRef]
36. Hänel, G. The properties of atmospheric aerosol particles as functions of the relative humidity at thermodynamic equilibrium with the surrounding moist air. In *Advances in Geophysics*; Landsberg, H.E., Miehem, J.V., Eds.; Elsevier: New York, NY, USA, 1976; Volume 19.
37. Shettle, E.P.; Fenn, R.W. *Models for the Aerosols of the Lower Atmosphere and the Effects of Humidity Variations on their Optical Properties*; Air Force Geophysics Laboratory, Hanscomb AFB: Bedford, MA, USA, 1979.
38. Yan, B.; Stamnes, K.; Li, W.; Chen, B.; Stamnes, J.J.; Tsay, S.C. Pitfalls in atmospheric correction of ocean color imagery: How should aerosol optical properties be computed? *Appl. Opt.* **2002**, *41*, 412–423. [CrossRef] [PubMed]
39. Du, H. Mie-scattering calculation. *Appl. Opt.* **2004**, *43*, 1951–1956. [CrossRef] [PubMed]
40. Wendisch, M.; Yang, P. *Theory of Atmospheric Radiative Transfer*; John Wiley & Sons: Hoboken, NJ, USA, 2012.
41. Segelstein, D.J. The Complex Refractive Index of Water. Mater's Thesis, Department of Physics, University of Missouri, Kansas City, MO, USA, 1981.
42. Smith, R.C.; Baker, K.S. Optical properties of the clearest natural waters (200–800 nm). *Appl. Opt.* **1981**, *36*, 177–184. [CrossRef] [PubMed]
43. Sogandares, F.M.; Fry, E.S. Absorption spectrum (340–640 nm) off pure water. I. Photothermal measurements. *Appl. Opt.* **1997**, *36*, 8699–8709. [CrossRef] [PubMed]
44. Pope, R.M.; Fry, E.S. Absorption spectrum (380–700 nm) of pure water, II Integrating cavity measurements. *Appl. Opt.* **1997**, *36*, 8710–8723. [CrossRef] [PubMed]
45. Kou, L.; Labrie, D.; Chylek, P. Refractive indices of water and ice in the 0.65 μm to 2.5 μm spectral range. *Appl. Opt.* **1993**, *32*, 3531–3540. [CrossRef] [PubMed]
46. Wiscombe, W.J.; Warren, S.G. A Model for the Spectral Albedo of Snow. I: Pure Snow. *J. Atmos. Sci.* **1980**, *37*, 2712–2733. [CrossRef]
47. Warren, S.G.; Wiscombe, W.J. A Model for the Spectral Albedo of Snow. II: Snow Containing Atmospheric Aerosols. *J. Atmos. Sci.* **1980**, *37*, 2734–2745. [CrossRef]
48. Grenfell, T.S.; Warren, S.G.; Mullen, P.C. Reflection of solar radiation by the Antarctic snow surface at ultraviolet, visible, and near?infrared wavelengths. *J. Geophys. Res.* **1994**, *99*, 18669–18684. [CrossRef]
49. Warren, S.G.; Brandt, R.E. Optical constants of ice from the ultraviolet to the microwave: A revised compilation. *J. Geophys. Res. Atmos.* **2008**, *113*. [CrossRef]

50. Jin, Z.; Stamnes, K.; Weeks, W.F.; Tsay, S.C. The effect of sea ice on the solar energy budget in the atmosphere-sea ice-ocean system: A model study. *J. Geophys. Res.* **1994**, *99*, 25281–25294. [CrossRef]
51. Hamre, B.; Winther, J.G.; Gerland, S.; Stamnes, J.J.; Stamnes, K. Modeled and measured optical transmittance of snow-covered first-year sea ice in Kongsfjorden, Svalbard. *J. Geophys. Res. Oceans* **2004**, *109*. [CrossRef]
52. Jiang, S.; Stamnes, K.; Li, W.; Hamre, B. Enhanced solar irradiance across the atmosphere–sea ice interface: A quantitative numerical study. *Appl. Opt.* **2005**, *44*, 2613–2625. [CrossRef] [PubMed]
53. Stamnes, K.; Hamre, B.; Stamnes, J.J.; Ryzhikov, G.; Birylina, M.; Mahoney, R.; Hauss, B.; Sei, A. Modeling of radiation transport in coupled atmosphere-snow-ice-ocean systems. *J. Quant. Spectrosc. Radiat. Transf.* **2011**, *112*, 714–726. [CrossRef]
54. Ackermann, M.; Ahrens, J.; Bai, X.; Bartelt, M.; Barwick, S.W.; Bay, R.C.; Becka, T.; Becker, J.K.; Becker, K.H.; Berghaus, P.; et al. Optical properties of deep glacial ice at the South Pole. *J. Geophys. Res.* **2006**, *111*. [CrossRef]
55. Fialho, P.; Hansen, A.D.A.; Honrath, R.E. Absorption coefficients by aerosols in remote areas: A new approach to decouple dust and black carbon absorption coefficients using seven-wavelength Aethalometer data. *J. Aerosol Sci.* **2005**, *36*, 267–282. [CrossRef]
56. Twardowski, M.S.; Boss, E.; Sullivan, J.M.; Donaghay, P.L. Modeling the spectral shape of absorption by chromophoric dissolved organic matter. *Mar. Chem.* **2004**, *89*, 69–88. [CrossRef]
57. Uusikivi, J.; Vähätalo, A.V.; Granskog, M.A.; Sommaruga, R. Contribution of mycosporine-like amino acids and colored dissolved and particulate matter to sea ice optical properties and ultraviolet attenuation. *Limnol. Oceanogr.* **2010**, *55*, 703–713. [CrossRef] [PubMed]
58. Ruddick, K.; Bouchra, N.; Collaborators. *Coastcoulor Round Robin—Final Report*; 2013. Available online: ftp://ccrropen@ftp.coastcolour.org/RoundRobin/CCRR_report_OCSMART.pdf (accessed on April 12, 2018).
59. Babin, M.; Stramski, A.D.; Ferrari, G.M.; Claustre, H.; Bricaud, A.; Obelesky, G.; Hoepffner, N. Variations in the light absorption coefficients of phytoplankton, nonalgal particles and dissolved organic matter in coastal waters around Europe. *J. Geophys. Res.* **2003**, *108*. [CrossRef]
60. Babin, M.; Morel, A.; Fournier-Sicre, V.; Fell, F.; Stramski, D. Light scattering properties of marine particles in coastal and open ocean waters as related to the particle mass concentration. *Limnol. Oceanogr.* **2003**, *28*, 843–859. [CrossRef]
61. Bricaud, A.; Morel, A.; Babin, M.; Allali, K.; Claustre, H. Mie-Scattering Calculation. *J. Geophys. Res.* **1998**, *103*, 31033–31044. [CrossRef]
62. Loisel, H.; Morel, A. Light scattering and chlorophyll concentration in case 1 waters: A re-examination. *Limnol. Oceanogr.* **1998**, *43*, 847–857. [CrossRef]
63. Morel, A.; Antoine, D.; Gentili, B. Bidirectional reflectance of oceanic waters: Accounting for Raman emission and varying particle scattering phase function. *Appl. Opt.* **2002**, *41*, 6289–6306. [CrossRef] [PubMed]
64. Mobley, C.P.; Sundman, L.K.; Boss, E. Phase function effects on oceanic light fields. *Appl. Opt.* **2002**, *41*, 1035–1050. [CrossRef] [PubMed]
65. Diehl, P.; Haardt, H. Measurement of the spectral attenuation to support biological research in a "plankton tube" experiment. *Oceanol. Acta* **1980**, *3*, 89–96.
66. McCave, I.N. Particulate size spectra, behavior, and origin of nephloid layers over the Nova Scotia continental rise. *J. Geophys. Res.* **1983**, *88*, 7647–7660. [CrossRef]
67. Fournier, G.R.; Forand, J.L. Analytic phase function for ocean water. *Proc. SPIE Ocean Opt. XII* **1994**, *2558*, 194–202.
68. Hu, Y.X.; Wielicki, B.; Lin, B.; Gibson, G.; Tsay, S.C.; Stamnes, K.; Wong, T. Delta-fit: A fast and accurate treatment of particle scattering phase functions with weighted singular-value decomposition least squares fitting. *J. Quant. Spectrosc. Radiat. Transf.* **2000**, *65*, 681–690. [CrossRef]
69. Werdell, P.; Bailey, S. An improved in-situ bio-optical data set for ocean color algorithm development and satellite data product validation. *Remote Sens. Environ.* **2005**, *98*, 122–140. [CrossRef]
70. Li, W.; Stamnes, K.; Spurr, R.; Stamnes, J.J. Simultaneous Retrieval of Aerosols and Ocean Properties: A Classic Inverse Modeling Approach. II. SeaWiFS Case Study for the Santa Barbara Channel. *Int. J. Rem. Sens.* **2008**, *29*, 5689–5698. [CrossRef]
71. Garver, S.A.; Siegel, D. Inherent optical property inversion of ocean color spectra and its biogeochemical interpretation 1. Time series from the Sargasso Sea. *J. Geophys. Res.* **1997**, *102*, 18607–18625. [CrossRef]

72. Garver, S.A.; Siegel, D.; Peterson, A.R. Optimization of a semi-analytical ocean color model for global-scale applications. *Appl. Opt.* **2002**, *41*, 2705–2714.
73. Bohren, C.F.; Huffman, D.R. *Absorption and Scattering of Light by Small Particles*; John Wiley: New York, NY, USA, 1998.
74. Born, M.; Wolf, E. *Principles of Optics*; Cambridge University Press: Cambridge, UK, 1980.
75. Cox, C.; Munk, W. Measurement of the roughness of the sea surface from photographs of the sun's glitter. *J. Opt. Soc. Am.* **1954**, *44*, 838–850. [CrossRef]
76. Masuda, K. Effects of the speed and direction of surface winds on the radiation in the atmosphere—Ocean system. *Remote Sens. Environ.* **1998**, *64*, 53–63. [CrossRef]
77. Wang, M.; Bailey, S.W. Correction of sun glint contamination on the SeaWiFS ocean and atmosphere products. *Appl. Opt.* **2001**, *40*, 4790–4798. [CrossRef]
78. Nakajima, T.; Tanaka, M. Algorithms for radiative intensity calculations in moderately thick atmospheres using a truncation approximation. *J. Quant. Spectrosc. Radiat. Transf.* **1988**, *40*, 51–69. [CrossRef]
79. Lin, Z.; Stamnes, S.; Jin, Z.; Laszlo, I.; Tsay, S.C.; Wiscombe, W.; Stamnes, K. Improved discrete ordinate solutions in the presence of an anisotropically reflecting lower boundary: Upgrades of the DISORT computational tool. *J. Quant. Spectrosc. Radiat. Transf.* **2015**, *157*, 119–134. [CrossRef]
80. Lin, Z.; Li, W.; Gatebe, C.; Poudyal, R.; Stamnes, K. Radiative transfer simulations of the two-dimensional ocean glint reflectance and determination of the sea surface roughness. *Appl. Opt.* **2016**, *55*, 1206–1215. [CrossRef] [PubMed]
81. Gordon, H.R. Atmospheric correction of ocean color imagery in the Earth Observation System era. *J. Geophys. Res.* **1997**, *102*, 17081–17106. [CrossRef]
82. Rodgers, C.D. *Inverse Methods for Atmospheric Sounding: Theory and Practice*; World Scientific: London, UK, 2000.
83. Jin, Z.; Stamnes, K. Radiative transfer in nonuniformly refracting layered media: Atmosphere-ocean system. *Appl. Opt.* **1994**, *33*, 431–442. [CrossRef] [PubMed]
84. Zhai, P.W.; Hu, Y.; Chowdhary, J.; Trepte, C.R.; Lucker, P.L.; Josset, D.B. A vector radiative transfer model for coupled atmosphere and ocean systems with a rough interface. *J. Quant. Spectrosc. Radiat. Transf.* **2010**, *111*, 1025–1040. [CrossRef]
85. Chowdhary, J.; Cairns, B.; Travis, L.D. Case studies of aerosol retrievals over the ocean from multiangle, multispectral photopolarimetric remote sensing data. *J. Atmos. Sci.* **2002**, *59*, 383–397. [CrossRef]
86. Chowdhary, J.; Cairns, B.; Mishchenko, M.I.; Hobbs, P.V.; Cota, G.F.; Redemann, J.; Rutledge, K.; Holben, B.N.; Russell, E. Retrieval of aerosol scattering and absorption properties from photopolarimetric observations over the ocean during the CLAMS experiment. *J. Atmos. Sci.* **2005**, *62*, 1093–1117. [CrossRef]
87. Chowdhary, J.; Cairns, B.; Waquet, F.; Knobelspiesse, K.; Ottaviani, M.; Redemann, J.; Travis, L.; Mishchenko, M. Sensitivity of multiangle, multispectral polarimetric remote sensing over open oceans to water-leaving radiance: Analyses of RSP data acquired during the MILAGRO campaign. *Remote Sens. Environ.* **2012**, *118*, 284–308. [CrossRef]
88. Chami, M.; Santer, R.; Dilligeard, E. Radiative transfer model for the computation of radiance and polarization in an ocean–atmosphere system: Polarization properties of suspended matter for remote sensing. *Appl. Opt.* **2001**, *40*, 2398–2416. [CrossRef] [PubMed]
89. Min, Q.; Duan, M. A successive order of scattering model for solving vector radiative transfer in the atmosphere. *J. Quant. Spectrosc. Radiat. Transf.* **2004**, *87*, 243–259. [CrossRef]
90. Fischer, J.; Grassl, H. Radiative transfer in an atmosphere-ocean system: An azimuthally dependent matrix-operator approach. *Appl. Opt.* **1984**, *23*, 1032–1039. [CrossRef] [PubMed]
91. Ota, Y.; Higurashi, A.; Nakajima, T.; Yokota, T. Matrix formulations of radiative transfer including the polarization effect in a coupled atmosphere-ocean system. *J. Quant. Spectrosc. Radiat. Transfer* **2010**, *111*, 878–894. [CrossRef]
92. Kattawar, G.; Adams, C. Stokes vector calculations of the submarine light field in an atmosphere-ocean with scattering according to the Rayleigh phase matrix: Effect of interface refractive index on radiance and polarization. *Limnol. Oceanogr.* **1989**, *34*, 1453–1472. [CrossRef]
93. Lotsberg, J.; Stamnes, J. Impact of particulate oceanic composition on the radiance and polarization of underwater and backscattered light. *Opt. Express* **2010**, *18*, 10432–10445. [CrossRef] [PubMed]

94. Morel, A.; Gentili, B. Diffuse reflectance of oceanic waters. II. bidirectional aspect. *Appl. Opt.* **1993**, *32*, 2803–2804. [CrossRef] [PubMed]
95. Fan, Y.; Li, W.; Stamnes, K.; Gatebe, C. A neural network method to correct bidirectional effects in water-leaving radiance. *Appl. Opt.* **2016**, *55*, 10–421. [CrossRef] [PubMed]
96. Voss, K.J.; Chapin, A.L. An Upwelling Radiance Distribution Camera System, NURADS. *Opt. Express* **2005**, *13*, 4250–4262. [CrossRef] [PubMed]
97. Kay, S.; Hedley, J.D.; Lavender, S. Sun Glint Correction of High and Low Spatial Resolution Images of Aquatic Scenes: A Review of Methods for Visible and Near-Infrared Wavelengths. *Remote Sens.* **2009**, *1*, 697–730. [CrossRef]
98. Steinmetz, F.; Deschamps, P.Y.; Ramon, D. Atmospheric correction in presence of sun glint: Application to MERIS. *Opt. Express* **2011**, *19*, 9783–9800. [CrossRef] [PubMed]
99. Fukushima, H.; Suzuki, K.; Li, L.; Suzuki, N.; Murakami, H. Improvement of the ADEOS-II/GLI sun-glint algorithm using concomitant microwave scatterometer-derived wind data. *Adv. Space Res.* **2009**, *43*, 941–947. [CrossRef]
100. Wang, M. The Rayleigh lookup tables for the SeaWiFS data processing: accounting for the effects of ocean surface roughness. *Int. J. Remote Sens.* **2002**, *23*, 2693–2702. [CrossRef]
101. Ottaviani, M.; Spurr, R.; Stamnes, K.; Li, W.; Su, W.; Wiscombe, W. Improving the description of sunglint for accurate prediction of remotely sensed radiances. *J. Quant. Spectrosc. Radiat. Transf.* **2008**, *109*, 2364–2375. [CrossRef]
102. Gatebe, C.K.; King, M.D.; Lyapustin, A.I.; Arnold, G.T.; Redemann, J. Airborne spectral measurements of ocean directional reflectance. *J. Atmos. Sci.* **2005**, *62*, 1072–1092. [CrossRef]
103. Stamnes, K.; Tsay, S.C.; Wiscombe, W.; Jayaweera, K. Numerically stable algorithm for discrete-ordinate-method radiative transfer in multiple scattering and emitting layered media. *Appl. Opt.* **1988**, *27*, 2502–2509. [CrossRef] [PubMed]
104. National Oceanic and Atmospheric Administration (NOAA). *US Standard Atmosphere*; Technical Report; NOAA: Washington, DC, USA, 1976.
105. Jin, Z.; Charlock, T.P.; Rutledge, K.; Stamnes, K.; Wang, Y. Analytical solution of radiative transfer in the coupled atmosphere-ocean system with a rough surface. *Appl. Opt.* **2006**, *45*, 7443–7455. [CrossRef] [PubMed]
106. Mobley, C.D. Polarized reflectance and transmittance properties of windblown sea surfaces. *Appl. Opt.* **2015**, *54*, 4828–4849. [CrossRef] [PubMed]
107. Fan, Y.; Li, W.; Stamnes, K.; Stamnes, J.J.; Sørensen, K. Simultaneous Retrieval of AEROSOL and Marine Parameters in Coastal areas Using a Coupled Atmosphere-Ocean Radiative Transfer Model. In Proceedings of the Sentinel-3 for Science Workshop, Venice-Lido, Italy, 2–5 June 2015.
108. Spurr, R.; Stamnes, K.; Eide, H.; Li, W.; Zheng, K.; Stamnes, J. Simultaneous retrieval of aerosol and ocean color: A classic inverse modeling approach: I. Analytic Jacobians from the linearized CAO-DISORT model. *J. Quant. Spectrosc. Radiat. Transf.* **2007**, *104*, 428–449. [CrossRef]
109. Fan, Y.; Li, W.; Gatebe, C.K.; Jamet, C.; Zibordi, G.; Schroeder, T.; Stamnes, K. Atmospheric correction and aerosol retrieval over coastal waters using multilayer neural networks. *Remote Sens. Environ.* **2017**, *199*, 218–240. [CrossRef]
110. He, X.; Stamnes, K.; Bai, Y.; Li, W.; Wang, D. Effects of Earth curvature on atmospheric correction for ocean color remote sensing. *Remote Sens. Environ.* **2018**, *209*, 118–133. [CrossRef]
111. Stamnes, S.; Hostetler, C.; Ferrari, R.; Burton, S.; Lui, X.; Hair, J.; Hu, Y.; Wasilewski, A.; Martin, W.; Van Diedenhoven, B.; et al. Simultaneous polarimeter retrievals of microphysical aerosol and ocean color parameters from the MAPP algorithm with comparison to high spectral resolution lidar aerosol and ocean products. *Appl. Opt.* **2018**, *57*, 2394–2413. [CrossRef] [PubMed]
112. Stamnes, S.; Fan, Y.; Chen, N.; Li, W.; Tanikawa, T.; Lin, Z.; Liu, X.; Burton, S.; Omar, A.; Stamnes, J.; et al. Advantages of measuring the Q Stokes parameter in addition to the total radiance I in the detection of absorbing aerosols. *Front. Earth Sci.* **2018**, *6*, 1–11. [CrossRef]
113. Dahlback, A.; Stamnes, K. A new spherical model for computing the radiation field available for photolysis and heating at twilight. *Planet. Space Sci.* **1991**, *39*, 671–683. [CrossRef]

114. Spurr, R.J.D. VLIDORT: A linearized pseudo-spherical vector discrete ordinate radiative transfer code for forward model and retrieval studies in multilayer multiple scattering media. *J. Quant. Spectrosc. Radiat. Transf.* **2006**, *102*, 316–342. [CrossRef]
115. Rozanov, V.; Rozanov, A.; Kokhanovsky, A.; Burrows, J. Radiative transfer through terrestrial atmosphere and ocean: Software package SCIATRAN. *J. Quant. Spectrosc. Radiat. Transf.* **2014**, *133*, 13–71. [CrossRef]
116. Davis, A.; Marshak, A. *3D Radiative Transfer in Cloudy Atmospheres*; Springer: Berlin, Germany, 2005.
117. Chandrasekhar, S. On the diffuse reflection of a pencil of radiation by a plane-parallel atmosphere. *Proc. Natl. Acad. Sci. USA* **1958**, *44*, 933–940. [CrossRef] [PubMed]
118. Shiina, T.; Yoshida, K.; Ito, M.; Okamura, Y. Long-range propagation of annular beam for lidar application. *Opt. Commun.* **2007**, *279*, 159–167. [CrossRef]
119. Habel, R.; Christensen, P.H.; Jarosz, W. Photon Beam Diffusion: A Hybrid Monte Carlo Method for Subsurface Scattering. In Proceedings of the 24th Eurographics Symposium on Rendering, Zaragoza, Spain, 19–21 June 2013; Volume 32.
120. Barichello, L.; Siewert, C. The searchlight problem for radiative transfer in a finite slab. *J. Comput. Phys.* **2000**, *157*, 707–726. [CrossRef]
121. Kim, A.D.; Moscoso, M. Radiative transfer computations for optical beams. *J. Comput. Phys.* **2003**, *185*, 50–60. [CrossRef]
122. Mobley, C.D.; Gentili, B.; Gordon, H.R.; Jin, Z.; Kattawar, G.W.; Morel, A.; Reinersman, P.; Stamnes, K.; Stavn, R.H. Comparison of numerical models for computing underwater light fields. *Appl. Opt.* **1993**, *32*, 7484–7504. [CrossRef] [PubMed]
123. Mitra, K.; Churnside, J.H. Transient radiative transfer equation applied to oceanographic lidar. *Appl. Opt.* **1999**, *38*, 889–895. [CrossRef] [PubMed]
124. Stamnes, K.; Lie-Svendsen, Ø.; Rees, M.H. The linear Boltzmann equation in slab geometry: Development and verification of a reliable and efficient solution. *Planet. Space Sci.* **1991**, *39*, 1453–1463. [CrossRef]
125. De Beek, R.; Vountas, M.; Rozanov, V.; Richter, A.; Burrows, J. The Ring effect in the cloudy atmosphere. *Geophys. Res. Lett.* **2001**, *28*, 721–724. [CrossRef]
126. Landgraf, J.; Hasekamp, O.; Van Deelen, R.; Aben, I. Rotational Raman scattering of polarized light in the Earth atmosphere: a vector radiative transfer model using the radiative transfer perturbation theory approach. *J. Quant. Spectrosc. Radiat. Transf.* **2004**, *87*, 399–433. [CrossRef]
127. Spurr, R.; de Haan, J.; van Oss, R.; Vasilkov, A. Discrete-ordinate radiative transfer in a stratified medium with first-order rotational Raman scattering. *J. Quant. Spectrosc. Radiat. Transf.* **2008**, *109*, 404–425. [CrossRef]
128. Ge, Y.; Gordon, H.; Voss, K. Simulation of inelastic scattering contributions to the irradiance field in the oceanic variation in Fraunhofer line depths. *Appl. Opt.* **1993**, *32*, 4028–4036. [CrossRef] [PubMed]
129. Kattawar, G.; Xu, X. Filling-in of Fraunhofer lines in the ocean by Raman scattering. *Appl. Opt.* **1992**, *31*, 1055–1065. [CrossRef] [PubMed]
130. Voss, K.J.; Fry, E.S. Measurement of the Mueller matrix for ocean water. *Appl. Opt.* **1984**, *23*, 4427–4439. [CrossRef] [PubMed]
131. Werdell, J.P.; Franz, B.A.; Bailey, S.W.; Feldman, G.C.; Boss, E.; Brando, V.E.; Dowell, M.; Hirata, T.; Lavender, S.; Lee, Z.P.; et al. Generalized ocean color inversion model for retrieving marine inherent optical properties. *Appl. Opt.* **2013**, *52*, 2019–2037. [CrossRef] [PubMed]
132. Lee, Z.P.; Carder, K.L.; Arnone, R. Deriving inherent optical properties from water color: A multi-band quasi-analytical algorithm for optically deep waters. *Appl. Opt.* **2002**, *41*, 5755–5772. [CrossRef] [PubMed]

6

Estimation of Suspended Matter, Organic Carbon and Chlorophyll-a Concentrations from Particle Size and Refractive Index Distributions

Jacopo Agagliate [1,*], Rüdiger Röttgers [2], Kerstin Heymann [2] and David McKee [1]

1 Department of Physics, University of Strathclyde, 107 Rottenrow, Glasgow G4 0NG, UK; david.mckee@strath.ac.uk

2 Helmholtz-Zentrum Geesthacht, Max-Planck-Straße 1, 21502 Geesthacht, Germany; rroettgers@hzg.de (R.R.); kerstin.heymann@hzg.de (K.H.)

* Correspondence: jacopo.agagliate@strath.ac.uk

Abstract: Models of particle density and of organic carbon and chlorophyll-a intraparticle concentration were applied to particle size distributions and particle real refractive index distributions determined from flow cytometry measurements of natural seawater samples from a range of UK coastal waters. The models allowed for the estimation of suspended particulate matter, organic suspended matter, inorganic suspended matter, particulate organic carbon, and chlorophyll-a concentrations. These were then compared with independent measurements of each of these parameters. Particle density models were initially applied to a simple spherical model of particle volume, but generally overestimated independently measured values, sometimes by over two orders of magnitude. However, when the same density models were applied to a fractal model of particle volume, successful agreement was reached for suspended particulate matter and both inorganic and organic suspended matter values (RMS%E: 57.4%, 148.5%, and 83.1% respectively). Non-linear organic carbon and chlorophyll-a volume scaling models were also applied to a spherical model of particle volume, and after an optimization procedure achieved successful agreement with independent measurements of particulate organic carbon and chlorophyll-a concentrations (RMS%E: 45.6% and 51.8% respectively). Refractive index-based models of carbon and chlorophyll-a intraparticle concentration were similarly tested, and were also found to require a fractal model of particle volume to achieve successful agreement with independent measurements, producing RMS%E values of 50.2% and 45.2% respectively after an optimization procedure. It is further shown that the non-linear exponents of the volume scaling models are mathematically equivalent to the fractal dimensionality coefficients that link cell volume to mass concentration, reflecting the impact of non-uniform distribution of intracellular carbon within cells. Fractal models of particle volume are thus found to be essential to successful closure between results provided by models of particle mass, intraparticle carbon and chlorophyll content, and bulk measurements of suspended mass and total particulate carbon and chlorophyll when natural mixed particle populations are concerned. The results also further confirm the value of determining both size and refractive index distributions of natural particle populations using flow cytometry.

Keywords: forward modeling; suspended matter; marine particles; fractal structure; organic carbon; chlorophyll-a

---

1. Introduction

The determination of suspended particulate mass concentrations (minerogenic mass, biomass, chlorophyll content) in marine particle populations is a matter of particular interest to ocean sciences, and one of the key aspects of the characterization of the properties of marine particles. The relationship between particulate mass properties and optical properties of seawater is important for understanding the formation of optical remote sensing signals and their interpretation.

In previous work carried out on a set of seawater samples collected in UK coastal waters (UKCW dataset), a Mie-based flow cytometric method (FC method) was developed to determine particle size distributions (PSDs) and real refractive index distributions (PRIDs), and its results used as inputs for Mie theory forward modelling to reconstruct not only bulk inherent optical properties (IOPs), but also individual fractions and optical contributions from inorganic, organic, and fluorescent particle subpopulations [1,2]. Although flow cytometric determination of particle physical properties and the subsequent modelling of IOPs have some precedent [3–5], no true effort has been devoted to extending the procedure to the reconstruction of particulate mass concentrations from flow cytometric data. Indeed, coupled with models of particle density and carbon intraparticle concentration, particle size, and real refractive index distributions offer the chance to explore the biogeochemical properties of a particle population from a new perspective.

In this study, models of organic and inorganic particle density are adapted from literature and applied to the UKCW dataset to produce modelled values of suspended particulate matter (SPM), organic suspended matter (OSM), and inorganic suspended matter (ISM). Furthermore, cell volume scaling models and refractive index-based models are also adapted from literature to allow estimation of intraparticle carbon (C_i) and intraparticle chlorophyll-a (Chl_i) content, ultimately producing modelled values of particulate organic carbon (POC) and chlorophyll-a concentrations (ChlA) from the particle data of the UKCW dataset. The modelled values thus obtained are then compared against the results of actual biogeochemical measurements, and the parameters used to assess carbon and chlorophyll concentrations are optimized on a dataset-wide basis to explore the physiology of the cells encountered during the He442 research cruise. A summary of the abbreviations and notations used throughout the study is given in Table 1.

Table 1. Abbreviations and notations used in this study.

Notation	Definition
ChlA	Chlorophyll-a concentration, mg m^{-3}
FC	Flow cytometer; flow cytometry
IOP	Inherent optical property
ISM	Inorganic suspended matter, g m^{-3}
OSM	Organic suspended matter, g m^{-3}
SPM	Suspended particulate matter, g m^{-3}
POC	Particulate organic carbon, mg m^{-3}
PRID	Particle real refractive index distribution
PSD	Particle size distribution
RMS%E	Root-mean-square percentage error
RMSE	Root-mean-square error
UKCW	UK coastal waters (dataset)
a, a_C, a_{chl}	Slopes of the refractive index-based models and of the of the C_i and Chl_i optimized refractive index-based models respectively, kg m^{-3}
b	y-intercepts of the refractive index-based models, kg m^{-3}
$F(r)$	Fractal dimension, dimensionless
h_1, h_2	Power law exponents of the optimized POC and ChlA volume scaling functions respectively, dimensionless
k	Particle size distribution scaling coefficient, mL^{-1}
k_1, k_2	Scaling coefficients of the optimized POC and ChlA volume scaling functions respectively, pg µm^{-3}
m_{tot}	Total particle mass, mg
C_i	Intraparticle carbon concentration, kg m^{-3}
Chl_i	Intraparticle chlorophyll-a concentration, kg m^{-3}

Table 1. *Cont.*

Notation	Definition
$N(D)$	Number concentration of particles within particle size bin corresponding to particle diameter D, mL^{-1}
$N'(D)$	Density function of the particle size distribution, mL^{-1} μm^{-1}
N_r	Number of particles within particle size bin corresponding to particle radius r, dimensionless
$N_{D,nr}$	Number of particles within particle bin corresponding to particle diameter D and real refractive index n_r, dimensionless
n_o	Real refractive index of the dry matter fraction of the particle, dimensionless
n_r	Real refractive index of the particle, dimensionless
n_i	Imaginary refractive index of the particle, dimensionless
$\overline{n}_{r,1}, \overline{n}_{r,2}$	Average real refractive indices at the upper and lower extremes of the particle size distribution respectively, dimensionless
r, D	Particle radius and particle diameter, μm
r_o	Primary particle radius, μm
V_D	Particle volume, μm^3
V_o	Volume of the dry matter fraction of the particle, μm^3
$y(r)$	Volume scaling function
B	Fractal dimension exponent, dimensionless
γ	Power law slope, dimensionless
P	Particle density, g/m^3
ρ_{nr}	Density of a particle with real refractive index n_r, g/m^3
ρ_o	Density of the dry matter fraction of the particle, g/m^3

2. Materials and Methods

2.1. Theory

2.1.1. Particle Mass Modelling from Apparent Density of Hydrated Matter

Calculations for modelled values of SPM, OSM, and ISM were made following the technique presented by Zhang et al. [6]. Building on the approach presented by Morel & Ahn [7] and Babin et al. [8] the technique estimates a density value for the particulate matter which is dependent on the real part of the refractive index of the particles and is designed to account for their water content. Since this value is neither the value of the dry matter fraction of the particle nor that of water, but rather a combination of the two, this global density is also known as "apparent" density. The equation takes the form

$$\rho = \rho_o V_o = \rho_o \frac{n_r - 1}{n_o - 1}, \tag{1}$$

where n_r is the real refractive index of the whole particle and n_o, ρ_o, and V_o are respectively the real refractive index, density, and fractional volume of the dry matter fraction of the particle. All refractive index values are given relative to water.

Values for the $\rho_o/(n_o - 1)$ ratio were defined following those employed by Zhang et al. [6]. For organic particles (defined as the fraction of the particle population with $n_r < 1.1$) the mean value of the ratio was set at $(8.56 \pm 1.1) \times 10^6$ g/m^3. These are particles with high water content, as high as ~$80 \pm 10\%$ for algal cells [9]. For mineral particles (defined as the fraction of the particle population with $n_r \geq 1.1$) the mean value of the ratio was instead set at $(15.52 \pm 1.84) \times 10^6$ g/m^3. These particles have low water content; when the fractional volume of dry matter reaches unity (i.e., water content within the particles is zero) the apparent density of the particle becomes equal to the density of the dry mineral matter and can be calculated accordingly. Zhang et al. [6] find $n_r = 1.16$ as the threshold above which $V_o = 1$, and give $\rho = [(6.42 \pm 0.85)n_r - (4.86 \pm 0.99)] \times 10^6$ g/m^3 as the corresponding density based on a linear regression of literature values of density and refractive index for a number of mineral species. Overall, the final expression of Equation (1) used in practice was

$$\rho = \begin{cases} 8.56 \times 10^6 (n_r - 1) & n_r < 1.1 \\ 15.52 \times 10^6 (n_r - 1) & 1.1 \leq n_r < 1.16 \\ 6.42 \times 10^6 n_r - 4.86 \times 10^6 & n_r \geq 1.16 \end{cases} \tag{2}$$

with all density values given as g/m^3.

2.1.2. Particulate Organic Carbon and Chlorophyll-a Cell Volume Scaling

Organic carbon and chlorophyll-a concentrations within a cell are not linear functions of the cell volume (also defined as biovolume by some authors); C_i and Chl_i values can be instead derived using empirical relationships defined by volume scaling exponents, which can be then summed over organic and fluorescent PSDs to obtain POC and ChlA values respectively, i.e.,

$$C = \sum_r y(r) N_r \tag{3}$$

where C represents either POC or ChlA, $y(r)$ is the corresponding size-dependent total carbon or chlorophyll concentration per cell and N_r is the number of particles within each size bin. A number of these empirical conversion relationships can be found in the literature for the modelling described here: four sets of parameters for carbon [10–12] and two sets of parameters for chlorophyll-a [11,13] were employed. These are presented in Table 2.

Table 2. Particulate organic carbon and chlorophyll-a cell volume scaling models used in this study

POC & ChlA Cell Volume Scaling	
Particulate Organic Carbon	**Source**
$y(r) = 0.433V(r)^{0.863}$	[10]
$y(r) = 0.109V(r)^{0.991}$	[11]
$y(r) = 0.288V(r)^{0.811}$	[12] (diatoms)
$y(r) = 0.216V(r)^{0.939}$	[12] (non-diatom mixed protists)
Chlorophyll-a	**Source**
$y(r) = 0.00429V(r)^{0.917}$	[11]
$y(r) = 0.0398V(r)^{0.863}$	[13]

2.1.3. Refractive Index-Based Estimation of Particulate Organic Carbon and Chlorophyll-a

Research carried out in the 1990s demonstrated that cell volume is not the only parameter that can be used to estimate C_i and Chl_i values. In a series of studies [14–18], a number of empirical relationships were established for various phytoplankton species between the real refractive index n_r and C_i and between the imaginary refractive index (n_i) and Chl_i. Expanding on this premise, Stramski [19] established refractive index-based linear models for the estimation of C_i and Chl_i based on data from two phytoplankton species

$$C_i = 3441.055 n_r(660\,\text{nm}) - 3404.99 \tag{4}$$

$$Chl_i = 996.86 n_i(675\,\text{nm}) + 1.17. \tag{5}$$

A follow-up work by DuRand et al. [20] established slightly modified relationships with the inclusion of data from additional phytoplankton species

$$C_i = 3946 n_r(650\,\text{nm}) - 3922 \tag{6}$$

$$Chl_i = 1244n_i(675\,\text{nm}) - 0.32. \tag{7}$$

These can then be associated with a particle volume model and with particle sizes as provided by PSDs to determine POC and ChlA values. Both sets of equations were employed in this work. The n_r values contained in the PRIDs were originally determined by the FC method for λ = 488 nm, i.e., the wavelength of the laser source used within the flow cytometer [1]; however, n_r values are only weakly dependent on the wavelength (e.g., [21]), and Equations (4) and (6) are thus likely to be usable as is for the UKCW PRIDs as well. The FC method does not provide any information on the n_i values; as will be described in Section 2.2.4, these were derived from literature. The relevant n_i value (i.e., for organic particles at λ = 675 nm) was adapted from Babin et al. [8] as $n_i = 1.620 \times 10^{-3}$.

2.2. *Methods*

The particle density, carbon and chlorophyll cell volume scaling, and C_i and Chl_i refractive-index based estimation models were applied to the PSDs and PRIDs of the UKCW dataset as determined by the FC method, which can be found described in detail in [1]. A description of the dataset and of the measurement protocols (particularly those relative to SPM, ISM, OSM, POC, and ChlA measurements) is summarised below. Resulting mass concentrations obtained by modelling from FC data were then compared with corresponding suspended matter, organic carbon, and chlorophyll concentration values determined from traditional sample analysis. Cumulative contributions from different size classes were also calculated for SPM, ISM, OSM, and POC.

2.2.1. UK Coastal Waters (UKCW) Dataset

The UKCW dataset consists of natural water samples obtained during the He442 research cruise in UK waters (4–21 April 2015) on board the R/V Heincke (Alfred-Wegener-Institute, Bremerhaven, Germany). Sixty-two stations were sampled across a variety of optical water conditions around the coast of the UK (Figure 1), supplying a total of 50 samples with complete sets of FC data and matching data from other instruments and independent measurements. This included SPM, ISM, OSM, POC, and ChlA values obtained from lab analysis of the water samples retrieved during the research cruise. Wind conditions were favourable throughout, ranging from calm to moderate gale, and did not hamper the measurement process at any point during the cruise. Day-to-day weather ranged widely from clear sky conditions to heavy rain, although good weather was generally prevalent. Of particular note was the very high particle load found in Bristol Channel waters, which resulted in particularly large values of SPM, ISM, and OSM, as will be described in the following.

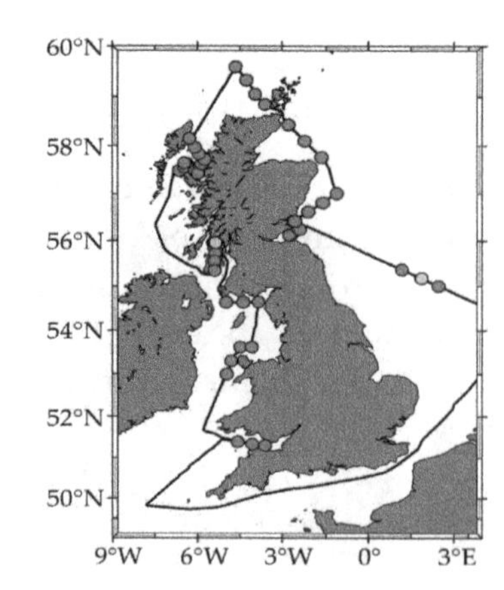

Figure 1. Track of the He442 research cruise, which took place in April 2015 in UK coastal waters aboard R/V Heincke. Out of the 62 measurement stations visited a total of 50 complete sets of data were retrieved, matching flow cytometric data, and ancillary measurements (blue circles). Yellow circles denote stations where two samples were taken. The figure was adapted and modified with permission from Figure 1 of Agagliate et al. [2].

2.2.2. Depth Profiling

Main depth profiling was done via an instrument frame equipped with Niskin bottles for sample retrieval. The frame was lowered through the water column at each of the stations, kept at a maximum depth for a first round of sampling, then raised to near surface depth to retrieve further samples. The samples were taken from the Niskin bottles on the frame as quickly as possible after the frame was back on deck and filled into 10-L plastic containers. In waters with high turbidity the Niskin bottles were flushed twice to avoid settling out of particulate matter. Forty-eight out of the 50 samples of the UKCW dataset are surface samples (depth: 5–7 m), with further two samples taken from bottom depths instead. The prevalence of surface samples within the dataset is due to the focus of the cruise, which was on developing data sets in support of the Sentinel remote sensing missions.

2.2.3. Flow Cytometry Measurement Protocol

All samples were measured by a CytoSense flow cytometer (CytoBuoy b.v., Woerden, The Netherlands) once for each of four sensitivity settings of the side scattering photomultiplier tube (50, 60, 70, 80), for 6 min and at a flow rate of 0.5 μL/s. Side scattering was used as the trigger channel in all cases. The reader is directed to [1] for a detailed description of the CytoSense flow cytometer and its operation, particularly in the context of the UKCW dataset. Additional measurements of standard polymer beads necessary for calibration of the FC method were taken daily across the whole sampling period. A detailed description of the FC method and of the procedure followed to reconstruct PSDs and PRIDs can also be found in [1], where the application of the method to the UKCW dataset is also discussed specifically.

2.2.4. PSD Extrapolations

The mass concentration modelling will require the entire optically relevant particle distribution to be included as the input, or the output will not be comparable with independently measured mass concentration values. The FC method was found to reliably retrieve particle diameters between ~0.5–10 μm [1]. This range covers a large fraction of the contribution to scattering and backscattering, but the whole optically relevant range spans from tens of nanometers to a few millimeters [22,23]. The undetectable fraction of the particle population has to be accounted for using an approximation of the PSD to extend the range of the distribution over the whole relevant range.

Ever since pioneering work in the '60s and '70s found that the number of particles suspended in the ocean increased continuously and monotonically towards smaller scales [24,25], power law distributions of the type used by Junge [26] for aerosols have been the most common form of approximation for natural seawater particle populations [27,28]. The PSDs determined by the FC method for the UKCW dataset broadly conformed to this model, and consequently, following in the steps of Green et al. [5], a least squares best fit of the UKCW PSDs through power law distributions as defined by

$$N(D) = N'(D)dD = kD^{-\gamma}dD \tag{8}$$

was used to extend the range of the measured PSDs (Figure 2). Here $N(D)$ is the number concentration of particles within the size bin corresponding to diameter D, $N'(D)$ the density function of the PSD, dD the width of the size bin, k the scaling coefficient of the PSD, and γ its slope. The form given in Equation (8) is necessary because the FC PSDs have bin-like nature; accordingly, the extrapolations need to be bin-like as well.

Values for the real refractive index n_r in the Junge extensions must also be accounted for using some approximation of the PRID to extend the range of known refractive indices; this was done by averaging n_r at the extremes of the measured PSD fraction (last four bins on either side) and using these averaged values on the respective arms of the extension (Figure 2). Since the FC method does not offer any information on the imaginary part of the refractive indices, n_i values are unknown both in the available FC PSDs and in their extrapolations. Typical values for n_i were therefore adapted from literature (Figure 8 of Babin et al. [8]), for both organic and inorganic particles. These were then assigned to the particles according to the value of the real refractive index of each bin, both directly determined by the FC method and extrapolated.

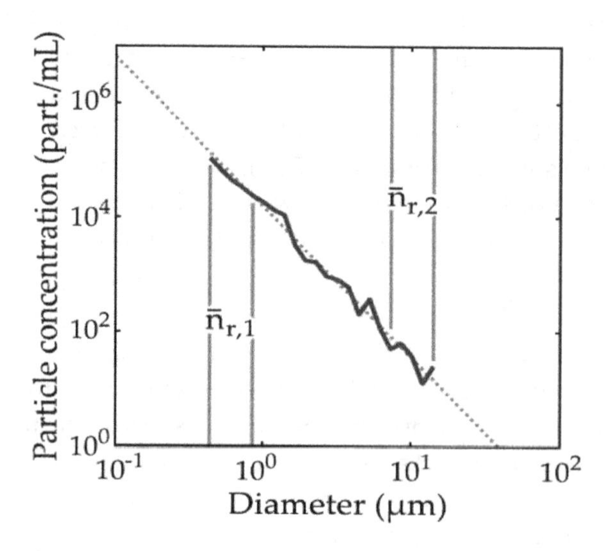

Figure 2. Power law best fit and real refractive index approximation in a typical natural particle population sample. Independent n_r values obtained by averaging the n_r of particles at the extremes of the PSD ($\overline{n}_{r,1}$, $\overline{n}_{r,2}$) were used to approximate the real refractive index within the respective ends of the PSD extrapolation (dotted line). The figure was adapted and modified with permission from Figure 3 of Agagliate et al. [2].

2.2.5. Suspended Particulate Matter and Inorganic/Organic Suspended Matter

Suspended particulate matter (SPM) was obtained from each sample following procedures detailed by Röttgers et al. [29]. The sample was filtered through filter pads under low vacuum (47-mm Whatman GF/F glass-fiber filters), then immediately placed in a petri dish after filtration and put to dry in a vacuum desiccator: SPM values were subsequently obtained by weighing the mass of dried sample and dividing it by the sample volume used. SPM values ranged between 45.519 and 0.113 g/m^3, with highest values found in the Bristol Channel and lowest values found in the North Sea, although variance in this latter area was large. Further separation into organic suspended matter and inorganic suspended matter fractions (OSM and ISM, also found in literature respectively as particulate organic matter and particulate inorganic matter, POM and PIM) was obtained by volatization of organics at 500 °C. ISM values ranged between 38.373 and 0.038 g/m^3, and followed a geographic pattern similar to that of SPM. OSM values ranged between 4.487 and 0.053 g/m^3, and were found to be more evenly distributed across the cruise track. The lowest value was found in the Irish Sea, while the highest was found once again in the Bristol Channel, due to the very large particle load of its turbid waters.

2.2.6. Particulate Organic Carbon

Particulate organic carbon (POC) concentrations were determined by catalytic combustion using a Vario TOC Cube instrument (Elementar, Langenselbold, Germany). Between 0.5 and 1 L of collected water for each sample were initially filtered onto 25-mm, combusted Whatman GF/F glass-fiber filters. The filters were then frozen and transported to the home laboratory. Once there, the filters were dried at 55 °C, packed into tin capsules and analyzed for their carbon content. Separation or removal of inorganic carbon was not considered, as concentrations of inorganic carbon (calcite or carbonates) are assumed low for these waters. Calibration of the POC/TOC analyzer was done regularly using sulphanilamide as the calibration standard. POC values were found to range between 2.307×10^3 and 6.783×10^1 mg/m^3, and followed a geographic pattern similar to that of OSM. Lowest values were found around Skye and the Hebrides, and highest values in the Bristol Channel.

2.2.7. Chlorophyll-a

Chlorophyll-a concentration (ChlA) was determined using high-performance liquid chromatography (HPLC) following Zapata et al. [30]: specifically, ChlA was determined from fluorescence values using excitation at 440 nm and emission at 650 nm and by comparison with standards of known chlorophyll concentration. ChlA values ranged between 7.620 and 0.096 mg/m^3, with lowest values found in the Firth of Clyde and highest values found in the North Sea.

3. Results

3.1. PSDs and PRIDs

The PSDs retrieved by the FC method for the UKCW dataset were found to broadly follow power law distributions, with the main difference between stations being the overall concentration of the particle population (Figure 3a). Two obvious outliers are present, corresponding to samples from the turbid waters of the Bristol Channel; close inspection reveals structures that may be closer in nature to models such as the double gamma distribution proposed by Risović [31], and that indeed may be identified to a lesser degree in the other samples as well. Nonetheless, the power law approach remains a reasonable approximation for a large majority of the dataset, and was used for PSD extrapolations accordingly for all samples including the Bristol Channel ones.

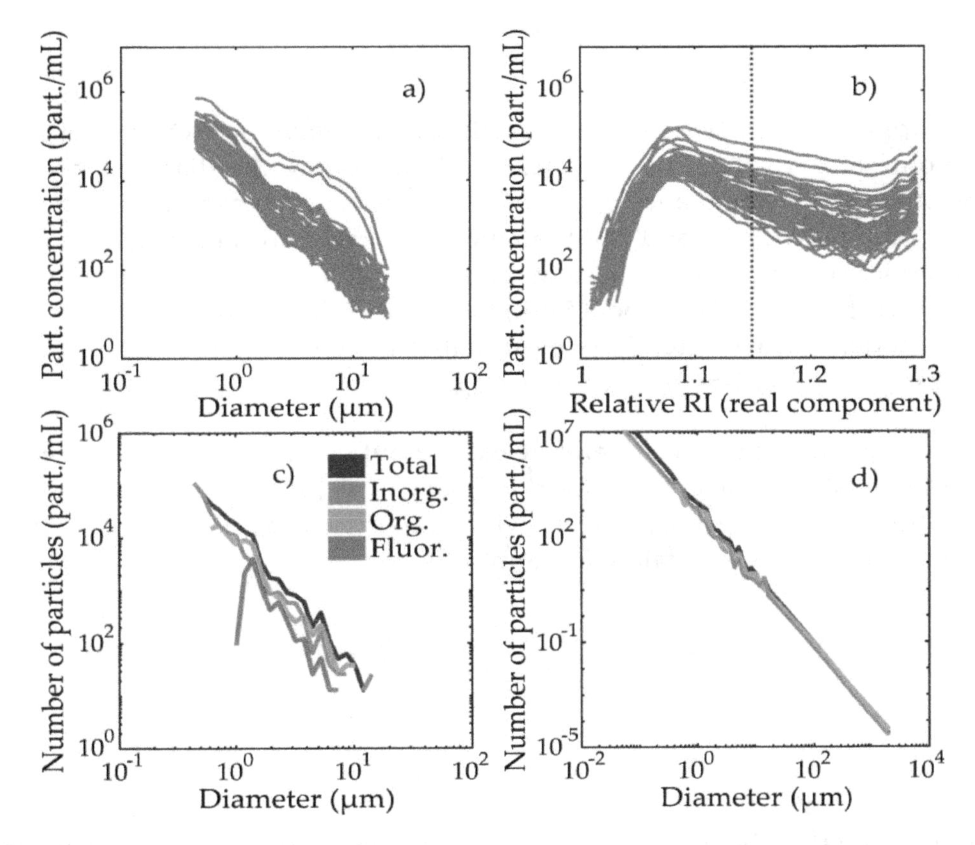

Figure 3. Collective view of (**a**) all 50 UKCW PSDs and (**b**) UKCW PRIDs produced by the FC method. Note that real refractive index values above 1.15 are not precise, but still indicate high refractive indices [1]. (**c**) Total, organic, inorganic, and fluorescent PSDs for a typical sample of the UKCW dataset and (**d**) power law extension of the total, organic and inorganic PSDs. Note that the extended organic and inorganic PSDs intersect the extended total PSD; therefore the sum of the extended organic and inorganic PSDs is not exactly equal to the extended total SPM. To evaluate the error thus introduced, SPM values are modelled both from the total PSD and by summing model ISM and OSM values. Panels (**a**,**b**) of the figure were adapted and modified with permission from Figure 2 of Agagliate et al. [2].

Particle refractive index distributions were found to be fairly homogeneous across all samples (Figure 3b), with distribution peaks found between 1.05–1.15 and within expectations for the real refractive index of the most common components of marine particle populations [9,32]. By exploiting the particle composition information given by the PRIDs, the total PSD of each UKCW sample was further separated into an inorganic PSD, an organic PSD, and a fluorescent fraction, which is itself a sub-fraction of the organic PSD (Figure 3c). These were used to calculate ISM, OSM & POC, and ChlA respectively, while SPM was calculated both from the total PSD and as the sum of the model ISM and OSM. To account for the fraction of the particle population undetected by the FC method, the total organic and inorganic PSDs were extended between 0.05 and 2000 μm as suggested by Davies et al. [23], following Equation (8) (Figure 3d); note that since the PSD extrapolations only approximate the particle populations outside the FC method detection range, the sum of the extended organic and inorganic PSDs generally is not exactly equal to the extended total PSD, making the dual calculation of SPM both from total PSD and as the sum of model ISM and OSM values a useful check of the error so introduced.

3.2. Particle Mass Modelling

Keeping with the assumption of particle sphericity used in the Mie-based models employed by the FC method, a first attempt of total particle mass calculation for the total, organic and inorganic fractions of the particle population was made as a simple bin-by-bin summation of spherical masses, i.e.,

$$m_{tot} = \sum_{D,nr} \rho_{nr} V_D N_{D,nr}, \tag{9}$$

where ρ_{nr} is the apparent density of a particle with real refractive index n_r as defined in Equation (2), V_D is the volume of a sphere of diameter D, D and n_r are the diameter and real refractive index corresponding to each bin, and $N_{D,nr}$ is the number concentration of particles within each bin. Given Equation (9) SPM, ISM, and OSM can then be obtained respectively by summing over the entire range of refractive indices or by limiting the summation to real refractive index values above or below the $n_r = 1.1$ threshold. The model SPM, ISM, and OSM values produced using this simple particle volume model however grossly overestimated the corresponding UKCW measurements, in certain cases by over two orders of magnitude (Figure 4).

A second calculation attempt was therefore carried out using a slightly modified version of the particle volume and total mass model employed in Zhang et al. [6]. This model is designed to account for the fractal nature of some marine particles, which can exist as aggregates of smaller units rather than as individuals exclusively, and takes the form

$$m_{tot} = \sum_{r,nr} \frac{4\pi}{3} \left(\frac{r}{r_o}\right)^{F(r)} r_o^3 \rho_{nr} N_{r,nr}, \tag{10}$$

where

$$F(r) = 3\left(\frac{r}{r_o}\right)^{\beta}. \tag{11}$$

Here it is assumed that the flow cytometer has measured the radius r of an aggregate particle, which is itself constructed from primary particles of radius r_o. $F(r)$ is known as the fractal dimension of the aggregate. The value of r_o and of exponent β are given as 0.5 μm and -0.0533 respectively [33], and for $r < r_o$ the value of F is fixed at 3. Crucially, the implementation of the model used here substitutes summation for the original integration to reflect the nature of the FC PSDs, extends the original range of 0.25–1000 μm to the 0.05–2000 μm range suggested by Davies et al. [23] and implements the effective radius r directly as half the particle diameter D determined by the FC method (rather than as the geometric formulation $4/3 \times V/A$ used by Zhang et al. [6], where A is the average projected area). This usage of Mie-derived parameters within a fractal model of mass highlights the dual nature of such modelling procedure: Mie theory is initially employed within the FC method to retrieve an optical size for the particles; the fractal model then reconciles this value with their physical size.

Over the extended size range the value of F was found to vary between 3 (its maximum possible value) and 1.998. The SPM, ISM, and OSM values produced using this fractal procedure were found to model the corresponding UKCW measurements much better than those produced using the simple spherical model, with RMS%E values 57.4%, 148.5%, and 83.1% for SPM, ISM, and OSM respectively (Figure 5). SPM values obtained as the sum of ISM and OSM were found to be close to those derived from the total PSD (RMS%E value 65.2%), indicating that the error introduced by the PSD extension is small. The two Bristol Channel samples, which deviated most obviously from the power law PSD model, produced clear outliers on all three accounts and were not included in the analysis. Median cumulative distributions of SPM, ISM, and OSM were also produced, showing that in a majority of samples 90% of the contribution to all three parameters is from particles between 0.2 and 200 μm (Figure 6).

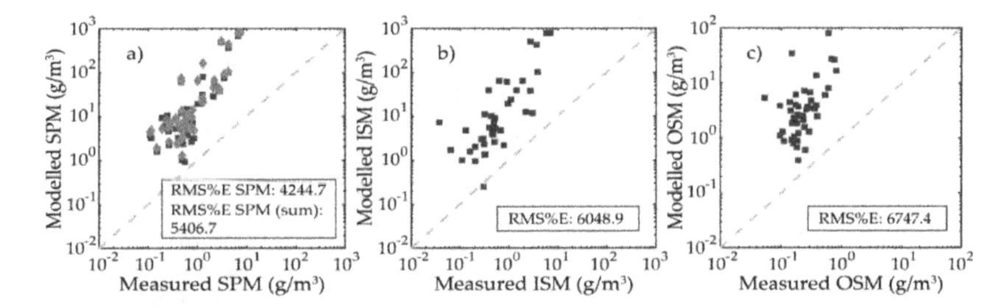

Figure 4. Comparison of modelled vs. measured (**a**) SPM, (**b**) ISM, and (**c**) OSM values for a simple spherical volume model. SPM values derived from the total PSD are represented as dark grey squares, while SPM values calculated as the sum of ISM and OSM are represented as light grey diamonds. The dashed grey lines indicate the 1:1 relationship.

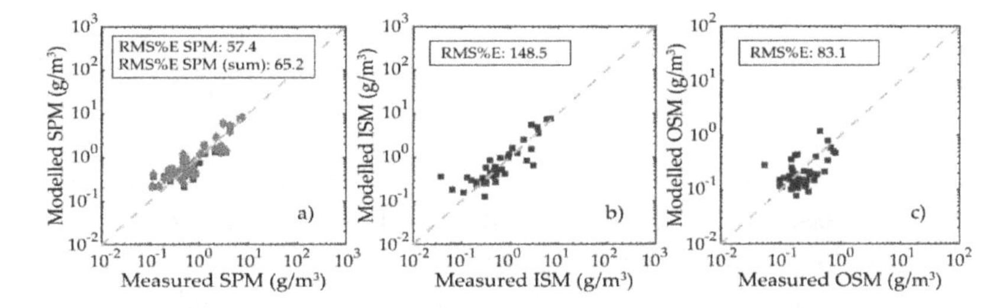

Figure 5. Comparison of modelled vs. measured (**a**) SPM, (**b**) ISM, and (**c**) OSM values for the fractal volume model. SPM values derived from the total PSD are represented as dark grey squares, while SPM values calculated as the sum of ISM and OSM are represented as light grey diamonds. The dashed grey lines indicate the 1:1 relationship.

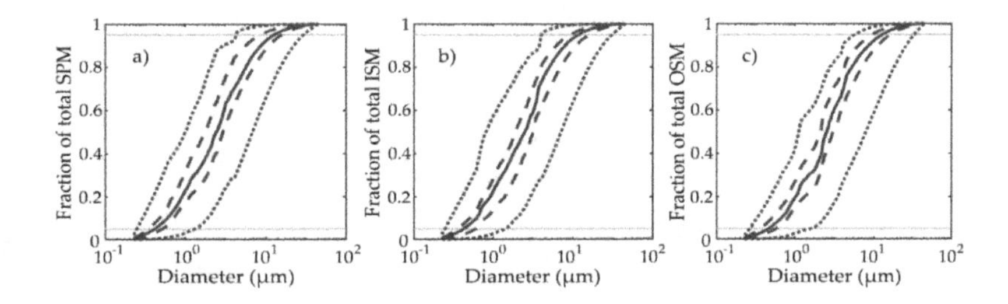

Figure 6. Cumulative distributions of modelled (**a**) SPM, (**b**) ISM, and (**c**) OSM values for the fractal volume model. The SPM curves refer to SPM values calculated from the total PSD. Solid, dashed, and dotted lines represent median, upper/lower quartiles, and maximum/minimum values respectively. The light grey horizontal lines mark the middle 90% of the contribution (i.e., from 5% to 95%) to the total value of each parameter.

3.3. Particulate Organic Carbon and Chlorophyll-a Concentration Modelling

For POC and ChlA modelling, the particle volume V was once again defined using a simple spherical model. The first to be applied were the cell volume scaling models (Table 2). Of the four models used to calculate POC values, only the diatom model given by Menden-Deuer and Lessard [12] produced results compatible with POC measurements (RMS%E: 92.9%, Figure 7a). This possibly reflects the taxonomical composition of the algal populations encountered during the He442 research cruise being mainly composed of diatom species typically associated with the spring bloom. The median cumulative distribution of POC for the diatom model shows an almost linear contribution from all size classes in a majority of samples, although results are shown to range widely from cases where the contribution is dominated by small particles to cases where, oppositely, the largest particles contributed the most (Figure 7b). This is likely the result of the interaction between the model parameters and the slope of the PSDs, and may also indicate that the parameters of the model work well for a majority, but not all of the samples. The two chlorophyll-a models both produced unsatisfactory results, with one data set underestimating ChlA, and the other over-estimating ChlA (Figure 8). The RMS%E values for the two sets were found to be 64.2% and 212.3% respectively. Although the overall quality of the match-up was low, the underlying structure of the data suggests that the form of the relationship might be useful subject to appropriate optimization.

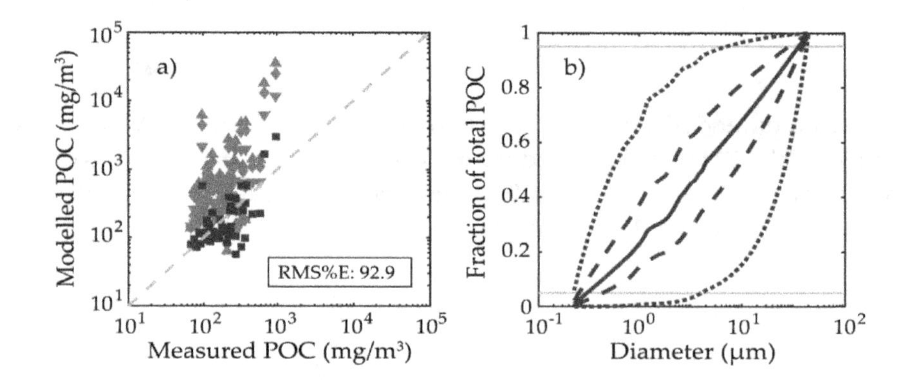

Figure 7. (**a**) Comparison of modelled vs. measured POC. POC values calculated using the diatom model (Menden-Deuer & Lessard, 2000) are represented by dark grey squares; the RMS%E value refers to these. POC values calculated using the other three models are represented by light grey diamonds and triangles; (**b**) Cumulative distribution of modelled POC for the diatom model. Solid, dashed, and dotted lines represent median, upper/lower quartiles, and max./min. values respectively. The light grey horizontal lines mark the middle 90% of the contribution (i.e., from 5% to 95%) to the total POC value. The dashed grey line indicates the 1:1 relationship.

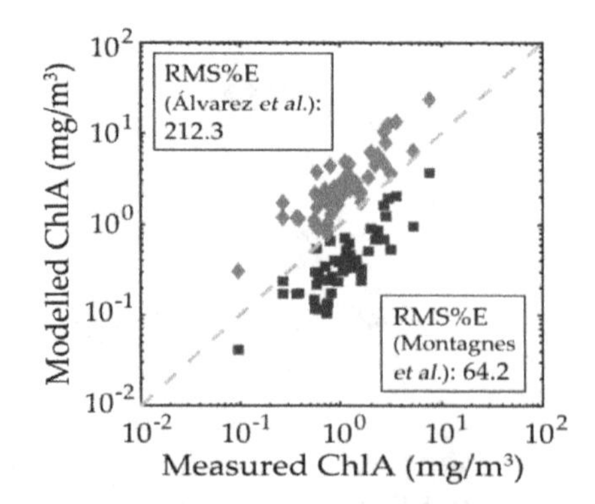

Figure 8. Comparison of modelled vs. measured ChlA. ChlA values calculated using the Montagnes et al. [11] model and the Álvarez et al. [13] model are represented by dark grey squares and light grey diamonds respectively. The dashed grey line indicates the 1:1 relationship.

When the refractive index-based C_i and Chl_i estimation models (Equations (4)–(7)) were applied to POC and ChlA modelling, the comparison between modelled and measured values of POC produced results (Figure 9a) which echo those found for SPM, ISM, and OSM when simple spherical particle volumes are used (see Figure 4). As was the case for particle mass values, the model POC values grossly overestimated the corresponding UKCW measurements, in certain cases by over two orders of magnitude. In contrast, the comparison between modelled and measured values of ChlA (Figure 9b) produced results which are similar to those found with the Montagnes et al. [11] cell volume scaling model (see Figure 8): both the Stramski and the DuRand models i.e., Equations (5) and (7) underestimated the measured ChlA values, producing RMS%E values of 63.3% and 76.5% respectively. Following the successful application of a fractal model of particle volume to the modelling of SPM, ISM, and OSM values, the same fractal model was applied to the refractive index-based C_i and Chl_i estimation as well by simple substitution of the ρ term of Equation (10) with either C_i or Chl_i. The POC values produced using the fractal procedure were found to model the corresponding UKCW measurements much better than those produced using the simple spherical model, with RMS%E values of 51.4% and 49.2% for the Stramski [19] and Durand et al. [20] models respectively (Figure 9c). However, the modelled values of ChlA were driven to further underestimate the measured values by the adoption of fractal volumes. RMS%E values for this new ChlA comparison were found to be 80.7% and 88.1% for the Stramski [19] and Durand et al. [20] models respectively (Figure 9d).

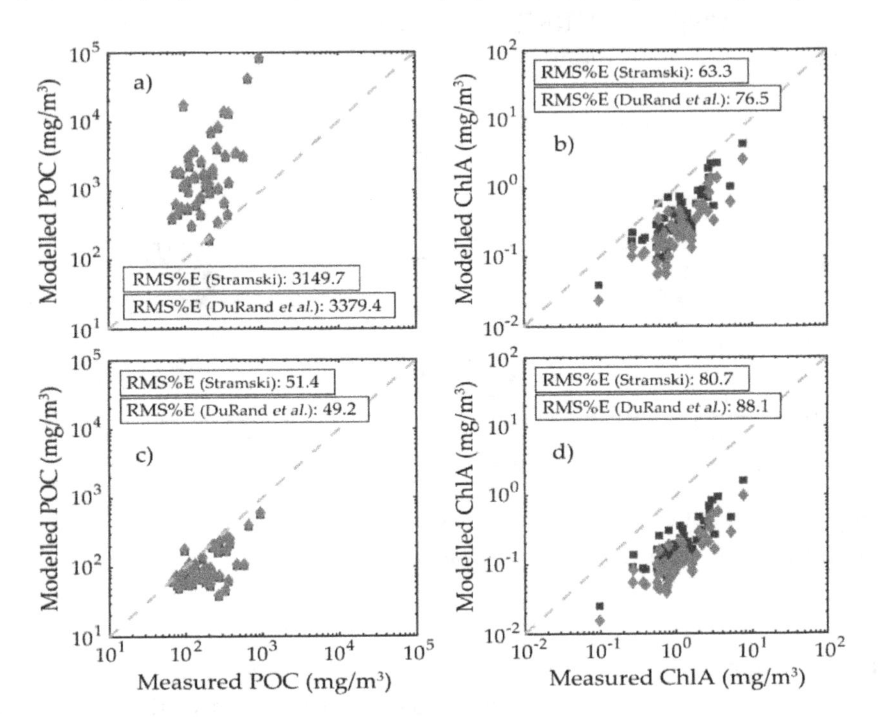

Figure 9. Comparison of (**a**) modelled vs. measured POC and (**b**) modelled vs. measured ChlA when a spherical model of particle volume is employed, and comparison of (**c**) modelled vs. measured POC and (**d**) modelled vs. measured ChlA when a fractal model of particle volume is employed instead. POC and ChlA values calculated using the Stramski [19] and Durand et al. [20] models are represented by dark grey squares and light grey diamonds respectively. The dashed grey lines indicate the 1:1 relationship.

3.4. Particulate Organic Carbon and Chlorophyll-a Concentration Modelling Optimization

A simple inversion of the procedure used to calculate POC and ChlA values allows for the empirical optimization of the parameter pairs used in the cell volume scaling models. Maintaining the general form

$$y(r) = kV(r)^h, \tag{12}$$

arrays of values for the parameters k and h can be generated, combined, and substituted in Equation (12), and the results compared and fitted against available measurements to identify the best parameter combinations, respectively (h_1, k_1) and (h_2, k_2) for POC and chlorophyll-a intraparticle concentrations. The optimization was initially applied to the whole UKCW dataset. Three parameter arrays were generated, one shared by exponents h_1 and h_2 plus one each for factors k_1 and k_2. The ranges were designed to encompass the parameter values of the models used thus far: specifically, 201 linearly spaced values for exponents h_1 and h_2 in a 0.6–1 range, 301 linearly spaced values for factor k_1 in a 0.05–0.65 range and 461 linearly spaced values for factor k_2 in a 0.004–0.05 range. Each (h_1, k_1) and (h_2, k_2) combination was then applied dataset-wide, compared against measured POC and ChlA and evaluated using the goodness-of-fit of a forced linear fit of the 1:1 line. The best parameter combinations were selected as those that minimized the RMSE values of the forced fit. The best POC model for the UKCW dataset was found as

$$y(r) = 0.442V(r)^{0.720} \tag{13}$$

while the best chlorophyll-a model was found as

$$y(r) = 0.029V(r)^{0.736}. \tag{14}$$

Comparisons of the optimized model results with the measured POC and ChlA values of the UKCW dataset are shown in Figure 10. RMS%E values for the comparisons were found to be 45.6% and 51.8% for POC and ChlA respectively.

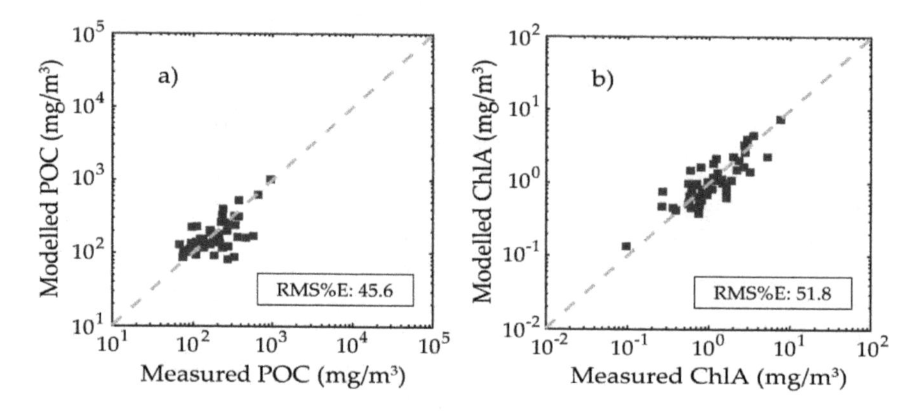

Figure 10. Comparison of (**a**) POC and (**b**) ChlA values as determined by the optimized models of Equations (13) and (14) vs. their respective measured values. The dashed grey lines indicate the 1:1 relationship.

An analogous optimization procedure can be applied to the refractive index-based C_i and Chl_i estimation models. Maintaining the use of fractal particle volumes and the general form

$$C_x = an_x - b \tag{15}$$

common to both models as described by Equations (4)–(7), where C_x is either C_i or Chl_i and n_x is respectively either n_r or n_i, arrays of values for the parameters a and b can be generated, combined and substituted in Equation (15), and the results compared and fitted against available measurements to identify the best parameter combinations. Interestingly, for both POC and ChlA the procedure could not identify single parameter combinations that minimized the RMSE; indeed, in both cases what was found was instead a continuum of parameter pairs which all minimized the RMSE to very similar values across the entire parameter ranges. We interpret this to be indicative of an excess of degrees of freedom in the relationship described by Equation (15) for the models to be properly constrained.

Returning to the physical basis of these relationships given in Stramski [19], the empirical linear relationship between C_i and n_r reflects a physical realization that increasing carbon content will generally increase real refractive index. Here we consider the implication of a natural boundary condition for this relationship: when $n_r = 1$ then $C_i = 0$. Including this constraint results in a slightly modified form of relationship

$$C_i = a_C(n_r(660\,\text{nm}) - 1) \tag{16}$$

We note that this is equivalent to forcing the Stramski [19] regression (Figure 2 therein) through $(n_r, C_i) = (1, 0)$. Given the experimental uncertainties noted in Stramski [19] and the similarity of the a and b components of the original regressions, we believe that the form proposed by Equation (16) is broadly comparable but better reflects the physical relationship under investigation. Similarly, in the case of Chl_i and n_i Stramski [19] provides a clear physical basis for a simple linear relationship based on earlier work by Morel and Bricaud [34]. A small offset is found in the resulting best-fit regressions, which is either small and positive (see Equation (5) here, Stramski [19]), or small and negative (see Equation (7) here, Durand et al. [20]). The initial model suggests that n_i(675 nm) ought to be a simple linear function of Chl_i, but in practice the small offset could represent residual absorption by pigments other than ChlA (Stramski [19]), or it could be a statistical artefact associated with limitations in data quality (the discrepancy in sign between the two aforementioned studies is possibly significant). In either case, a slightly modified version of the n_i and Chl_i relationship can be given as

$$Chl_i = a_{chl} n_i(675\,\text{nm}) + const. \tag{17}$$

However, it must be noted that the information contained in our data does not provide an indication of the appropriate value for the constant offset in Equation (17): indeed, the absence of a well-determined pair of best parameters when optimization is executed on both a and b terms of the linear model shows that minimal error for the dataset may be reached for any value of the offset. Therefore, the only likely way forward is to use the single well-defined scenario available i.e., the assumption of negligible residual absorption by pigments other than Chl_i. Under this assumption, the constant offset equals zero.

Once applied to Equations (16) and (17), the optimization procedure thus defined identified minimal RMSE values for relationships

$$C_i = 6880(n_r(660\,\text{nm}) - 1) \tag{18}$$

$$Chl_i = 8320 n_i(675\,\text{nm}). \tag{19}$$

Comparisons of the optimized model results with the measured POC and ChlA values of the UKCW dataset are shown in Figure 11. RMS%E values for the comparisons were found to be 50.2% and 45.2% for POC and ChlA respectively.

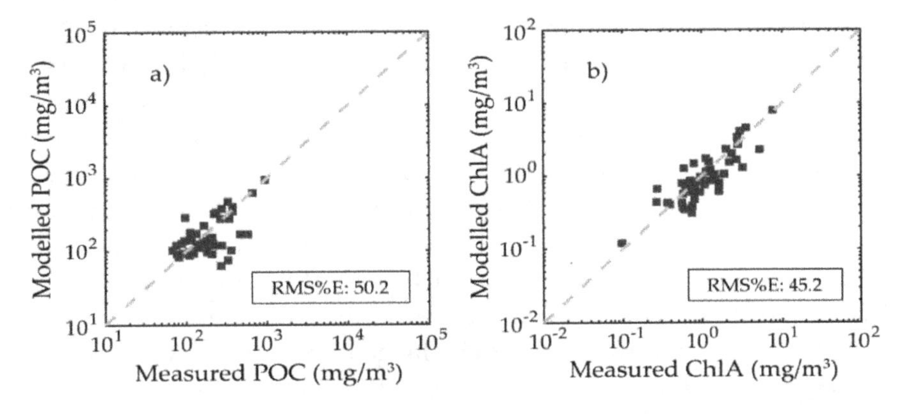

Figure 11. Comparison of (**a**) POC and (**b**) ChlA values as determined by the optimized models of Equations (17) and (18) vs. their respective measured values. The dashed grey lines indicate the 1:1 relationship.

4. Discussion

The modelling of suspended matter parameters for the samples of the UKCW dataset produced an interesting result: PSDs and PRIDs generated using a sphere-based, Mie-derived methodology produced SPM, ISM, and OSM values which compared poorly with their corresponding measured parameters when paired with an equally simple spherical volume model (Figure 4). However, the same PSDs and PRIDs produced comparable results when used directly as inputs for a fractal model of particle volume instead (Figure 5). This reflects two important aspects that should be considered carefully. The first is that the particle diameters found by the FC method are equivalent diameters rather than a direct measure of the physical dimension of the particles strictly [2]. The second is that natural particle populations are not solid spheres, so it is necessary to consider the impact of volume scaling and shape effects on apparent densities. The fractal model effectively rescales apparent particle densities across the size range. This is a feature of dealing with polydisperse natural particle populations that may not be encountered when dealing with effectively monodisperse algal cultures.

Application of volume scaling models (Figures 7a and 8) was moderately successful in determining POC and ChlA concentrations, although the variable rate of success between models mirrors the lack of consensus on a single model to accurately represent size–physiology relationships in marine phytoplankton. Indeed, a single set of model parameters is likely to be insufficient to adequately represent the metabolic complexity of all algal organisms [35]. This seems to substantiate the concerns expressed by Stramski [19] in noting the large variability between results obtained using different cell volume scaling models. Nevertheless, the further application of an empirical optimization procedure produced good match-ups between modeled and measured POC and ChlA values across the UKCW dataset (Figure 10).

Refractive index-based models of C_i and Chl_i were found to require a fractal model of particle volume to produce modelled values comparable with measured ones, particularly in the case of POC (Figure 9). This mirrors the results found for the particle density models for SPM, ISM, and OSM, and suggests that fractal models of volume are necessary to reconcile linear formulations of particle density and intraparticle carbon content with corresponding bulk measurements when natural particle populations with a wide size range and complex composition are involved. The matter is less clear in the case of ChlA, for which the refractive index-based models underestimated measured values in the case of spherical volumes, and even more so in the case of fractal volumes. Since the particle volume model must be consistent between POC and ChlA, we interpret this not as an issue with the fractal approach but rather as a sign that other elements within the overall procedure may be problematic. For example, it is important to remember that the imaginary refractive indices used in the Chl_i estimation represent an approximation based on a single n_i value derived from the literature. A simple optimization procedure shows that the modelled ChlA values of Figure 9b,d are reconciled with the corresponding measured values if $n_i(675\text{ nm}) = 12.3 \times 10^{-3}$ and $n_i(675\text{ nm}) = 11.1 \times 10^{-3}$ in the fractal volume case and if $n_i(675\text{ nm}) = 4.2 \times 10^{-3}$ and $n_i(675\text{ nm}) = 4.6 \times 10^{-3}$ in the spherical volume case (Stramski [19] and Durand et al. [20] models respectively in both cases). All four values are within the range shown in Table 1 of Stramski et al. [36], suggesting that a more dynamic range of n_i values may in fact improve the agreement between measurements and prediction. Furthermore, the fluorescent PSDs used in the ChlA calculations are not always easily reconciled with a Junge-like size distribution, and thus cannot be extended beyond the FC functional size range in a straightforward manner (Figure 3c). This can easily lead to unobserved particle fractions (especially for particle sizes above ~10 μm) and thus to underestimation of the actual total fluorescence.

This underestimation has a direct effect on the parameter values of the optimized refractive index-based models, which were found to be very large compared to those found in the literature (Equations (4)–(7), (18), and (19)). However, it should be noted that while the large value of the parameter a_{chl} can be explained by the issues mentioned above, the large value of the parameter a_C may instead be explained by the very different context of the model's application. The original C_i model was developed from the analysis of a small number of cultured phytoplankton species, which compared to the varied composition and extended size range of the He442 samples are essentially monotypic and probably log-normally distributed in their size. For both parameters, the good agreement obtained here (Figure 11) further confirms that fractal models of particle volume are appropriate and essential for successful closure between the results provided by linear formulations of biogeochemical parameters and their corresponding bulk measurements when natural mixed particle populations are concerned.

Given the results obtained for SPM, ISM, and OSM and for refractive index-derived values of POC and ChlA using a fractal model of particle volume, at a first glance the application of POC and ChlA volume scaling models to simple spherical volumes appears incongruous. However, close inspection of Equations (10)–(12) reveals that the POC and ChlA empirical volume scaling relationships hide proportionalities which are very close to those of the SPM, ISM, and OSM models. Specifically, from Equations (10) and (12) and for a single particle

$$\begin{aligned} m = r_o^{3-F(r)} \rho \left(\tfrac{4\pi}{3}\right)^{1-\frac{F(r)}{3}} \left(\tfrac{4}{3}\pi r^3\right)^{\frac{F(r)}{3}} = \\ = u(r) V^{\frac{F(r)}{3}} \propto y = kV^h. \end{aligned} \quad (20)$$

Of particular interest then is the comparison between the exponents h and $F(r)/3$. Over the 0.05–2000 μm size range considered here, the value of $F(r)$ was found to vary between 3 and ~2, producing values of $F(r)/3$ between 1 and 2/3. These not only cover the range of values of exponent h in literature-derived and UKCW optimized volume scaling models for both POC and ChlA, but also echo general results found in literature for the volume scaling coefficients of chlorophyll and organic carbon. A physical interpretation of these results is that intracellular carbon is not uniformly distributed within cells, and this non-uniform distribution of intracellular carbon causes the observed non-linear relationship with cell volume. Based on the analysis above, we have found that the observed exponent in the cell volume model reflects the non-uniform distribution of carbon within the cell and can be used to predict the subsequent impact on the fractal dimensionality of the particle. As shown earlier in the paper, the fractal dimension of the particles controls the relationship between particle size distribution and observed mass concentrations (Figure 5).

Non-uniform distribution of intracellular carbon is also closely linked with cellular metabolism. Metabolic rate under optimal growth conditions is seen to scale with volume following a 3/4 exponent for a large number of organisms in what is known as the 3/4 rule or Kleiber's rule [37], and in phytoplankton this relationship is directly tied to the photosynthetic rate, and ultimately to the intracellular chlorophyll-a concentration. In general, phytoplankton cells regulate their pigment concentration in response to environmental irradiance changes [37]. Values for the volume scaling coefficient are then variably predicted to range between 3/4 and 2/3 for optimal growth and light limited conditions respectively [37], or to reach ~1 when nutrients are abundant [35]. Álvarez et al. [13] report values between 3/4 and 1, and Mei et al. [38] modelled cellular growth rate scaling exponents using chlorophyll-a intraparticle concentration scaling exponents ranging between 2/3 and 1. Similarly, empirical estimates of the scaling between cell volume and particulate carbon also vary. Some authors find carbon intraparticle concentration to decrease proportionally with increasing cell size i.e., $h < 1$ [10,12], while other find it to be isometric to cell size i.e., h ~1 [11].

Ultimately, the results presented in this study seem to suggest that, while cellular metabolic rate can be logically expected to play an important part in defining chlorophyll and organic carbon concentrations within organic particles, structural characteristics of the organic particles as described by fractal models can also offer a complementary interpretation for the proportionalities observed in nature and described in existing microbiology literature. This opens up interesting avenues for future research. For example, a simple attempt at repeating the optimization procedure for parameters k_1 and k_2 while substituting $F(r)/3$ for volume scaling coefficient h produced values $k_1 = 0.232$ for POC and $k_2 = 0.100$ for ChlA. The latter value is much larger than corresponding parameters found within the literature-derived ChlA volume scaling models employed in this work; furthermore, in both cases, modelled values of POC and ChlA resulted in a larger RMS%E when compared with their respective measured values from the UKCW dataset, respectively 61.2% and 93.6% for POC and ChlA. This may suggest that the application of fractal volumes to organic carbon and chlorophyll-a concentrations determination via volume scaling models will require a different set of values for the primary radius r_0 and/or exponent β compared to those used in the estimation of suspended matter concentrations. Coincidentally, this might also be further explanation for the underestimation found in modelled POC and especially ChlA values derived from refractive index based models of C_i and Chl_i when fractal volumes are applied. Most certainly, further in-depth research will be needed to answer these questions.

5. Conclusions

The results obtained by the FC method for the UKCW dataset were combined with models of particle density and of organic carbon and chlorophyll-a intraparticle concentrations to investigate the biogeochemical properties of the particle populations. The success of the resulting SPM, ISM, OSM and (after empirical optimization) POC and ChlA estimations lends further credibility to the PSD and PRID determination capabilities of the FC method, and further supports the usefulness of flow cytometry and of the FC method as a tool to complement other established techniques.

The fact that a fractal model of particle structure was key to ensure the quality of the SPM, ISM, OSM and refractive index-derived POC and ChlA match-ups suggests that the FC method observes particles as equivalent spheres, and has therefore the potential to be to some extent resilient both to particles which violate Mie-compatible aspect ratios and to the break-up of flocs and aggregates which derives from the flow cytometric measurement technique. Furthermore, this characteristic makes FC method results more readily comparable with those of other more common marine optics instruments, which for the most part observe bulk seawater IOPs and PSDs. It should be also noted that while the dataset was chiefly composed of surface water samples, no particular limitation exists a priori for the application of the FC method in its present form to samples from any water depth.

The volume scaling models used to calculate POC and ChlA were successfully applied to simple spherical particles instead. This may appear incongruous at first; however, these models too are revealed to hide proportionalities analogous to those caused by fractal structures, shining interesting new light on the volume scaling coefficients described in marine microbiology literature. We have shown here that the volume scaling models provide a route to better understand the impact of carbon distribution within particles and resulting fractal dimensionality, while the successful optimization of refractive index-based models reinforces previous findings that cell composition in the form of intracellular carbon concentration is well represented by corresponding changes in real refractive index. Finally, the size discrimination offered by FC also allows for a better understanding of the contribution of different size classes to the bulk biogeochemical properties, as it previously did for the IOPs of natural particle populations [2]. Taken in combination, these results represent a significant demonstration of the quality of size and refractive index information that is provided by FC data.

Author Contributions: J.A. and D.M. conceived and outlined the concept for this study; R.R. coordinated work during the HE442 research cruise; D.M. and R.R. supervised the retrieval of water samples; J.A. carried out flow cytometry operations; R.R. and K.H. carried out SPM, ISM, OSM, POC, and ChlA measurements on the water samples; J.A. analyzed the data; R.R. provided the Methods paragraphs relative to SPM, ISM, OSM, POC, and ChlA measurements; J.A. wrote the paper.

Acknowledgments: Agagliate and McKee gratefully acknowledge financial support from the MASTS pooling initiative. The authors wish to thank the captain and the crew of RV Heincke for their support and help during the HE442 research cruise. The authors also duly thank D. Stramski and two anonymous reviewers, who all helped improve this manuscript with their comments and suggestions.

References

1. Agagliate, J.; Röttgers, R.; Twardowski, M.S.; McKee, D. Evaluation of a flow cytometry method to determine size and real refractive index distributions in natural marine particle populations. *Appl. Opt.* **2018**, *57*, 1705–1716. [CrossRef] [PubMed]
2. Agagliate, J.; Lefering, I.; McKee, D. Forward modelling of inherent optical properties from flow cytometry estimates of particle size and refractive index. *Appl. Opt.* **2018**, *57*, 1777–1788. [CrossRef] [PubMed]
3. Ackleson, S.G.; Spinrad, R.W. Size and refractive index of individual marine particulates: A flow cytometric approach. *Appl. Opt.* **1988**, *27*, 1270–1277. [CrossRef] [PubMed]
4. Green, R.E.; Sosik, H.M.; Olson, R.J.; DuRand, M.D. Flow cytometric determination of size and complex refractive index for marine particles: Comparison with independent and bulk estimates. *Appl. Opt.* **2003**, *42*, 526–541. [CrossRef] [PubMed]
5. Green, R.E.; Sosik, H.M.; Olson, R.J. Contributions of phytoplankton and other particles to inherent optical properties in New England continental shelf waters. *Limnol. Oceanogr.* **2003**, *48*, 2377–2391. [CrossRef]
6. Zhang, X.; Stavn, R.H.; Falster, A.U.; Gray, D.; Gould, R.W., Jr. New insight into particulate mineral and organic matter in coastal ocean waters through optical inversion. *Estuar. Coast. Shelf Sci.* **2014**, *149*, 1–12. [CrossRef]
7. Morel, A.; Ahn, Y.-H. Optical efficiency factors of free-living marine bacteria: Influence of bacterioplankton upon the optical properties and particulate organic carbon in oceanic waters. *J. Mar. Res.* **1990**, *48*, 145–175. [CrossRef]
8. Babin, M.; Morel, A.; Fournier-Sicre, V.; Fell, F.; Stramski, D. Light scattering properties of marine particles in coastal and open ocean waters as related to the particle mass concentration. *Limnol. Oceanogr.* **2003**, *48*, 843–859. [CrossRef]
9. Aas, E. Refractive index of phytoplankton derived from its metabolite composition. *J. Plankton Res.* **1996**, *18*, 2223–2249. [CrossRef]
10. Verity, P.G.; Robertson, C.Y.; Tronzo, C.R.; Andrews, M.G.; Nelson, J.R.; Sieracki, M.E. Relationships between cell volume and the carbon and nitrogen content of marine photosynthetic nanoplankton. *Limnol. Oceanogr.* **1992**, *37*, 1434–1446. [CrossRef]
11. Montagnes, D.J.S.; Berges, J.A.; Harrison, P.J.; Taylor, F.J.R. Estimating carbon, nitrogen, protein, and chlorophyll *a* from volume in marine phytoplankton. *Limnol. Oceanogr.* **1994**, *39*, 1044–1060. [CrossRef]
12. Menden-Deuer, S.; Lessard, E.J. Carbon to volume relationships for dinoflagellates, diatoms, and other protist plankton. *Limnol. Oceanogr.* **2000**, *45*, 569–579. [CrossRef]
13. Álvarez, E.; Nogueira, E.; López-Urrutia, Á. In Vivo Single-Cell Fluorescence and Size Scaling of Phytoplankton Chlorophyll Content. *Appl. Environ. Microbiol.* **2017**, *83*, e03317-16. [CrossRef] [PubMed]
14. Stramski, D.; Morel, A. Optical properties of photosynthetic picoplankton in different physiological states as affected by growth irradiance. *Deep Sea Res. Part A Oceanogr. Res. Pap.* **1990**, *37*, 245–266. [CrossRef]
15. Stramski, D.; Reynolds, R.A. Diel variations in the optical properties of a marine diatom. *Limnol. Oceanogr.* **1993**, *38*, 1347–1364. [CrossRef]
16. Stramski, D.; Shalapyonok, A.; Reynolds, R.A. Optical characterization of the oceanic unicellular cyanobacterium *Synechococcus* grown under a day-night cycle in natural irradiance. *J. Geophys. Res.* **1995**, *100*, 13295–13307. [CrossRef]
17. Reynolds, R.A.; Stramski, D.; Kiefer, D.A. The effect of nitrogen limitation on the absorption and scattering properties of the marine diatom *Thalassiosira pseudonana*. *Limnol. Oceanogr.* **1997**, *42*, 881–892. [CrossRef]
18. Durand, M.D.; Olson, R.J. Diel patterns in optical properties of the chlorophyte *Nannochloris* sp.: Relating individual-cell to bulk measurements. *Limnol. Oceanogr.* **1998**, *43*, 1107–1118. [CrossRef]
19. Stramski, D. Refractive index of planktonic cells as a measure of cellular carbon and chlorophyll a content. *Deep Sea Res. Part I Oceanogr. Res. Pap.* **1999**, *46*, 335–351. [CrossRef]
20. Durand, M.D.; Green, R.E.; Sosik, H.M.; Olson, R.J. Diel Variations in Optical Properties of *Micromonas Pusilla* (Prasinophyceae). *J. Phycol.* **2002**, *38*, 1132–1142. [CrossRef]
21. Stramski, D.; Morel, A.; Bricaud, A. Modeling the light attenuation and scattering by spherical phytoplanktonic cells: A retrieval of the bulk refractive index. *Appl. Opt.* **1988**, *27*, 3954–3956. [CrossRef] [PubMed]
22. Reynolds, R.A.; Stramski, D.; Wright, V.M.; Woźniak, S.B. Measurements and characterization of particle size distributions in coastal waters. *J. Geophys. Res.* **2010**, 115. [CrossRef]

23. Davies, E.J.; McKee, D.; Bowers, D.; Graham, G.W.; Nimmo-Smith, W.A.M. Optically significant particle sizes in seawater. *Appl. Opt.* **2014**, *53*, 1067. [CrossRef] [PubMed]
24. Bader, H. The hyperbolic distribution of particle sizes. *J. Geophys. Res.* **1970**, *75*, 2822–2830. [CrossRef]
25. Sheldon, R.W.; Prakash, A.; Sutcliffe, W.H., Jr. The size distribution of particles in the ocean. *Limnol. Oceanogr.* **1972**, *17*, 327–340. [CrossRef]
26. Junge, C.E. *Air Chemistry and Radioactivity*; Academic Press: New York, NY, USA, 1963; ISBN-13 9780123921505.
27. Stramski, D.; Kiefer, D.A. Light scattering by microorganisms in the open ocean. *Prog. Oceanogr.* **1991**, *28*, 343–383. [CrossRef]
28. Ulloa, O.; Sathyendranath, S.; Platt, T. Effect of the particle-size distribution on the backscattering ratio in seawater. *Appl. Opt.* **1994**, *33*, 7070. [CrossRef] [PubMed]
29. Röttgers, R.; Heymann, K.; Krasemann, H. Suspended matter concentrations in coastal waters: Methodological improvements to quantify individual measurement uncertainty. *Estuar. Coast. Shelf Sci.* **2014**, *151*, 148–155. [CrossRef]
30. Zapata, M.; Rodríguez, F.; Garrido, J.L. Separation of chlorophylls and carotenoids from marine phytoplankton: A new HPLC method using a reversed phase C8 column and pyridine-containing mobile phases. *Mar. Ecol. Prog. Ser.* **2000**, *195*, 29–45. [CrossRef]
31. Risović, D. Two-component model of sea particle size distribution. *Deep Sea Res. Part I Oceanogr. Res. Pap.* **1993**, *40*, 1459–1473. [CrossRef]
32. Twardowski, M.S.; Boss, E.; Macdonald, J.B.; Pegau, W.S.; Barnard, A.H.; Zaneveld, J.R.V. A model for estimating bulk refractive index from the optical backscattering ratio and the implications for understanding particle composition in case I and case II waters. *J. Geophys. Res.* **2001**, *106*, 14129–14142. [CrossRef]
33. Khelifa, A.; Hill, P.S. Models for effective density and settling velocity of flocs. *J. Hydraul. Res.* **2006**, *44*, 390–401. [CrossRef]
34. Morel, A.; Bricaud, A. Theoretical results concerning light absorption in a discrete medium, and application to specific absorption by phytoplankton. *Deep Sea Res. Part A Oceanogr. Res. Pap.* **1981**, *28*, 1375–1393. [CrossRef]
35. Marañón, E.; Cermeño, P.; Rodríguez, J.; Zubkov, M.V.; Harris, R.P. Scaling of phytoplankton photosynthesis and cell size in the ocean. *Limnol. Oceanogr.* **2007**, *52*, 2190–2198. [CrossRef]
36. Stramski, D.; Bricaud, A.; Morel, A. Modeling the inherent optical properties of the ocean based on the detailed composition of the planktonic community. *Appl. Opt.* **2001**, *40*, 2929–2945. [CrossRef] [PubMed]
37. Finkel, Z.V.; Irwin, A.J.; Schofield, O. Resource limitation alters the 3/4 size scaling of metabolic rates in phytoplankton. *Mar. Ecol. Prog. Ser.* **2004**, *273*, 269–279. [CrossRef]
38. Mei, Z.P.; Finkel, Z.V.; Irwin, A.J. Light and nutrient availability affect the size-scaling of growth in phytoplankton. *J. Theor. Biol.* **2009**, *259*, 582–588. [CrossRef] [PubMed]

7

Influence of Three-Dimensional Coral Structures on Hyperspectral Benthic Reflectance and Water-Leaving Reflectance

John D. Hedley [1,*], Maryam Mirhakak [2], Adam Wentworth [3] and Heidi M. Dierssen [2]

[1] Numerical Optics Ltd., Tiverton EX16 8AA, UK
[2] Department of Marine Sciences, University of Connecticut, Groton, CT 06340, USA; ma.mirhakak@gmail.com (M.M.); heidi.dierssen@uconn.edu (H.M.D.)
[3] Materials Science and Engineering Department, University of Connecticut, Storrs, CT 06269, USA; awentworth2@bwh.harvard.edu
* Correspondence: j.d.hedley@numopt.com

Featured Application: Hyperspectral remote sensing of coral reefs. We present simplifying factors and guidance on handling or avoiding the potential variability in spectral reflectances caused by the structural nature of corals and illumination conditions. Results presented here can be implemented in remote sensing algorithms for bottom mapping of coral reefs, and guide data collection practice.

Abstract: Shading and inter-reflections created by the three-dimensional coral canopy structure play an important role on benthic reflectance and its propagation above the water. Here, a plane parallel model was coupled with a three-dimensional radiative transfer canopy model, incorporating measured coral shapes and hyperspectral benthic reflectances, to investigate this question under different illumination and water column conditions. Results indicated that a Lambertian treatment of the bottom reflectance can be a reasonable assumption if a variable shading factor is included. Without flexibility in the shading treatment, nadir view bottom reflectances can vary by as much as ±20% (or ±9% in above-water remote sensing reflectance) under solar zenith angles (SZAs) up to 50°. Spectrally-independent shading factors are developed for benthic coral reflectance measurements based on the rugosity of the coral. In remote sensing applications, where the rugosity is unknown, a shading factor could be incorporated as an endmember for retrieval in the inversion scheme. In dense coral canopies in clear shallow waters, the benthos cannot always be treated as Lambertian, and for large solar-view angles the bi-directional reflectance distribution functions (BRDF) hotspot propagated to above water reflectances can create up to a 50% or more difference in water-leaving reflectances, and discrepancies of 20% even for nadir-view geometries.

Keywords: remote sensing; hyperspectral; shallow water; coral; derivative; radiative transfer; canopy

---

1. Introduction

Remote sensing of coral reefs is an important complementary survey technique for science, monitoring, and management, being able to cover substantially larger areas than in-situ surveys albeit at lower accuracy [1,2]. A well-established paradigm in coral reef remote sensing is that hyperspectral data offers the best results for discrimination of both benthic habitats and specific benthic types, such as corals, algae, and sand [3,4]. Applied studies have shown that benthic habitat classification accuracy increases with the number of available spectral bands [3,5], specifically in the 400–700 nm range since the opacity of water beyond 700 nm limits use of NIR wavelengths to only the shallowest waters

(<2 m). Habitats are broad mapping categories, whereas mapping of specific types, e.g., corals vs. algae, is very challenging since their spectral reflectances can be similar. Modelling and sensitivity analyses based on reflectance spectra measured, either in-situ using diver-operated spectroradiometers, or ex-situ in the lab, imply that spectral separation of benthic types by hyperspectral reflectances is possible [4,6,7], and there are increasing numbers of applied demonstrations of discriminations of benthic types (e.g., coral, sand, and algae) from airborne hyperspectral imagery [8–14].

These studies imply that capability for discrimination of coral reef types is reliant on the discrimination of spectral features of the pigments that are present, such as peridinin in the coral symbionts, and accessory pigments of algae [8]. Pigment-based reflectance features are present at the surface of the corals and other benthic types, but the coral reef environment is structurally complex, with multiple benthic types typically present within even the highest spatial resolution imagery (e.g., pixels 1 m). Vertical structure in the reef canopy and the benthic types themselves gives rise to shading, inter-reflections, and a generally complex interaction with the light environment [15,16]. Most sensitivity analyses to date work with empirically measured reflectances of individual types, at a scale roughly equivalent to their morphology, and use a linear mixing model, which essentially assumes that the individual types can be treated in the same way as a spatially flat Lambertian patchwork [4,7]. However, even in a carefully defined lab experiment, the linear mixing model does not always work well with structural components [17]. Given the complexity of light interaction with a mixed structural canopy, it is not immediately obvious to what extent a pigment based hyperspectral reflectance feature present at the scale of a coral surface translates to an above water measurement at remote sensing scales.

The purpose of the study presented here was to investigate propagation of hyperspectral reflectances from the surfaces of structural benthos (corals and surrounding substrate) to water leaving reflectance. The study combined a three-dimensional canopy model [18], with a plane parallel water-column model (HydroLight [19]) (Figure 1). The model was parameterized by 3D reconstructions of actual coral shapes and surface reflectances derived from hyperspectral images of the same corals. A variety of investigations were conducted (Table 1), to track the main factors that mitigate the transmission of hyperspectral features, and to justify practical simplifications to provide useful results that can be immediately applied in other contexts. Investigations included the effects of shading, and water attenuation and scattering, both in the region of individual corals and within mixtures of corals and surrounding substrates. The three-dimensional model was used to generate fully populated bi-directional reflectance distribution functions (BRDFs) of mixed canopies, and these were input to HydroLight to model above water reflectance for different depths, water inherent optical properties (IOPs), and solar-view geometries.

In summary, specific objectives were:

1. Investigate how the spectral reflectance at the scale of coral macro-morphology (shape) relates to coral surface-scale spectral reflectance.
2. Quantify the error arising from making simplifying modelling assumptions, such as using a single reflectance spectrum to represent a coral under different light environments, or excluding water attenuation and scattering between coral structures within the canopy.
3. Characterize the BRDF of assemblages of corals in different densities.
4. Apply those BRDFs as the bottom boundary in HydroLight, to model BRDF effects in the remote sensing reflectance arising from canopy BRDF properties in different depths and conditions.

The results indicated that whilst structural factors can introduce substantial variability in spectral reflectances, this variability can be avoided or accommodated by simplifying factors. The major variations in spectral reflectance over corals, due to structure and illumination, can be captured by incorporating a spectrally flat shading factor or black shade endmember into linear mixing models. Above-water BRDF effects on reflectance can be 20% or more for dense shallow canopies when viewing close to the solar plane in typical remote sensing solar-view geometries, but are less apparent if the solar-view geometry is at 90° to the solar plane.

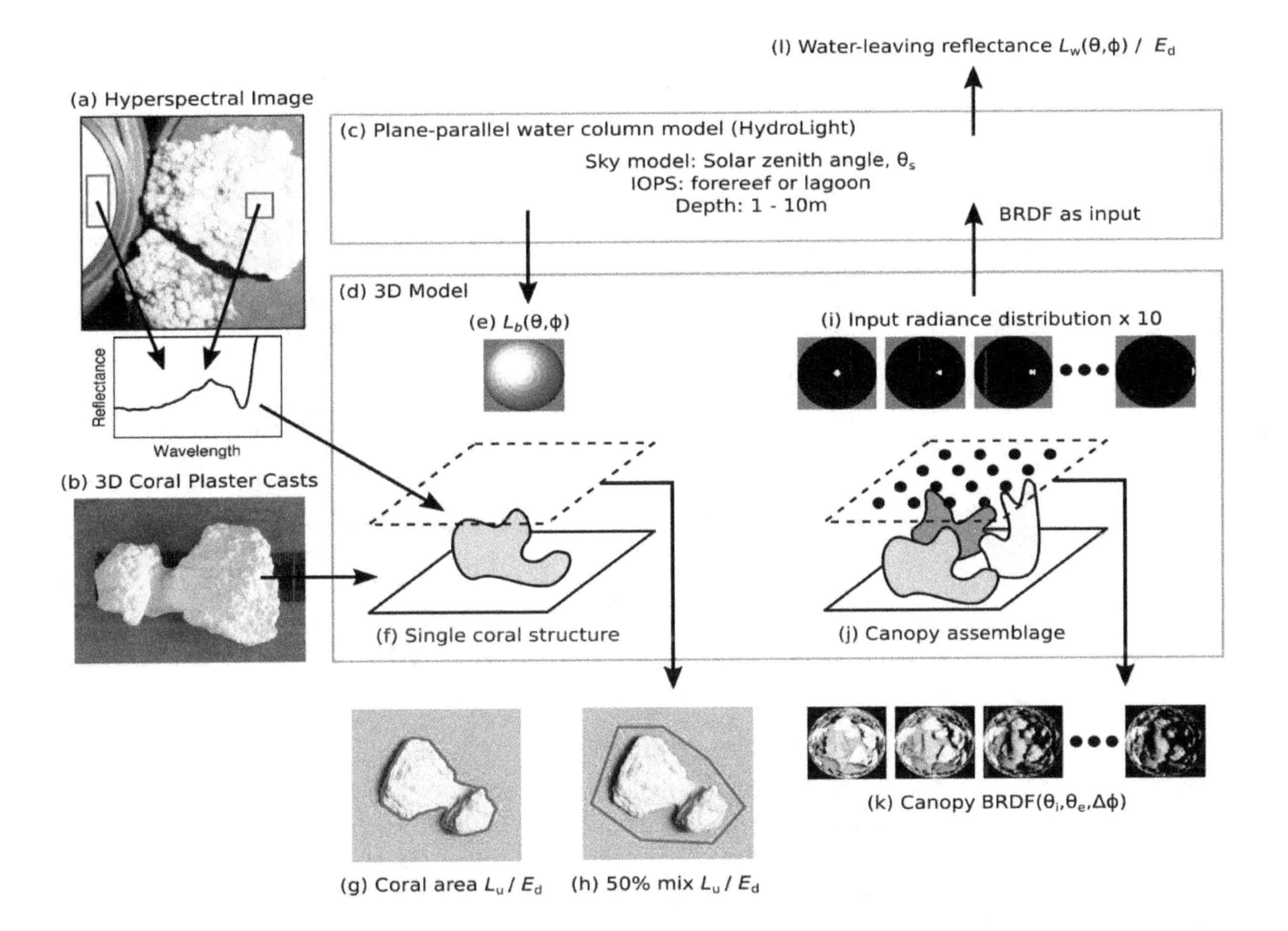

Figure 1. Model setup: (**a**) Coral surface reflectances were derived from hyperspectral image, and (**b**) shape from 3D reconstructions from plaster casts; (**c**) A plane parallel water column model was coupled with (**d**) a 3D canopy model in two ways: (1) (**e**) To model bottom of water column light fields over (**f**) single structures and estimate (**g**) reflectance over the coral and (**h**) of a 50% mix with surrounding substrate; and (2) (**i**) Directional incident radiance at different angles over (**j**) assemblages of structures were used to characterize (**k**) the bi-directional reflectance distribution function (BRDF). The BRDF was then input to the water column model to give (**l**) water-leaving reflectances.

2. Materials and Methods

2.1. Overview

Toward the aims listed above, three-dimensional models of actual coral shapes were input to a 3D canopy radiative transfer model [18] coupled to a plane-parallel water column model, HydroLight [19]. Six different modelling exercises were conducted (Table 1, Figure 1). These were structured to progressively incorporate complexity and justify the design of subsequent modelling steps: from reflectance over a single coral shape only, to then including substrate, then include assemblages of corals, and finally the water column. Intermediate results from these activities also provided results of interest, for example, on the validity of using a linear mixing model for bottom of water column reflectance. The following sections give details on the input data and model set up.

Table 1. Different properties and treatments modelled in this study and the aim of each investigation. "Res." indicates if the high- or low-resolution coral structure models were used. "I." indicates if interstitial scattering and attenuation by the water around the coral shapes was included (N-No, Y-Yes). "Position" indicates the notional location of the estimated reflectance.

<table>
<tr><th></th><th>Modelled Property</th><th>Res.</th><th>I.</th><th>Derived Reflectance and Purpose of Activity</th><th>Position</th></tr>
<tr><td>1</td><td rowspan="2">Reflectance averaged over each of the 16 coral shapes, for the coral area only (substrate masked out). Bottom of water column incident radiance distribution generated by HydroLight with 32 treatments, combining: Solar zenith angles 10° and 50°; depths 1 m and 10 m; forereef and lagoon IOPs; and four rotational positions of the coral shape.</td><td>high</td><td>N</td><td>$\pi L_u/E_d$
Effect of shape on reflectance over coral area only.</td><td>above canopy</td></tr>
<tr><td>2</td><td>low</td><td>N</td><td>$\pi L_u/E_d$
Effect of resolution of 3D models.</td><td>above canopy</td></tr>
<tr><td>3</td><td rowspan="2">Reflectance averaged over a single coral shape, plus in each case, surrounding substrate to give a 50% mix in terms of areal cover. Incident radiance distributions as above.</td><td>low</td><td>N</td><td>$\pi L_u/E_d$
Validity of linear mixing model for reflectance over coral and surrounding substrate.</td><td>above canopy</td></tr>
<tr><td>4</td><td>low</td><td>Y</td><td>$\pi L_u/E_d$
Effect of ignoring water scattering and attenuation within canopy.</td><td>above canopy</td></tr>
<tr><td>5</td><td>Azimuthally averaged BRDF function for six canopy assemblages ranging from 39% to 75% coral cover. BRDF tabulated in HydroLight standard directional discretization.</td><td>low</td><td>N</td><td>$\pi \, BRDF(\theta_i, \theta_e, \Delta\varphi)$
Demonstrate canopy BRDF effects and generate function for input to HydroLight.</td><td>above canopy</td></tr>
<tr><td>6</td><td>Above-water remote sensing reflectance as a function of solar-view geometry, depth, and IOPs, for six canopy BRDFs generated as above.</td><td>low</td><td>N</td><td>$\pi L_w(\theta_s, \theta_v, \Delta\varphi)/E_d$
Demonstrate above water BRDF effects under typical solar-view geometries.</td><td>above water</td></tr>
</table>

2.2. Coral Surface Reflectances and Shape

The three-dimensional shape and surface reflectance of 16 corals were used as input to the model. Live corals of various species, including *Porites compressa, Porites evermanni*, and *Montipora capitate*, were sampled from Mahukona Beach Park, Potters, and Pauko Bay Boat Ramp in Big Island, Hawaii, in February 2017. For the purposes of this paper, corals were grouped based on sampling depth, referred to as shallow 'S' (<8 m) or deep 'D' (10 to 20 m) (Table 2). For practical reasons the corals were relatively small (<20 cm diameter). Corals were stored in open buckets filled with natural salt water and kept in a cool water bath that was shaded from the direct sun. Coral samples were imaged in the air, immediately after being removed from the water, using a tripod-mounted 710 hyperspectral imager (Surface Optics Corporation). This instrument records an image cube of 520 × 696 pixels with spectral information at 128 spectral bands, and with 5 nm spacing from 380 to 1040 nm. All images were made under natural illumination at noon ±2 h, to minimize any shading effects on the coral surfaces. Images were obtained within 24 h of collection, and before any noticeable degradation. A Labsphere Spectralon™ Diffuse Reflectance Standard with a reflectance value of 20% was placed in the image frame during each measurement (Figure 1a). Using the reference, for each coral, a single surface reflectance, $R(\lambda)$, i.e., the reflectance at the scale of the coral surface, assuming the surface is locally flat, was estimated by taking the mean over one or more small areas of the image where the coral surface was relatively horizontal (Figure 1a). While in reality, the reflectance over a coral surface is typically variable, in the scope of this study, we were interested in the effect of coral macro-morphology and canopy structure on reflectance. Hence, each coral was treated as if it had a uniform surface reflectance. The availability of horizontal areas for extracting surface reflectance possibly biased the measured reflectances slightly, but the sample areas where sufficiently large to mitigate small scale influences, such as low pigmented apical polyps (Figure 1a). For the underlying substrate, a spectral reflectance of dead coral rock was used, this data was from a previous study, collected with a GER 1500 spectroradiometer as described in Reference [17]. All spectra were resampled to 168 bands in 2 nm intervals from 400 nm to 736 nm, as this was the spectral resolution of the modelling software.

Table 2. Coral shapes used in this study. "Group" indicates grouping based on depth of collection, shallow (S, <8 m) or deep (D, 10–20 m). Coral 03 was from a site subjected to terrestrial run-off and was not placed in either group. Polygon counts for the high- and low-resolution models are given. The final column is discussed in the results and is the maximum percentage difference that choice of resolution leads to in reflectance averaged over the coral area in any band.

ID	Species		Polygon Count		Rugosity	Max L_u/E_d Difference
		Group	Low Res.	High Res.		Low vs. High
03	*Montipora capitata*	n/a	2306	152,185	1.98	0.36
08	*Porites evermanni*	D	1712	69,945	2.74	0.51
10	*Porites evermanni*	D	1832	75,524	2.16	0.28
11	*Montipora capitata*	D	1830	103,394	1.77	0.29
12	*Porites evermanni*	D	2150	51,954	2.16	0.24
16	*Porites compressa*	D	1454	89,703	5.43	1.33
17	*Porites evermanni*	S	2079	124,092	2.05	0.27
18	*Porites compressa*	D	1898	103,430	5.82	0.47
19	*Porites evermanni*	S	1804	183,283	1.61	0.22
31	*Porites compressa*	D	1226	56,726	2.96	0.43
32	*Montipora capitata*	S	2461	100,886	2.01	0.31
33	*Porites compressa*	S	2160	98,502	2.90	0.47
35	*Pocillopora* sp.	S	2966	240,353	3.06	0.21
38	*Porites evermanni* or *lutea*	S	1908	95,265	1.72	0.17
39	*Montipora patula*	S	2264	133,594	1.81	0.39
40	*Porites evermanni* or *lutea*	S	1403	56,696	1.45	0.17

To capture the shape of each field-sampled coral, they were impressed into silicone rubber putty after imaging and the moulds were allowed to air dry. These moulds were transported back to the University of Connecticut; each mold was filled with plaster to recreate the coral shape (Figure 1b). Digital 3D models of the plaster cast of natural corals were created using Autodesk ReCapTM photogrammetry software (version 5.0.1.30). The process involved taking between 20 and 40 overlapping photographs of the plaster corals and converting them into 3D digital models consisting of a mesh of vertices and triangular surface polygons. Then, the 3D models were edited in Autodesk® MeshmixerTM (version 3.4.35) to isolate the coral shape from the background. For each coral, two models were produced: High resolution, ranging from 50,000 to 200,000 triangles, and low-resolution, typically ~2000 triangles (Table 1). The low-resolution models are computationally easier to handle, and the first intended test was to determine if low-resolution models would be sufficient. For interpretation of results, a surface rugosity measure [20] was calculated for each coral shape, being the ratio of the coral surface area to its projected area (i.e., nadir view areal extent) in the high-resolution models.

2.3. Radiative Transfer Modelling

The radiative transfer modelling was split into two components: (1) a three-dimensional canopy model designed to evaluate optical properties just above the canopy or above individual coral structures (Figure 1d) [18,21], and (2) a plane-parallel water column model (Figure 1c), HydroLight [19]. All modelling was conducted hyperspectrally in 168 bands of 2 nm step, from 400 to 736 nm.

The 3D model has been described and used in a number of previous publications [18,21–23]. The solution method involves breaking down all surfaces and volumes into discrete elements, triangular surface polygons (as already provided by the 3D digital models), and cubic voxels, for scattering and attenuating media (the water). In this application, the coral shapes were placed on a flat underlying substrate that was decomposed into 100 × 100 squares (each being two triangles) over an area of approximately 20 cm × 20 cm. The model domain can either have light incident from above with horizontally periodic boundaries (i.e., the model set-up repeats in all horizontal directions), or be embedded into a "far-field" radiance distribution characterized from all directions (appropriate for modelling an isolated structure). The model is solved by calculating light transfer between all pairs of elements and propagating the incident light through the system by iteration, for a predetermined number of passes or until convergence is detected. Surfaces are treated as locally Lambertian, but volumetric elements embody full directional scattering, according to a scattering coefficient combined with a phase function [18]. As an update to previous work, this model is now implemented on modern Graphics Processing Units (GPUs), which facilitates improved performance on computationally demanding applications with many bands and elements.

Since the coral structures were small (~20 cm max. height), it would be inefficient to use the 3D model for the full water column, since depths up to 10 m were of interest. More efficiently, the 3D model was coupled to HydroLight in two ways:

1. HydroLight was used to generate a bottom of water column radiance distribution, and this was used to illuminate the individual coral structures, standing on flat dead coral substrate, by being used as the input radiance boundary condition (Figure 1e,f). From these model runs, the reflectance over the coral structure was determined under different illumination conditions to deduce the average reflectance and its variability due to shading and other effects, across a range of illumination conditions (Figure 1g,h). Treatments were applied corresponding to the range limits of interest: Depths of 1 m and 10 m; solar zenith angles of 10° and 50°, azimuth angles of 0°, 90°, 180°, 270°, and two sets of Inherent Optical Properties (IOPs) denoted "forereef" and "lagoon" (described below). Water surface roughness was set to correspond to a wind speed of 5 ms^{-1} but note that only water leaving radiance was used in the results, so surface reflectance was excluded. This method was used for tests 1 to 4, listed in Table 1.
2. Mixtures of 3D living coral structures were assembled on flat dead coral rock substrate, in areal cover densities from 39% to 75%. The 3D model was used with horizontal periodic boundaries to generate a canopy BRDF function, which was then used as a bottom boundary condition in HydroLight (Figure 1i,j). By this method, spectral water-leaving reflectances derived as water-leaving radiance at a given wavelength (λ) normalized to downwelling planar irradiance ($L_w(\theta, \Delta\varphi, \lambda)/E_d(\lambda)$), could be derived above the water for different view directions under different solar zenith angles and water column conditions (Figure 1l). Incorporating the water column this way spatially averages the results, and scope of the results corresponds to pixels larger than the coral structures, i.e., pixels $\geq$ 1 m, since the coral structures were less than 20 cm across. This modelling covered the same range of depths, solar zenith angles, and the two IOP treatments mentioned above. This aspect of the work corresponds to activities 5 and 6 in Table 1.

One important test was to determine if water attenuation and scattering within the canopy, i.e., at the scale of, and in between, coral shapes was at all important with respect to the derived reflectances. The 3D model can incorporate volumetric absorption and scattering but it is computationally slow, especially for computations in 168 bands. The "interstitial" IOPs were incorporated by embedding the shape in a mesh of 16 × 16 × 16 voxels in activity 4 (Table 1), with IOPs consistent with the corresponding HydroLight modelling (see below). In the other modelling activities, interstitial IOPs were not included. Since the coral shapes were small, within the canopy path lengths of only a few centimeters through the water were achieved. Scattering is predominantly in the forward direction so on short path lengths it has very little effect; the primary issue was absorption on the paths between surface patches. An absorption coefficient of 0.4 (approximately as at 700 nm) gives losses of just 4% on a 10 cm path, so whilst it was not expected that excluding interstitial IOPs would have a substantial effect, a specific test was conducted, and this is evaluated in the results section.

2.4. Inherent Optical Properties

The two IOP treatments "forereef" and "lagoon" (Figure 2) were configured using the HydroLight 5.3 New Case 2 model [19] to produce total absorption, $a(\lambda)$, and attenuation, $c(\lambda)$, values very similar to representative AC-S measurements taken in forereef and lagoon locations at Glovers Reef in Belize and in Palau, Micronesia in 2006 [7]. Both treatments had chlorophyll set at 0.12 $mg \cdot m^{-3}$, but the lagoon treatment had more scattering and more coloured dissolved organic matter (CDOM). The independent

CDOM $a_g(440)$ (not associated with phytoplankton) was 0.008 m^{-1} for forereef and 0.04 m^{-1} for lagoon, scattering was introduced by a calcareous sand component of 0.01 and 0.4 $mg \cdot m^{-3}$, respectively. These component concentrations were derived by trial and error to produce bulk IOPs close to the measured values, which represented the limit of the range for high clarity (forereef) and high scattering (lagoon) in the AC-S dataset (see figures in [7]). Under the scattering phase function, Petzold's phase function was used throughout, as described in Reference [24]. These IOPs were within the range of those measured across different reefs of the Pacific [25].

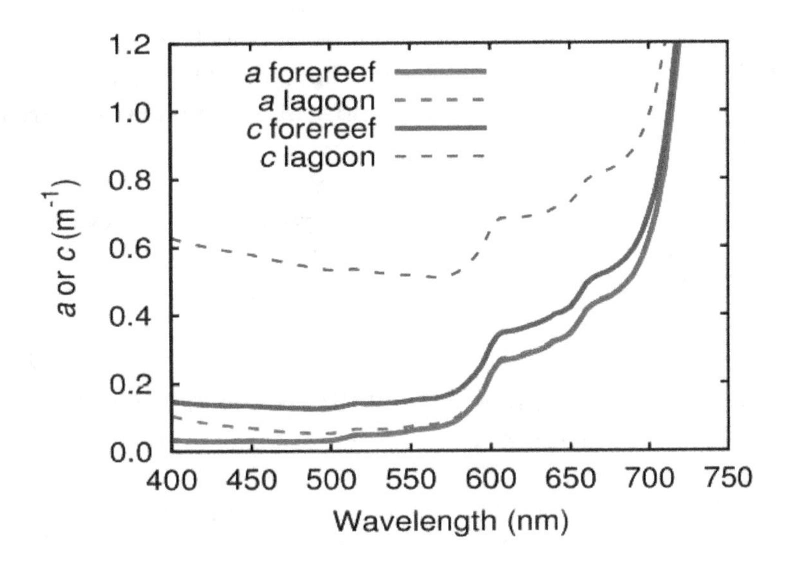

Figure 2. Total absorption, $a(\lambda)$, and attenuation, $c(\lambda)$, as used in the two water column inherent optical property (IOP) treatments, forereef and lagoon (includes the contribution of pure water itself).

2.5. Coral Canopy Assemblages

Incorporating individual corals into a water column model would correspond to a single coral surrounded by an infinite extent of bare substrate, which is not a situation of practical interest for remote sensing. Propagation of reflectance through a plane-parallel water column model, such as HydroLight, necessitates constructing a "pseudo plane-parallel" canopy to characterize a horizontally invariant bottom boundary BRDF. To this aim, six assemblages of corals on flat substrate were constructed by randomly manually placing and rotating the individual low-resolution coral models in an area of 20 cm $\times$ 20 cm (periodically repeating). Three assemblages were generated using corals from the shallow group, and three from the deep group. Viewed from above, the areal cover of the corals versus the substrate varied from 39% to 75%. Assemblages were named based on the group and percentage cover, specifically: S46, S58, S75, D39, D46, D74. There was no intention to specifically compare or construct any hypothesis concerning the shallow and deep groups, but it was considered prudent to be consistent and not to mix deep corals with shallow. The assemblages contained between 7 and 15 individual coral shapes, where some shapes appeared twice or three times in specific assemblages, but in different rotational positions. These randomly constructed canopies may not be strictly representative of a real canopy, but they do incorporate a mixture of typical coral morphologies and surface reflectances and enable an estimate of the optical properties of assemblages of such structures. In these models the total number of surface polygons, including substrate, varied from 35,000 to 68,000.

2.6. Coral Model Outputs and Processing

For activities 1 to 4 (Table 1), the main output of interest was reflectance just above the canopy under naturalistic illumination conditions, for a typical range of situations, i.e., the kind of reflectance that could be used as a simple bottom reflectance boundary condition in a remote sensing model (Figure 1g,h). The key questions were if this reflectance was consistent under different illumination conditions (hence robust), and the extent of the effect of shading and other spectral mixing processes due to structure. Toward this aim, for each coral shape and illumination treatment, the model generated a 512 × 512 pixel hyperspectral nadir-view image of the coral in orthographic projection (no perspective) (Figure 1g,h). For activities 1 to 3, the upward radiance directly above the coral shape only (no substrate) was extracted from this image and averaged to give an average nadir view upward radiance, $L_u(\lambda)$, over the coral shape. Reflectance was calculated as $L_u(\lambda)/E_d(\lambda)$ and multiplied by π, for the purpose of comparing to the surface scale diffuse reflectance. For activity 4, which was concerned with spectral mixing with the substrate reflectance (Table 1), the upward radiance was averaged over the shape plus enough surrounding substrate to make a 50% areal cover mix between the coral shape and substrate (Figure 1h).

The extent of this area was manually determined as a roughly consistent border around the shape. Activities 1 to 4 illuminated corals with azimuth angles of 0°, 9 0 °, 180°, and 270°, so shadows and other effects would be variable between these repeats, corresponding to the same shape illuminated from different directions.

The evaluation of BRDFs required that the full BRDF function be tabulated in a form suitable to use as input to HydroLight. HydroLight requires azimuthally averaged BRDFs, i.e., as a function of relative azimuth only, so the form of the function is BRDF (θ_i, θ_e, $\Delta\varphi$) (units sr^{-1}) in terms of quad-averaged radiance in the HydroLight directional discretization (directional segments of 10° × 15°, [19]). The incident and exitant zenith angles are denoted by θ_i and θ_e, and only the relative azimuth angle, $\Delta\varphi$, is relevant. The 3D model can work directly with the input and output quad-averaged radiances, where a single run of the model illuminates the canopy with radiance from a single directional quad, and the full hemisphere of exitant quad-averaged radiances is averaged over a 5 × 5 grid of points over the canopy (Figure 1k), see Reference [21] for more details.

A minimum of 10 model runs can populate the azimuthally averaged function, one run for each incident zenith angle quad position, θ_i (Figure 1i). Here, three runs per zenith angle were conducted with the canopy rotated by 120° each time. Mirror image symmetry was assumed BRDF (θ_i, θ_e, $\Delta\varphi$) = BRDF (θ_i, θ_e,$-\Delta\varphi$), and the reciprocity condition BRDF (θ_i, θ_e, $\Delta\varphi$) = BRDF (θ_e, θ_i, $-\Delta\varphi$) was checked to ensure there were no major errors and then enforced. Each value in the final function was the average of 12 values in general, although not all values were independent.

3. Results and Discussion

3.1. *Effect of Coral Shape on Nadir View Reflectance*

Pseudo-color renderings of the 512 × 512 pixel images used to derive the upward radiance over the corals are shown in Figure 3. These model outputs are for high resolution 3D models, for one of the 32 bottom of water column illumination treatments. The RGB images were derived from the 168 band model output by forming the product of the spectrum in each pixel with red, green, and blue visual tristimulus functions [24].

The surface reflectance for some corals was darker than the substrate reflectance, e.g., coral 03 (Figure 3), but for most, the coral surface reflectance was lighter. The necessity of obtaining the reflectance from a horizontal surface area in the hyperspectral image (Figure 1a) may bias the reflectance compared to the true total surface average, but this is of minor consequence in this study, since the overall brightness of the reflectance is of secondary importance to the spectral shape.

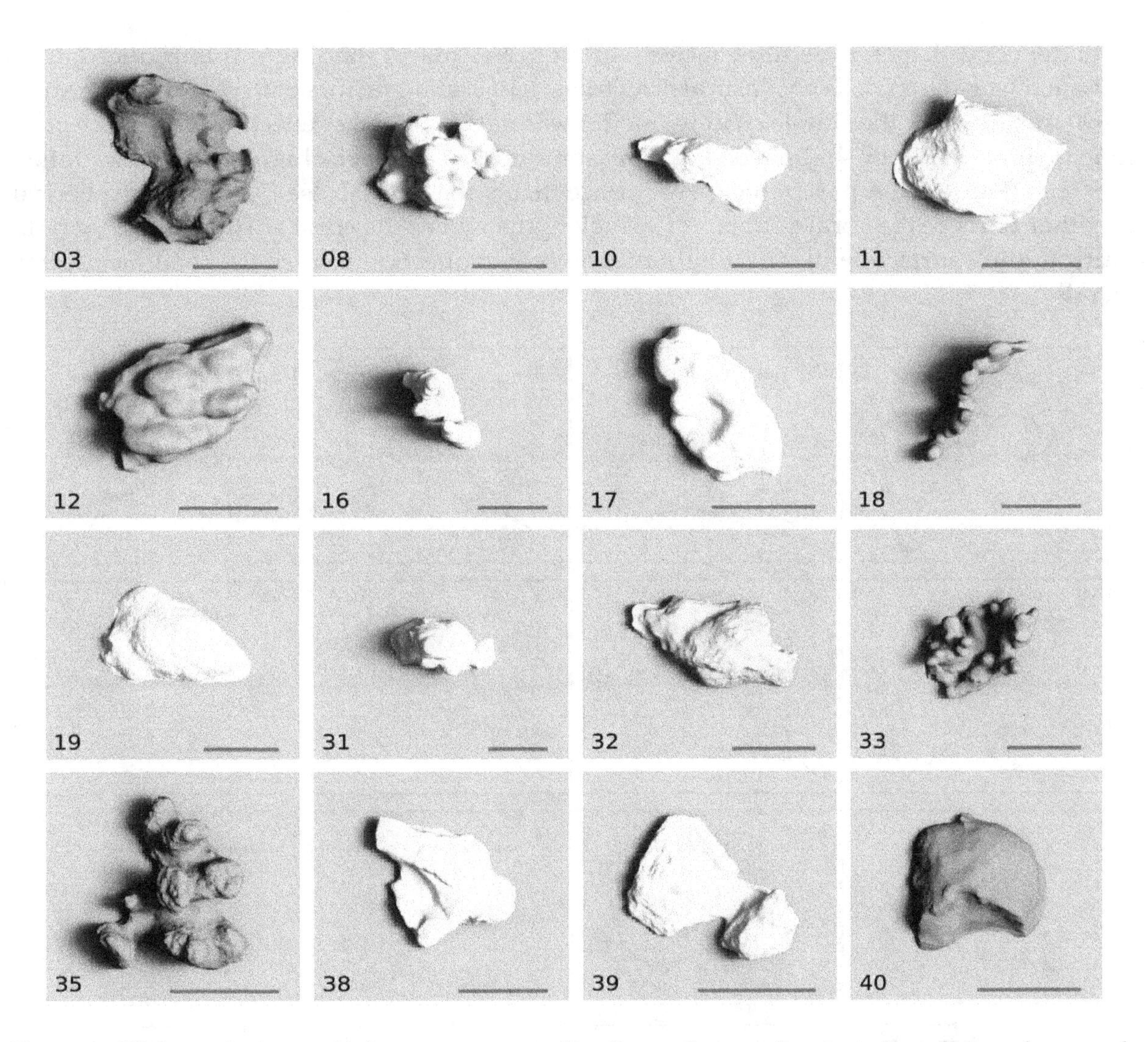

Figure 3. High resolution model outputs, as a nadir-view orthogonal projected rendering above each coral, converted to RGB using the tristimulus functions [24]. Illumination conditions are solar zenith $\theta_s = 50°$, forereef IOPs, depth 1 m. These are the high-resolution models without interstitial IOPs (activity 1, Table 1). Note: Corals are not shown in the same relative scale, bar is 5 cm.

With respect to the modelled reflectance over the coral area (Figure 1g), for all corals, the spectra averaged over the coral area had a very similar shape to the coral surface reflectance but was darker due to shading within the coral structure (Figure 4). The different illumination conditions introduced some variability in the reflectance but in general, this variability was small and certainly smaller than the shading effect. Coral 16 had the highest variability in reflectance under different illuminations, as this coral was a relatively vertical structure with an overall irregular shape when viewed from above; therefore the level of shading was quite dependent on azimuth angle of the incident light. Regarding corals shapes that were flatter (e.g., coral 39), or with numerous protrusions (coral 35), different lighting conditions produced little variation in the average reflectance (Figure 4).The first useful observation from these results was that in general, the average reflectance over the coral shape up to 690 nm can be represented by a scaled version of the surface reflectance, where the scaling factor < 1 effectively introduces shading as a black endmember in a linear mixing model, i.e., the transformation of surface reflectance $R(\lambda)$ is of the form $R'(\lambda) = R(\lambda) \times f$, where $0 \leq f \leq 1$ and for a flat Lambertian surface $f = 1$. Shading endmembers have been used in mineral applications [26]. For each coral, a single scaling factor was deduced as the median value of the fit from 400 to 690 nm to each of the 32 treatments (Table 3, Figure 4). These scaling factors varied from 0.50 (coral 16) to 0.84 (coral 39), and for flatter coral shapes, the scaling factor was higher since less shading was introduced. Corals 18 and 31, which were also relatively vertical structures, had scaling factors of 0.59 and 0.54, respectively. Above 690 nm, the fit to the simple shadow model was not as good, as reflectances over the coral area tended to be higher (Figure 4). It is likely the reason for this is the very high coral surface reflectance above 690 nm (Figure 4). A part of the coral surface that is shaded will "see" in its hemispherical field of view other parts of the coral surface. In wavelengths where the coral surface is highly reflective, those surrounding surfaces will reflect light to the point in question, and so the shading effect will be less overall.

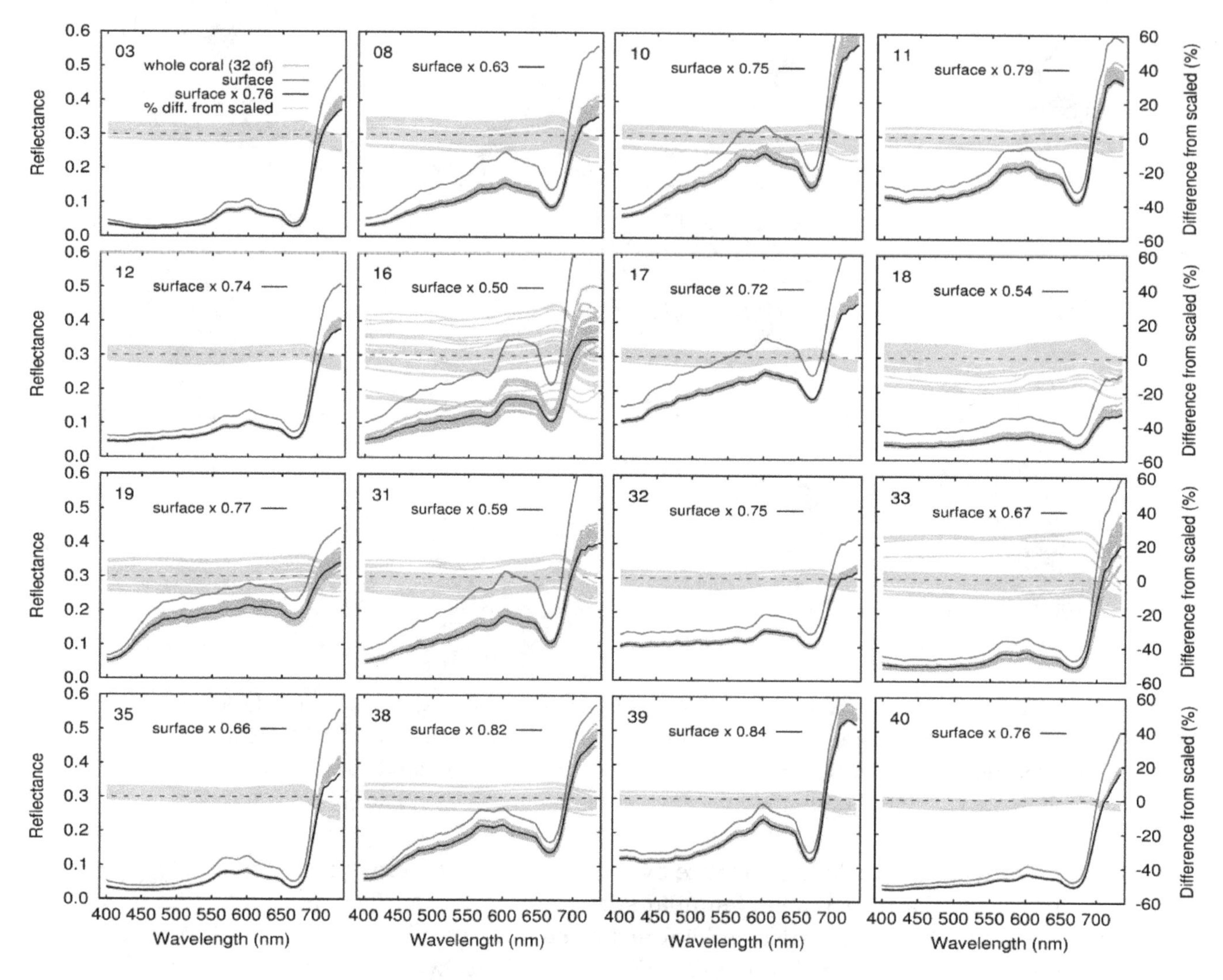

Figure 4. Reflectance over coral shape area only (no substrate) at the bottom of the water column with no interstitial water scattering or attenuation included. Plots show surface reflectance (red), reflectance over coral area under 32 treatments (grey), surface reflectance scaled by shading factor (black), and % difference of each treatment from scaled surface reflectance (light blue, right hand y-axis).

In each coral, most of the reflectances under different illumination conditions were within ±20% of the surface reflectance scaled by the mean shade factor (Figure 4). This shading factor, which in our data ranges from 0.50 to 0.84, could be of immediate use in approaches that use diffuse reflectance as a bottom boundary condition, for example, model inversion techniques for image processing [27]. If basic inputs are surface reflectances, a variable shading factor (effectively a shade endmember) could be included as an additional parameter to be estimated in the inversion, or fixed at a reasonable value, e.g., ~0.7, or certainly less than one. A variable factor would not only accommodate the difference between the coral shapes shown here but would also accommodate the residual error in each coral, due to variation under illumination conditions (Figure 4). However, this also points to the importance of the design of in-situ methodologies used for collection of spectral reflectance data [10,11,28,29]. Protocols which include multiple sampling over benthos will likely include a certain amount of the shading factor already, but this is scale-dependent, i.e., for small branching corals it may be completely included, whereas for larger massive morphologies the measurements may be closer to surface reflectances. Moreover, protocols that use shading or artificial light sources may or may not include the relevant amount of shading for a remote sensing context. Careful consideration of the relative scale of measurements in the context of the application is required. From our results, the maximum error in the brightness of the endmember reflectance, when using a surface scale reflectance instead of macro-morphology scale, could be a factor of as much as 0.5 (or 2). It may be worthwhile to consider if spectral libraries should be standardized or normalized in this respect, to either include or not include the shading factor. With this information, there would be the possibility to apply a post-hoc factor, at least as a rough estimate (between 0.5 and 0.84 in our data, e.g., ~0.67), or as a species-specific value (Table 3). Another consideration is that for very high spatial resolution imagery (pixels $\leq$ 0.5 m), where benthic structure is at a similar scale to the pixels, shading effects may vary from pixel to pixel, over for example large massive corals, where the sides and top may be imaged in different pixels.

Table 3. Shading factors for coral area only and coral with surrounding substrate in 50% mix. Final column shows the maximum difference in the reflectance of any band up to 690 nm, when interstitial attenuation and scattering is included in the canopy model.

ID	Species	Shading Scale Factor Coral only		Shading Scale Factor 50% Coral-Substrate Mix		Max L_u/E_d diff. with Interstitial IOPs ($\leq$690 nm)
		Range	Median	Range	Median	
03	*Montipora capitata*	0.71–0.79	0.76	0.83–0.88	0.85	2.6%
08	*Porites evermanni*	0.57–0.69	0.63	0.68–0.77	0.74	2.8%
10	*Porites evermanni*	0.71–0.82	0.75	0.75–0.84	0.81	1.6%
11	*Montipora capitata*	0.75–0.84	0.79	0.80–0.85	0.84	1.0%
12	*Porites evermanni*	0.71–0.77	0.74	0.79–0.85	0.82	1.7%
16	*Porites compressa*	0.39–0.66	0.50	0.51–0.64	0.59	3.1%
17	*Porites evermanni*	0.69–0.75	0.72	0.75–0.79	0.78	1.0%
18	*Porites compressa*	0.50–0.65	0.54	0.55–0.68	0.65	3.0%
19	*Porites evermanni*	0.69–0.85	0.77	0.78–0.86	0.83	1.9%
31	*Porites compressa*	0.54–0.65	0.59	0.64–0.71	0.70	2.8%
32	*Montipora capitata*	0.72–0.78	0.75	0.79–0.83	0.81	1.2%
33	*Porites compressa*	0.51–0.74	0.67	0.71–0.81	0.78	2.6%
35	*Pocillopora* sp.	0.62–0.68	0.66	0.70–0.81	0.74	2.1%
38	*Porites evermanni* or *lutea*	0.76–0.88	0.82	0.82–0.87	0.86	1.5%
39	*Montipora patula*	0.81–0.88	0.84	0.83–0.88	0.87	0.7%
40	*Porites evermanni* or *lutea*	0.75–0.79	0.76	0.87–0.89	0.88	1.7%

3.2. *Effect of 3D Model Resolution*

To determine if the low-resolution 3D coral models were sufficient, the modelling of reflectance over coral shape area (Figure 1g, results in Figures 3 and 4, and Table 3) was duplicated using the low-resolution versions of the coral 3D structures (activity 2, Table 1). Across all corals and illumination treatments, the difference in any band when using the low-resolution 3D models was not greater than 0.51%, apart from coral 16 which had a maximum difference of 1.33%. In most corals, the maximum difference was less than a third of a per cent (Table 2, final column). Visually the outputs were very similar (Figure 5). Therefore, in general, using the low-resolution structures made almost no difference to the reflectance averaged over the coral area, and there seemed little need to continue using the high-resolution models, which with up to ~100,000 polygons cause the 3D radiative transfer model to run substantially slower. For this reason, the remaining modelling activities used only the low-resolution models.

(a) high resolution (240353 triangles)

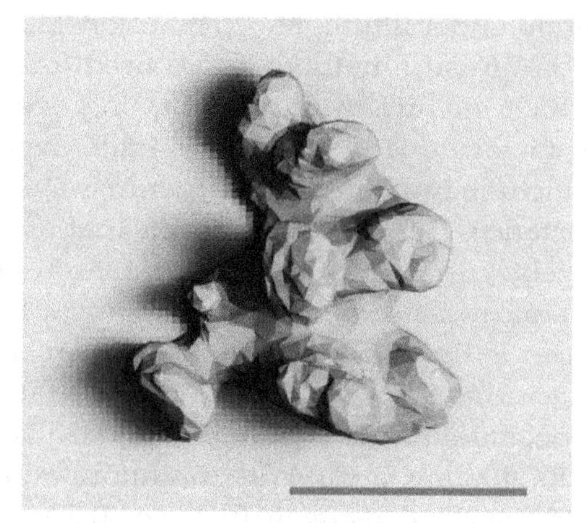

(b) low resolution (2966 triangles)

Figure 5. (**a**) High-resolution model versus; (**b**) low-resolution model of coral 35, scale bar is 5 cm.

3.3. Linear Mixing of Reflectance over Coral Shapes and Surrounding Substrate

As with the reflectance over the coral shape only, reflectance over a 50% areal mix of coral and surrounding substrate was consistently darker than a 50% linear mix of the surface reflectances, but could be well represented in wavelengths less than 690 nm by a uniformly scaled version of that linear mix (Figure 6). The shading factor when substrate was included was a consistently higher value (less shading) than the values over coral shape only, where the median for each coral ranged from 0.59 (coral 16) to 0.88 (coral 40) (Table 3). Individual values were from 4% to 20% higher than the corresponding values over the coral area only. This occurred because the flat substrate does not introduce any shading itself but is shaded by the coral structure (Figure 3). Including surrounding substrate introduces relatively less shading than would occur within the area of the structure. Therefore, both in the experimental set-up and the physical world, the shading factor is likely dependent on the ratio of coral cover to substrate.

The previous suggestion of including a variable black shade endmember in a linear mixing model for bottom reflectance would seem to be further supported by these findings. The shading factor is dependent on the areal cover of flat substrate included, and the difference was variable from 4% to 20%. A-priori inclusion of the shading factor, even by careful collection of in-situ spectra to include shadows, cannot account for the variability due to structural context. However, again the figures presented here assume that the basic input reflectances are surface reflectances, where for flat substrates such as sand or dead coral rock this is likely to be the case, whilst for corals some a-priori inclusion of shading is likely. The possible range of shading factors would have to be modified accordingly.

Figure 7 shows that median shading factors were a clear function of surface rugosity (coral surface area divided by projected top-down area) for the reflectance over the coral area only (Figure 7a), and for coral shape and substrate mixes (Figure 7b). Intuitively, the shading factor should be 1 when rugosity is 1 (flat surface), and very high rugosities should tend to show some positive shading factor greater than zero. To express these constraints, a simple two-factor negative exponential model was fit to the plots in Figure 7, of which $y = (1 - A) \times (\exp[-S \times (x - 1)] + A$, fit for S (slope) and, A, the asymptotic shading factor. This function gives a good fit in both cases but is just indicative of the general shape of the relationship, since the validity of the concept of an asymptotic shading factor is debatable. The slightly reduced shading effect when flat substrate is included, can be seen in the smaller magnitude slope (S) of the best-fit line in Figure 7b. Note the surface rugosity value refers only to the coral shape. If rugosity in Figure 7b were calculated incorporating the flat substrate, the rugosity values would be lower and the relationship would be closer to Figure 7a. Solar zenith angle of 10° vs.

50° only had a small effect on the shading factor (Figure 7b), but as expected the effect increased with higher rugosity. Several factors were relevant to the observed small effect of the solar zenith angle, i.e., due to refraction the corresponding sub-surface zenith angles are less, at approximately 7°–35°, and side-ward illumination reduces the irradiance on some surface facets, but at the same time increases it on others. Considering Figure 2, produced with $\theta_s = 50°$, for most corals, only a small part of the area is actually visibly in shadow. Those with the most shaded area correspond to those where solar zenith angle has the most effect, e.g., corals 08, 33, and 35, corresponding to the group with rugosity ~3 in Figure 7b. Overall Figure 7 indicates that rugosity, or equivalently species morphology, can be a robust indicator of the magnitude of the appropriate shading factor. This is consistent with empirical studies on BRDFs, which have shown that morphology and shadowing between branches are key determiners of spectral reflectance [16]. If a variable shading factor could be reliably estimated as a black endmember contribution in a model inversion image analysis [27], this estimate would carry information on benthic rugosity and benthic type.

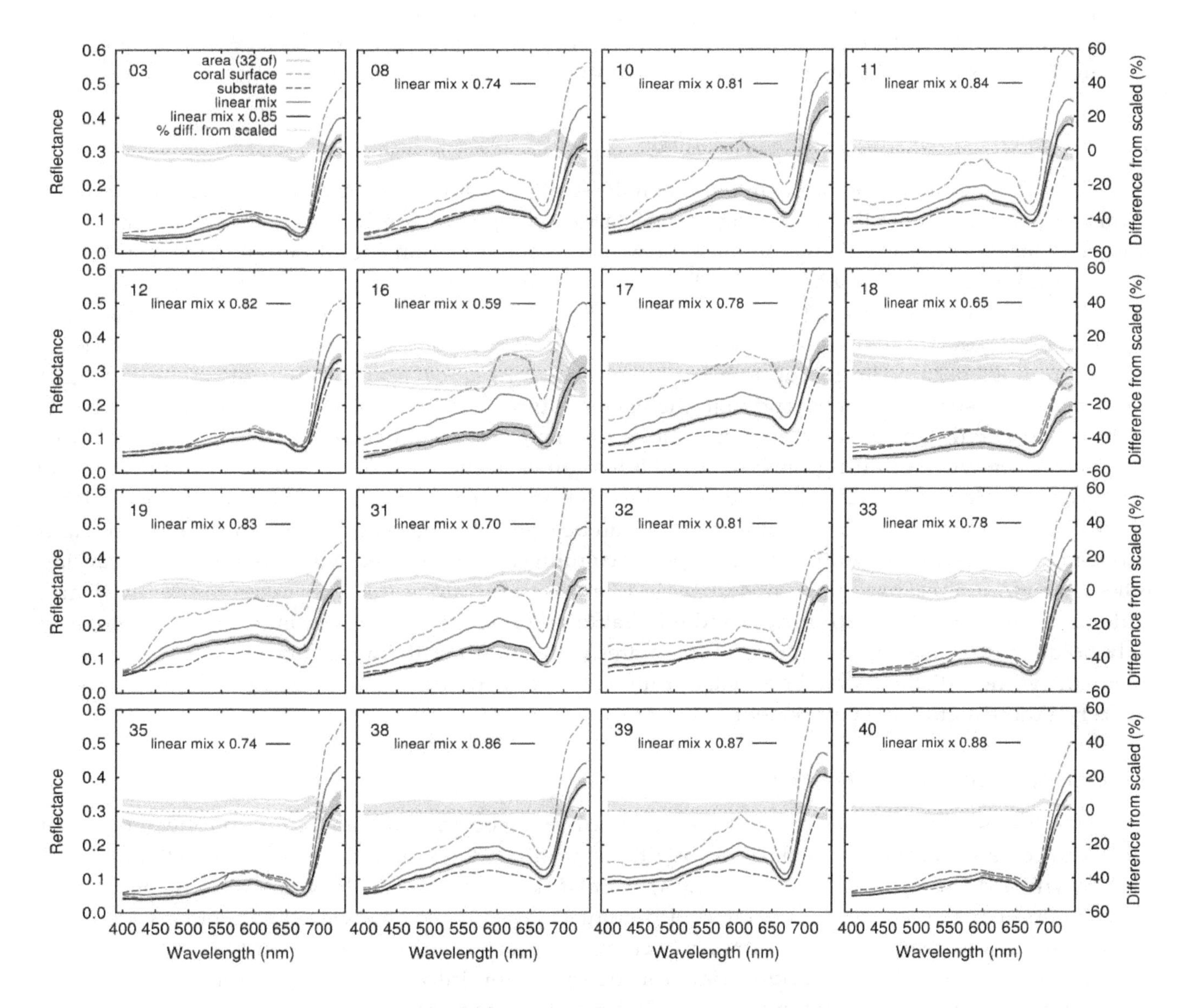

Figure 6. Reflectance over coral structure and surrounding substrate to give a 50% mix in areal cover. Plots show the coral surface and substrate surface reflectances (thin and thick green lines), and a 50% linear mix of those reflectances (red), reflectance over the coral and substrate area under 32 treatments (grey), surface reflectance scaled by shading factor (black), and % difference of each treatment from scaled surface reflectance (light blue, right hand y-axis).

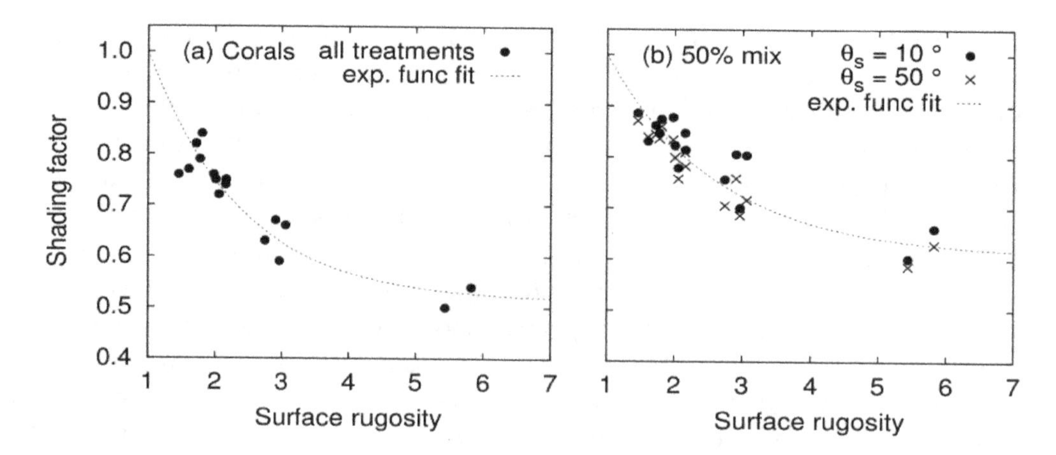

Figure 7. Median shading factor as a function of surface rugosity: (**a**) Over coral area only (Figure 4), and; (**b**) over coral area and surrounding substrate in a 50% areal mix (Figure 6) with data separated into solar zenith angles of 10° and 50°. Lines are of the form $y = (1 - A) \times \exp[-S \times (x - 1)] + A$ and are the least squares best fit, giving (**a**) $A = 0.51$, $S = 0.72$; (**b**) $A = 0.61$, $S = 0.61$.

3.4. Effect of Interstitial Scattering and Absorption

The difference when including interstitial IOPs (scattering and absorption in the water around and within the coral shape) in the modelling of reflectance averaged over a 50% mix of coral and substrate was in general, less than 3% in any band for wavelengths less than 690 nm (Table 3), and for most coral shapes of the order of 2% or less. This showed that the previous back-of-envelope calculation that suggested a 4% error at 700 nm, based on absorption over path lengths of 10 cm, was quite reasonable. Beyond 700 nm, absorption by pure water is high and the discrepancy when omitting interstitial IOPs became larger, up to 5%. However, in a practical application, unless corals almost touch the water surface, the absorption in the water column above the canopy will dominate over the within-canopy effect, so the discrepancies above 690 nm are of minor consequence. The cost of 2% or 3% in accuracy when omitting the interstitial IOPs comes at the advantage, not only of substantially faster computation in the 3D model, but also the BRDF of the canopy can be decoupled from the IOPs of the overlying water column. That is, a BRDF can be computed based on coral canopy structure and surface reflectance, and then the overlying water column can be specified with independently varying IOPs. Therefore, this answers point 4 in Table 1, and is the strategy that was taken in the BRDF and water column modelling activities (5 and 6 in Table 1). However, this result is scale dependent and only holds in this case because the coral structures were small. With larger corals and reef structures, where the vertical distances can be a meter or more, the effect of interstitial water scattering absorption is likely a more significant component of the reflectance.

3.5. Canopy Assemblage BRDF Effects

The discussion so far has concentrated on nadir view bottom of water column reflectance over individual corals under naturalistic light conditions, calculated based on the upward directed radiance and downward irradiance, i.e., $\pi L_u(\lambda)/E_d(\lambda)$. Whilst the previous results were fairly insensitive to illumination conditions, such as solar zenith angle (Figures 4 and 6), view angle effects were not considered. Direct use of these results as a bottom boundary requires the assumption of Lambertian reflectance. As such, they are appropriate for incorporation into models where this assumption is implicit, for example, in model inversion techniques for image analysis [27]. However, to correctly model propagation of light through the water column in a physically exact model, such as HydroLight, the full BRDF of canopy assemblages is required (activities 5 and 6, Table 1).

The BRDF function is calculated using light incident only for a specific direction in an otherwise black radiance field (Figure 1i), so in the model outputs, shading effects are very apparent (Figure 8). Visually, Figure 8 may appear to contradict Figure 7b, since shading as a function of incident zenith angle in Figure 8 appears stronger than implied by Figure 7b. To reconcile this first note that Figure 7b is expressed in terms of above-surface solar zenith angles of 10° and 50°, and so most closely corresponds to Figure 8a,b (sub-surface incident angles of 10° and 30°), Figure 8c is a more extreme example. Furthermore, Figure 8 arises from unidirectional illumination, whereas Figure 7b arises from a more diffuse illumination, being the sub-surface propagated solar and sky illumination. A final point is that the increased shadow from Figure 8a to 8b, although visually very apparent, is probably not more than 10% of the actual image area, so even under unidirectional illumination it likely would not affect the shading factor by more than 10%, which is comparable to the differences seen for rugosities ~3 in Figure 7b.

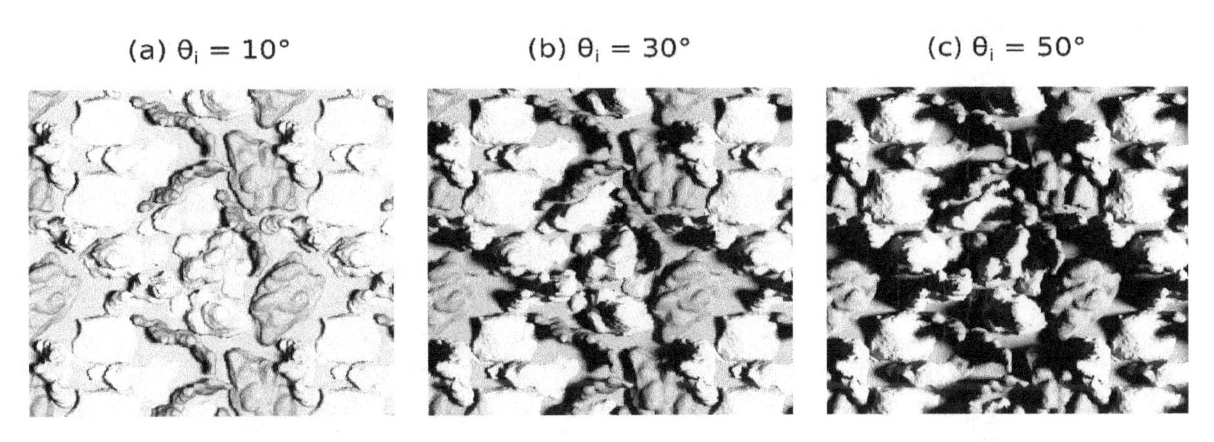

Figure 8. Examples from BRDF generation of canopy using reflectances of deep corals at 74% coral coverage (D74). Shown as nadir-view ($\theta_e = 0°$) orthogonal projected rendering above each canopy, corresponding to incident radiance from: (**a**) $\theta_i = 10°$; (**b**) $\theta_i = 30°$ and; (**c**) $\theta_i = 50°$, respectively, azimuth $\varphi s = 0°$. RGB images are created from hyperspectral data using the tristimulus functions and represent an area of 30 cm × 30 cm.

One of the densest canopy assemblages, D74 (74% coral cover), also tended to have vertically higher coral structures and consequentially showed the strongest BRDF effect, with an almost linear effect of view angle in the incident plane ($\Delta\varphi = 0$), when incident light was at a 50° zenith angle (Figure 9e). The reflected radiance decreased by a factor of seven when varying the view direction from almost horizontal in the same direction as the incident light, to horizontally toward the incident light (Figure 9e). In all canopy assemblages there is a clear hotspot effect [30], where brightness is at a maximum when the incident and view directions are the same. The BRDFs discussed here were spatially averaged, and the hot spot occurs because the viewing geometry determines the fraction of shadowed area detected. When the viewing and incident light directions are the same, shadows are maximally obscured by the illuminated surfaces. Interestingly, for the canopy structures modelled here, the hotspot effect is stretched out to larger view angles (Figure 9). This is likely a geometrical consequence of an increasingly horizontal view onto vertical structures illuminated from the side (Figure 8).

In all canopy assemblages, there were only small BRDF effects at 90° to the incident plane ($\Delta\varphi = 90°$) (Figure 9b,d,f). Typically reflectance decreased slightly as the view angle moved from nadir, although the pattern was stronger for denser canopies, such as D74 (Figure 9), and even included a slight increase in reflectance close to horizontal, being a pattern also seen in models of dense seagrass canopies [21].

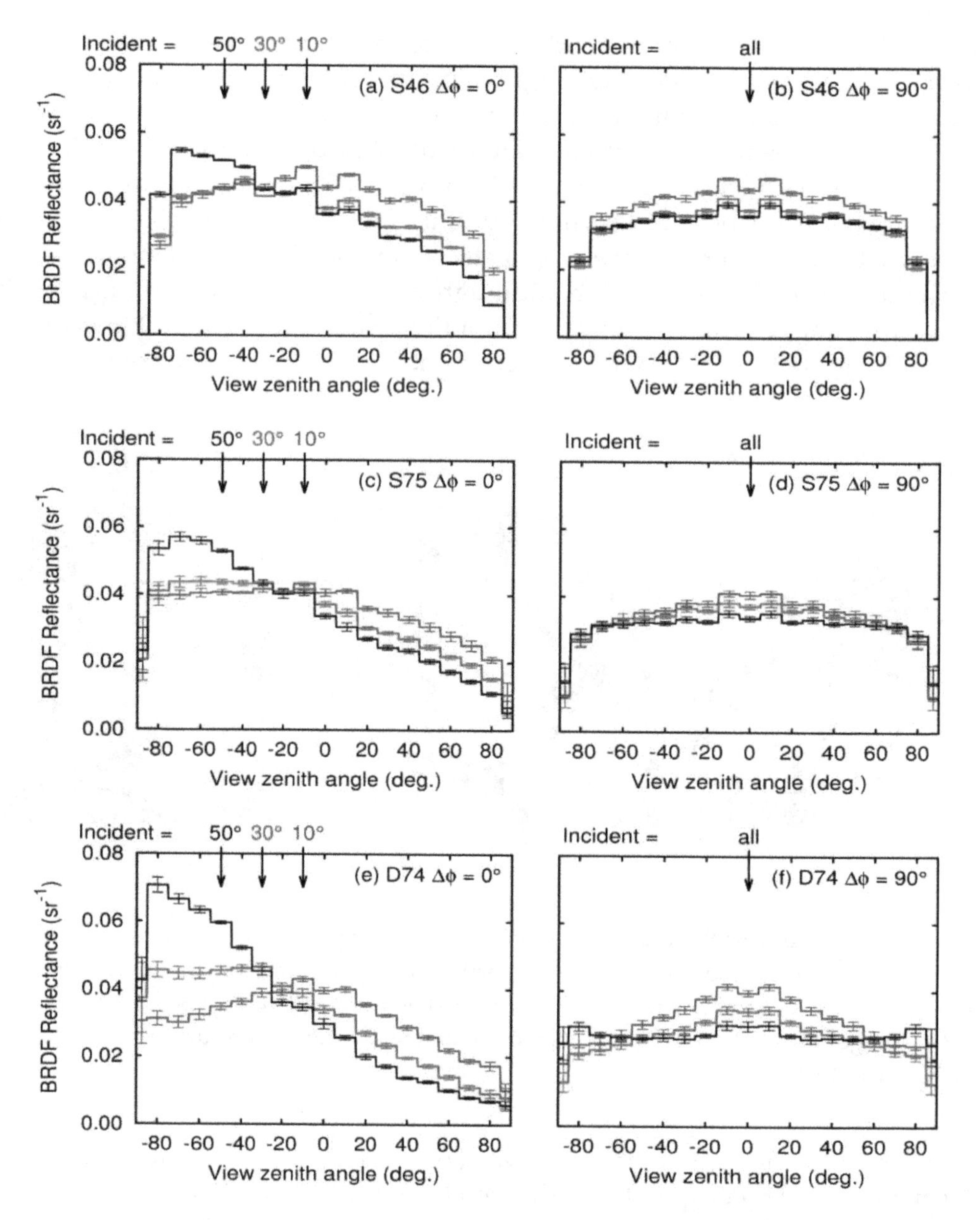

Figure 9. Bidirectional reflectance functions (BRDFs) at 550 nm, of three of the composite coral canopies for shallow (S, 0–10 m) and deep (D, 10–20 m) corals with areal cover: (**a**,**b**) 46%; (**c**,**d**) 75%; and (**e**,**f**) 74%, in the incident plane (a, c, e, $\Delta\varphi = 0°$) and at 90° to the incident plane (**b**,**d**,**f**, $\Delta\varphi = 90°$). Arrows show direction of incident light, each plot shows θi of 10°, 30° and 50°. Negative view zenith angle for $\Delta\varphi = 0°$ (**a**,**c**,**e**) means backward reflection (source and view point in the same hemisphere). Error bars are ± 1 standard error on 12 values from assumed reciprocity, rotational and mirror symmetries.

3.6. Above-Water BRDF Effects

Canopy BRDFs contributed to the directional pattern in water leaving radiance under naturalistic sky illumination primarily for dense canopies in shallow, clear water (Figure 10). For the 46% cover canopy, S46, a slight BRDF effect of increased retro-reflection in the incident plane, leading to effective propagation of the hotspot effect, was evident for depths less than 5 m with a low sun position, $\theta_s = 50°$ (Figure 10a). In deeper waters and at 90° to the incident plane L_W, it decreased with increased view angle (Figure 10b,d,f). In deeper waters over sparser canopies, the water column itself becomes a contributor to the above water BRDF, where at 10 m depth the BRDF response in the solar plane (Figure 10a) resembles the response at 90° to the solar plane (Figure 10b).

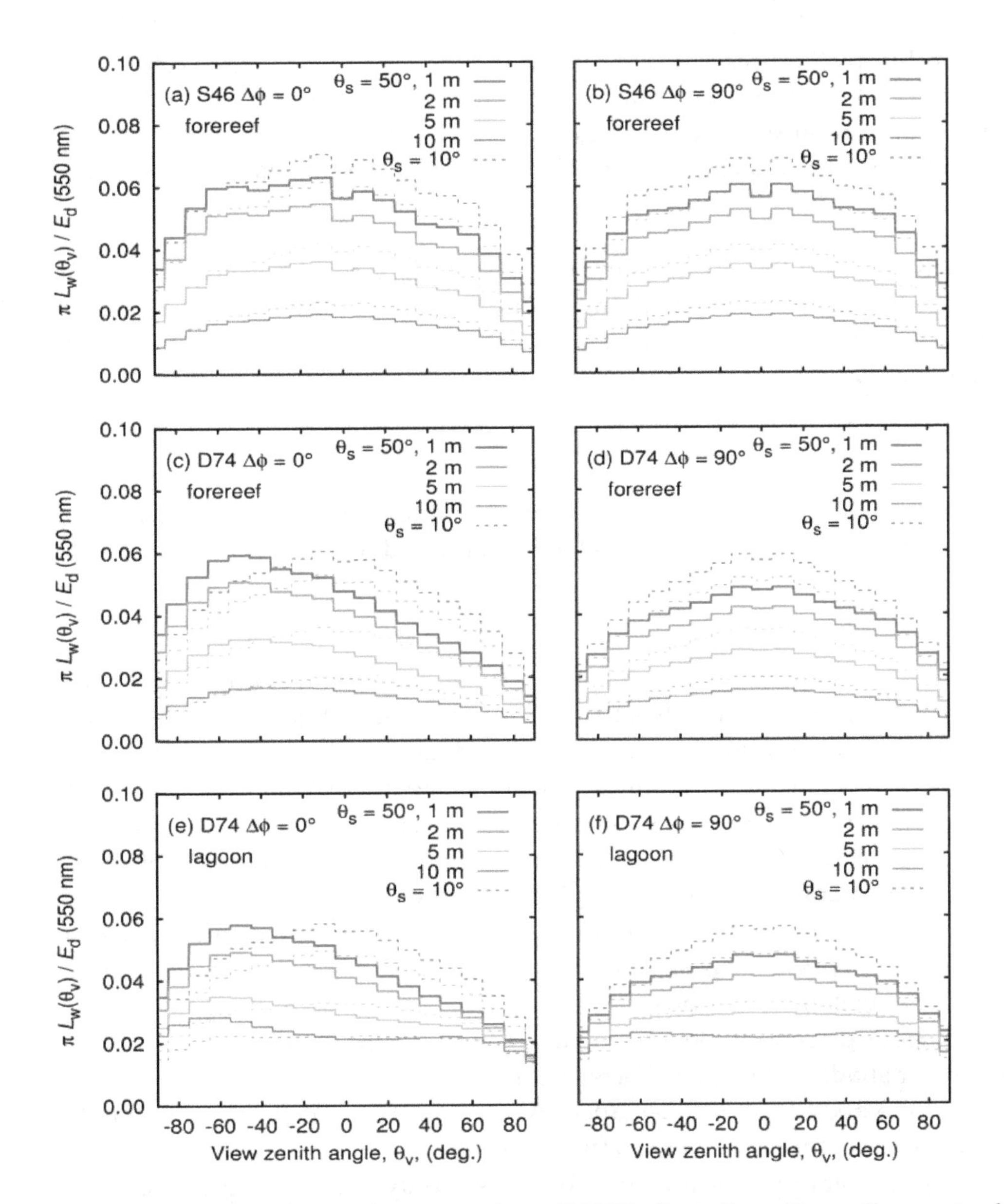

Figure 10. Modelled influence of three-dimensional coral BRDF effects (from Figure 9) on water-leaving reflectance (π Lw/Ed) at 550 nm for: (**a**,**b**) Composite canopies S46 (46% shallow coral cover) and; (**c**,**d**,**e**,**f**) D74 (74% deep coral cover), in the incident plane (**a**,**c**,**e**, $\Delta\varphi = 0°$) and at 90° to the incident plane (**b**,**d**,**f**, $\Delta\varphi = 90°$). Each plot shows results of incident solar zenith angles of 10° (dotted lines) and 50° (solid lines), and depths 1, 2, 5, and 10 m. Negative view zenith angle for $\Delta\varphi = 0°$ (**a**,**c**,**e**) means backward reflection (source and view point in the same hemisphere).

Regarding the dense 74% coral cover canopy D74, water-leaving BRDF effects in the incident plane were quite prominent for both the forereef IOP treatment and the higher scattering lagoon waters (Figure 10c,e). For example, when the water depth was 1 m, the difference in $L_W(550)$ between a view zenith angle of 50° toward and away from a sun at $\theta_s = 50°$ was about a factor of 1.5 (~0.06 vs. ~0.04, Figure 10c,e); whereas this factor was ~1.2 for view zenith angles at $\pm$ 20°. At 90° to the incident plane, even for dense canopies, BRDF effects are less apparent (Figure 10d,f). It is clear that for these modelled canopies, the assumption of a Lambertian bottom is primarily violated close to the retro-reflected (hotspot) direction for large zenith angles (>30°) but could be a more reasonable assumption under other solar-view geometries. The difference in nadir view water-leaving reflectance at 550 nm for canopy D74 at 1 m, for solar zenith angle of 50° vs. 10°, was 18% (Figure 10c,d, ~0.047 vs. ~0.057). This is equivalent to the variation previously shown in reflectance over individual corals under different illumination conditions (Figures 4 and 6), which for some corals was as much as ±20%, but here the variation was reduced by the water column, to effectively ±9%. Mobley et al. [31] estimated ignoring non-Lambertian bottom effects would in general, cause errors of less than 10% in remote sensing reflectance. While not strictly a comparison to a Lambertian bottom assumption, a discrepancy of 18% implies the potential for larger errors.

To avoid across-swath BRDF effects in airborne imagery, the advice is therefore broadly the same as for avoiding surface glint [32]; thus flying toward or away from the sun will ensure the cross track view is at 90° to the solar plane, and a solar zenith angle > 30° both avoids glint and slightly flattens the BRDF response (Figure 10d,f).

4. Conclusions

A number of conclusions from this work can be made that relate (1) to using Lambertian bottom boundary conditions and linear mixing models, and (2) to data collection of reflectances and imagery. In summary:

- If the input reflectances for different benthic types can be considered 'pure' surface reflectances, then structural complexity implies the introduction of a shading scale factor or a black shade endmember in mixing models is required. In our data, that shading scale factor could be anywhere from 0.5 to 0.9, whereas for a genuinely flat substrate it would be 1.0.
- The magnitude of the shading factor is inversely related to surface rugosity; hence the shade factor value could be set based on the expected bottom types, or if derivable in an image analysis, it would give additional information on benthic type.
- Nadir viewing reflectance over the bottom (L_u/E_d) for individual corals varies as much as ~20% (more only in exceptional cases) under the range of light environments found on reefs, under typical remote sensing conditions.
- Regarding view directions beyond a few 10s of degrees from nadir for dense canopies in shallow water shading, hotspot effects become relevant, leading to potentially a 50% difference (factor of 1.5) in above water reflectance. At narrower remote sensing geometries (view zenith angles < 20°), differences can be 20% (factor of 1.2).
- For all modelled canopy assemblages, there were only small BRDF effects at a relative viewing angle of 90° to the incident solar plane. The advice for minimizing cross-track BRDF effects in airborne imagery is therefore consistent with that for minimizing surface glint, i.e., fly in a direction close to the solar plane with a solar zenith angle greater than 30°.
- Collection of in-situ spectra of benthic types requires consideration of what level of shading is already included, or whether the data is close to being a surface reflectance measurement. However, including shading in field measurements cannot fully accommodate the influence of canopy interactions, because there are effects at canopy scale involving different benthic types.
- These modelling results provide useful concepts and parameter ranges, which can assist in the interpretation of empirical data and the development of image processing algorithms.

Author Contributions: Conceptualization J.D.H., H.M.D.; Methodology, J.D.H., M.M., A.W., H.M.D.; Writing-Original draft preparation, J.D.H.; Writing–Review & Editing J.D.H., M.M., A.W., H.M.D.

Acknowledgments: The authors extend thanks to Brandon Russell and Jeff Godfrey for assistance in data acquisition and field work. Part of Brandon Russell's time was funded by the NASA Ocean Biology and Biogeochemistry program and the COral Reef Airborne Laboratory (CORAL) project. Coral collection was conducted under special activity permit (SAP) 2018-02 issued on 2 February 2017, by the State of Hawaii Department of Land and Natural Resources.

References

1. Hedley, J.D.; Roelfsema, C.M.; Chollet-Ordaz, I.; Harborne, A.R.; Heron, S.F.; Weeks, S.; Skirving, W.J.; Strong, A.E.; Eakin, C.M.; Christensen, T.R.L.; et al. Remote sensing of coral reefs for monitoring and management: A review. *Remote Sens.* **2016**, *8*, 118. [CrossRef]
2. Mumby, P.J.; Skirving, W.; Strong, A.E.; Hardy, J.T.; LeDrew, E.F.; Hochberg, E.J.; Stumpf, R.P.; David, L.T. Remote sensing of coral reefs and their physical environment. *Mar. Pollut. Bull.* **2004**, *48*, 219–228. [CrossRef] [PubMed]
3. Mumby, P.J.; Green, E.P.; Edwards, A.J.; Clark, C.D. Coral reef habitat-mapping: How much detail can remote sensing provide? *Mar. Biol.* **1997**, *130*, 193–202. [CrossRef]
4. Hochberg, E.J.; Atkinson, M.J. Capabilities of remote sensors to classify coral, algae, and sand as pure and mixed spectra. *Remote Sens. Environ.* **2003**, *85*, 174–189. [CrossRef]
5. Capolsini, P.; Andréfouët, S.; Rion, C.; Payri, C. A comparison of Landsat ETM+, SPOT HRV, Ikonos, ASTER, and airborne MASTER data for coral reef habitat mapping in South Pacific islands. *Can. J. Remote Sens.* **2007**, *23*, 87–200. [CrossRef]
6. Lubin, D.; Li, W.; Dustan, P.; Mazel, C.H.; Stamnes, K. Spectral Signatures of coral reefs: Features from space. *Remote Sens. Environ.* **2001**, *75*, 127–137. [CrossRef]
7. Hedley, J.D.; Roelfsema, C.; Phinn, S.R.; Mumby, P.J. Environmental and sensor limitations in optical remote sensing of coral reefs: Implications for monitoring and sensor design. *Remote Sens.* **2012**, *4*, 271–302. [CrossRef]
8. Hedley, J.D.; Mumby, P.J. Biological and remote sensing perspectives of pigmentation in coral reef organisms. *Adv. Mar. Biol.* **2002**, *43*, 277–317. [PubMed]
9. Mumby, P.J.; Hedley, J.D.; Chisholm, J.R.M.; Clark, C.D.; Ripley, H.; Jaubert, J. The cover of living and dead corals from airborne remote sensing. *Coral Reefs* **2004**, *23*, 171–183. [CrossRef]
10. Goodman, J.A.; Ustin, S.L. Classification of benthic composition in a coral reef environment using spectral unmixing. *J. Appl. Remote Sens.* **2007**, *1*, 011501.
11. Hochberg, E.J.; Atkinson, M.J. Spectral discrimination of coral reef benthic communities. *Coral Reefs* **2000**, *19*, 164–171. [CrossRef]
12. Lesser, M.P.; Mobley, C.D. Bathymetry, water optical properties, and benthic classification of coral reefs using hyperspectral remote sensing imagery. *Coral Reefs* **2007**, *26*, 819–829. [CrossRef]
13. Leiper, I.A.; Phinn, S.R.; Roelfsema, C.M.; Joyce, K.E.; Dekker, A.G. Mapping Coral Reef Benthos, Substrates, and Bathymetry, Using Compact Airborne Spectrographic Imager (CASI) Data. *Remote Sens.* **2014**, *6*, 6423–6445. [CrossRef]
14. Garcia, R.A.; Lee, Z.; Hochberg, E.J. Hyperspectral Shallow-Water Remote Sensing with an Enhanced Benthic Classifier. *Remote Sens.* **2018**, *10*, 147. [CrossRef]
15. Joyce, K.E.; Phinn, S.R. Bi-directional reflectance of corals. *Int. J. Remote Sens.* **2002**, *23*, 389–394. [CrossRef]
16. Miller, I.; Foster, B.C.; Laffan, S.W.; Brander, R.W. Bidirectional reflectance of coral growth-forms. *Int. J. Remote Sens.* **2016**, *37*, 1553–1567. [CrossRef]
17. Hedley, J.D.; Mumby, P.J.; Joyce, K.E.; Phinn, S.R. Spectral unmixing of coral reef benthos under ideal conditions. *Coral Reefs* **2004**, *23*, 60–73. [CrossRef]
18. Hedley, J.D. A three-dimensional radiative transfer model for shallow water environments. *Optics Express* **2008**, *16*, 21887–21902. [CrossRef] [PubMed]
19. Mobley, C.D.; Sundman, L. Hydrolight 5.2 User's Guide. Sequoia Scientific. Available online: https://www.sequoiasci.com/wp-content/uploads/2013/07/HE52UsersGuide.pdf (accessed on 25 July 2018).
20. Young, G.C.; Dey, S.; Rogers, A.D.; Exton, D. Cost and time-effective method for multi-scale measures of rugosity, fractal dimension, and vector dispersion from coral reef 3D models. *PLoS ONE* **2017**, *12*, e0175341. [CrossRef] [PubMed]
21. Hedley, J.D.; Enríquez, S. Optical properties of canopies of the tropical seagrass *Thalassia testudinum* estimated by a three-dimensional radiative transfer model. *Limnol. Oceanogr.* **2010**, *55*, 1537–1550. [CrossRef]
22. Hedley, J.D.; McMahon, K.; Fearns, P. Seagrass canopy photosynthetic response is a function of canopy density and light environment: A model for Amphibolis griffithi. *PloS ONE* **2014**, *9*, e111454. [CrossRef] [PubMed]

23. Hedley, J.D.; Russell, B.; Randolph, K.; Dierssen, H. A physics-based method for the remote sensing of seagrasses. *Remote Sens. Environ.* **2015**, *174*, 134–147. [CrossRef]
24. Mobley, C.D. *Light and Water: Radiative Transfer in Natural Waters*; Academic Press: San Diego, CA, USA, 1994; ISBN 0-12-502750-8.
25. Russell, A.B.; Hochberg, E.; Dierssen, H.M. Comparison of water column optical properties across geomorphic zones of Pacific coral reefs. *Limnol. Oceanogr.* under review.
26. Adams, J.B.; Smith, M.O.; Johnson, P.E. Spectral mixture modelling: A new analysis of rock and soil types at the Viking Lander I site. *J. Geophys. Res.* **1986**, *91*, 8098–8112. [CrossRef]
27. Dekker, A.G.; Phinn, S.R.; Anstee, J.; Bissett, P.; Brando, V.E.; Casey, B.; Fearns, P.; Hedley, J.; Klonowski, W.; Lee, Z.P.; et al. Intercomparison of shallow water bathymetry, hydro–optics, and benthos mapping techniques in Australian and Caribbean coastal environments. *Limnol. Oceanogr. Methods* **2011**, *9*, 396–425. [CrossRef]
28. Roelfsema, C.M.; Marshall, J.; Phinn, S.R.; Joyce, K. *Underwater Spectrometer System 2006 (UWSS04)—Manual*; Biophysical Remote Sensing Group, Centre for Spatial Environmental Research, University of Queensland: Queensland, Australia, 2006.
29. Roelfsema, C.M.; Phinn, S.R. Spectral Reflectance Library of Healthy and Bleached Corals in the Keppel Islands, Great Barrier Reef. PANGAEA. Available online: https://doi.org/10.1594/PANGAEA.872507 (accessed on 25 July 2018).
30. Liang, S. *Quantitative Remote Sensing of Land Surfaces*; Wiley: Hoboken, NJ, USA, 2004; ISBN 978-0471281665.
31. Mobley, C.D.; Zhang, H.; Voss, K.J. Effects of optically shallow bottoms on upwelling radiances: Bidirectional reflectance distribution function effects. *Limnol. Oceanogr.* **2003**, *48*, 337–345. [CrossRef]
32. Kay, S.; Hedley, J.D.; Lavender, S. Sun glint correction of high and low spatial resolution images of aquatic scenes: A review of methods for visible and near-infrared wavelengths. *Remote Sens.* **2009**, *1*, 697–730. [CrossRef]

8

Concentrations of Multiple Phytoplankton Pigments in the Global Oceans Obtained from Satellite Ocean Color Measurements with MERIS

Guoqing Wang [1,*], Zhongping Lee [1] and Colleen B. Mouw [2]

[1] School for the Environment, University of Massachusetts Boston, Boston, MA 02125, USA; zhongping.lee@umb.edu
[2] Graduate School of Oceanography, University of Rhode Island, Narragansett, RI 02882, USA; cmouw@uri.edu
* Correspondence: guoqing.wang001@umb.edu or gqwang18@gmail.com

Abstract: The remote sensing of chlorophyll *a* concentration from ocean color satellites has been an essential variable quantifying phytoplankton in the past decades, yet estimation of accessory pigments from ocean color remote sensing data has remained largely elusive. In this study, we validated the concentrations of multiple pigments (Cpigs) retrieved from in situ and MEdium Resolution Imaging Spectrometer (MERIS) measured remote sensing reflectance ($R_{rs}(\lambda)$) in the global oceans. A multi-pigment inversion model (MuPI) was used to semi-analytically retrieve Cpigs from $R_{rs}(\lambda)$. With a set of globally optimized parameters, the accuracy of the retrievals obtained with MuPI is quite promising. Compared with High-Performance Liquid Chromatography (HPLC) measurements near Bermuda, the concentrations of chlorophyll *a*, *b*, *c* ([Chl-a], [Chl-b], [Chl-c]), photoprotective carotenoids ([PPC]), and photosynthetic carotenoids ([PSC]) can be retrieved from MERIS data with a mean unbiased absolute percentage difference of 38%, 78%, 65%, 36%, and 47%, respectively. The advantage of the MuPI approach is the simultaneous retrievals of [Chl-a] and the accessory pigments [Chl-b], [Chl-c], [PPC], [PSC] from MERIS $R_{rs}(\lambda)$ based on a closure between the input and output $R_{rs}(\lambda)$ spectra. These results can greatly expand scientific studies of ocean biology and biogeochemistry of the global oceans that are not possible when the only available information is [Chl-a].

Keywords: phytoplankton pigments; ocean color; remote sensing; MERIS; global oceans

---

1. Introduction

Ocean color remote sensing has been focused on phytoplankton due to the important role that they play in the global biogeochemical cycles and ocean food webs [1,2]. With the development of remote sensing technology, a variety of approaches have been developed to remotely obtain information about phytoplankton, such as their chlorophyll concentration [3–5], functional groups, and size classes [6–11]. The most widely used satellite-based product of phytoplankton is chlorophyll *a* concentration ([Chl-a], $mg \cdot m^{-3}$) [3–5,12,13]. Satellite retrieved [Chl-a] has been utilized in estimation of phytoplankton biomass, primary production, and detection of harmful algal blooms [14,15]. However, many studies have indicated that [Chl-a] alone is not a good indicator of phytoplankton biomass or physiological status [16–20]. Some accessory pigments have been recognized as biomarkers for phytoplankton groups or species [18,21–24]. These accessory pigments provide better estimation of the biomass of particular phytoplankton groups or species, such as phycocyanin (PC) for cyanobacteria [25,26]. The variation in the accessory pigment composition has been widely used in estimating different phytoplankton functional groups [27–42], and physiological status of the phytoplankton [43,44].

In an effort to obtain pigment concentrations beyond [Chl-a] from remote sensing reflectance ($R_{rs}(\lambda)$, sr^{-1}), phycocyanin concentration, instead of [Chl-a], has been retrieved and used as a better index for cyanobacteria biomass and potential toxicity for cyanobacteria bloom waters [25,26]. To obtain phycocyanin and [Chl-a], empirical and semi-analytical methods have been proposed and good results obtained in their application to bloom detection and monitoring [13,25,26]. Either empirical or analytical, these methods are based upon relationships between bio-optical information and one or two pigment concentrations.

Empirical approaches have been the most widely used to obtain information for two or more pigments. Similar to the empirical relationships used by NASA for the estimation of [Chl-a] from $R_{rs}(\lambda)$ [4], Pan et al. [45] developed empirical relationships between High-Performance Liquid Chromatography (HPLC) measured pigment concentrations and $R_{rs}(\lambda)$ for coastal waters in the northeast coast of the United States. Moissan et al. [46] directly used satellite-derived [Chl-a] as model input to retrieve other pigments in the Atlantic Ocean off the east coast of United States with the underlining assumption that all accessory pigments co-vary with chlorophyll *a*.

Semi-analytical models, which are based on mechanistic relationships derived from radiative transfer, allow the estimation of inherent optical properties (absorption and backscattering) from $R_{rs}(\lambda)$ measured by any radiometer [47–62]. These semi-analytical models make it possible to obtain optical properties of the water components simultaneously from measured $R_{rs}(\lambda)$. Taking advantage of this property of semi-analytical algorithms, Wang et al. [63] incorporated the Gaussian decomposition method proposed by Hoepffner and Sathyendranath [64] into a semi-analytical model, termed as multi-pigment inversion model (MuPI), and demonstrated the potential of obtaining Gaussian peak heights representing the absorption coefficients from various pigments. Chase et al. [65] also adopted a similar scheme and applied it to hyperspectral in situ $R_{rs}(\lambda)$ measurements from the open ocean for the estimation of accessory pigments. However, as demonstrated in many studies [37,39,64–66], the assumption that each Gaussian amplitude represents the light absorption of one specific pigment is not always feasible. This is further shown in Chase et al. [65] where mixed results were obtained when a Gaussian peak height was linked with a single pigment. In this study, with in situ data from the global oceans, a thorough examination between the Gaussian peak heights and pigment concentrations was conducted.

The purpose of this study is twofold: (1) to evaluate the updated MuPI in retrieving concentrations of multiple phytoplankton pigments across the global ocean from MEdium Resolution Imaging Spectrometer (MERIS) measurements, and (2) to present the spatial distributions of accessory pigments across the global ocean that were previously not available. Model parameters were updated, and its performance was evaluated with different datasets covering a large dynamic range of ocean water conditions. The model was then applied to satellite remote sensing data from MERIS to obtain the global distribution and variation of different pigment concentrations. Finally, limitations and future developments of the MuPI model are discussed.

2. Data and Methods

2.1. Datasets and Study Sites

The datasets used in this study can be broadly classified into six different categories: (a) phytoplankton absorption coefficients ($a_{ph}(\lambda)$) from the global oceans; (b) simultaneously collected $a_{ph}(\lambda)$ and HPLC; (c) simultaneously measured $R_{rs}(\lambda)$, $a_{ph}(\lambda)$ and/or HPLC; (d) HPLC time series; (e) $R_{rs}(\lambda)$ from MERIS imagery; and (f) HydroLight simulated $R_{rs}(\lambda)$, $a_{ph}(\lambda)$, particulate backscattering coefficients ($b_{bp}(\lambda)$) and absorption coefficients of colored dissolved and detrital matters ($a_{dg}(\lambda)$) (International Ocean-Color Coordinating Group (IOCCG) dataset [50]). Table 1 provides an overview of the different datasets, time, size, variables, [Chl-a] range, and the main usage. The description of each dataset and the data measurements are included in the following paragraphs.

Table 1. Datasets, time, variables, size and their usages in this study. Cpigs: pigment concentrations, N: the number of samples, NA: not applicable.

Datasets/Cruises	Time	Size (N)	Measurements	Chl-a ($mg\cdot m^{-3}$)	Usage
SeaBASS	2001–2012	1619	$a_{ph}(\lambda)$	NA	Gaussian curves
	1991–2007	430	$a_{ph}(\lambda)$, HPLC	0.02–13.2	$a_{Gau}(\lambda)$ vs. Cpigs relationships
IOCCG	NA	500	$R_{rs}(\lambda)$, $a_{ph}(\lambda)$, $a_{dg}(\lambda)$, $b_{bp}(\lambda)$	0.03–30	$a_{Gau}(\lambda)$ and Cpigs validation
Tara Oceans expedition	2010–2012	23	$R_{rs}(\lambda)$, $a_{ph}(\lambda)$, HPLC	0.02–0.95	
VIIRS cal/val	2014–2015	21	$R_{rs}(\lambda)$, $a_{ph}(\lambda)$, HPLC	0.15–1.5	
BIOSOPE	2004	31	$R_{rs}(\lambda)$, $a_{ph}(\lambda)$, HPLC	0.00036–3.06	
BATS	2002–2012	148	HPLC	0.002–0.486	Cpigs variation
MERIS	2002–2012	148	$R_{rs}(\lambda)$	0.037–0.325	

A series of $a_{ph}(\lambda)$ spectra measured with the quantitative filter technique (QFT [67]) were used to find the globally optimized Gaussian parameters and the relationships among them. This dataset was obtained by searching the SeaWiFS Bio-optical Archive and Storage System (SeaBASS), which covers 1619 stations across the global oceans observed during 2001–2011. A set of 430 observations that had $a_{ph}(\lambda)$ and HPLC measurements coincidently observed were obtained from this dataset and were further randomly separated into two equal subsets (N = 215). The Subset_1 was used for regression analysis between Gaussian peak height ($a_{Gau}(\lambda)$) and pigment concentrations (Cpigs) to obtain the relationships among them; and the Subset_2 together with the $a_{ph}(\lambda)$ and HPLC from Tara Oceans, BIOSOPE, and VIIRS cal/val cruises were used to validate the relationships obtained from Subset_1.

The Tara Oceans expedition contains 23 match-ups of $R_{rs}(\lambda)$, $a_{ph}(\lambda)$ and HPLC around the global ocean. The BIOSOPE dataset includes 31 match-ups of $R_{rs}(\lambda)$, $a_{ph}(\lambda)$ and HPLC which were collected in the southeastern Pacific Ocean (obtained from: http://www.obs-vlfr.fr/proof/vt/op/ec/biosope/bio.htm). The VIIRS cal/val dataset is composed of 21 $R_{rs}(\lambda)$, $a_{ph}(\lambda)$ and HPLC measurements obtained from cruises covering the coastal oceans in North Atlantic Ocean off the United States east coast. The BATS (Bermuda Atlantic Time-Series Study) dataset is composed of HPLC time series from 2002 to 2012 and was obtained from the Bermuda Atlantic Time-Series Study (near Bermuda) (http://bats.bios.edu/bats_measurements.html). All of the in situ measurements used are from the surface, defined as a depth $\leq$5 m. The sampling locations of all these measurements are shown in Figure 1.

The IOCCG dataset was simulated using HydroLight software version 5.1 [68]. It was designed to cover the dynamic range observed across the global ocean but is biased to coastal waters (http://www.ioccg.org/data/synthetic.html). The $R_{rs}(\lambda)$, $a_{ph}(\lambda)$, $a_{dg}(\lambda)$ and $b_{bp}(\lambda)$ obtained from this dataset were used to validate the MuPI retrievals.

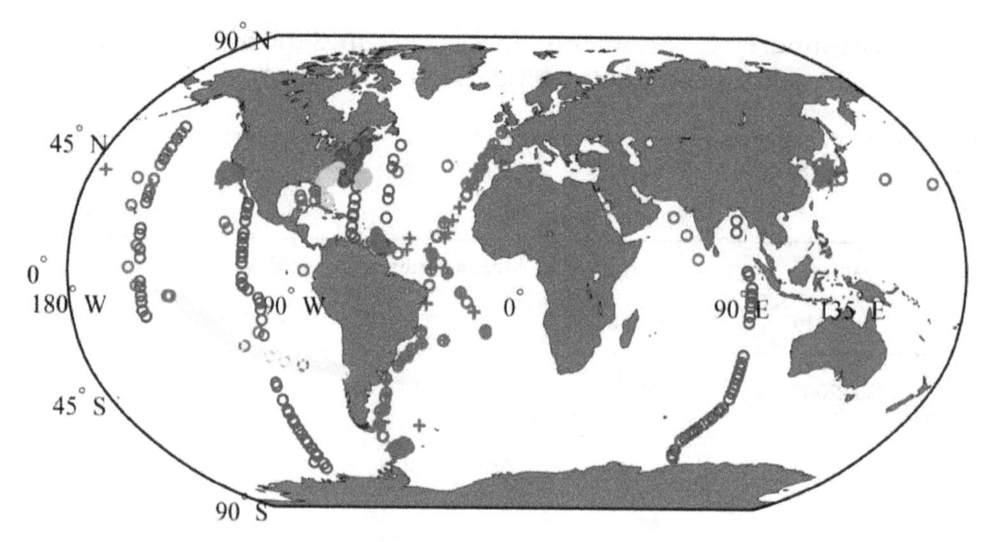

Figure 1. In situ data distribution, the (**o**) are the stations for quantitative filter technique (QFT) $a_{ph}(\lambda)$ from SeaBASS, (**o**) and (+) are the subset_1 and subset_2 stations of matchups of $a_{ph}(\lambda)$ and HPLC from SeaBASS, (**o**) is the HPLC location for BATS (Bermuda Atlantic Time-Series Study), (**o**) are the BIOSOPE $R_{rs}(\lambda)$, $a_{ph}(\lambda)$ and HPLC locations, (**o**) are locations of the $R_{rs}(\lambda)$, $a_{ph}(\lambda)$ and HPLC from VIIRS val/cal cruises in 2014 and 2015, and (**o**) are the locations of $R_{rs}(\lambda)$, $a_{ph}(\lambda)$ and HPLC from Tara Oceans expedition.

2.2. Radiometric Measurements

The in situ remote sensing reflectance, $R_{rs}(\lambda)$, was calculated based on the measurements of radiance and irradiance sampled with the Radiometer Incorporating the Skylight-Blocked Apparatus (RISBA) [69], Hyper Spectral Radiometer HyperPro free-fall profiler (Satlantic, Inc. Halifax, Nova Scotia, Canada), or above water radiometers [70]. The $R_{rs}(\lambda)$ spectra from 350–800 nm with different spectral increments were interpolated to 1 nm resolution.

Standard Level 3 MERIS $R_{rs}(\lambda)$ was acquired from the National Aeronautics and Space Administration (NASA) ocean color website (https://oceancolor.gsfc.nasa.gov). The HPLC data from BATS were matched to Level 3 MERIS 8-day products, at 4 km resolution and plus or minus one pixel (3 × 3 window). This criterion, although less restricting than NASA's 3-h window for data and algorithm validation [71], was adopted to maximize the number of match-ups.

The MuPI model was applied to MERIS $R_{rs}(\lambda)$ imagery from 2002–2012 to obtain the seasonal variation of chlorophyll *a*, *b*, *c*, photoprotective and photosynthetic carotenoids concentrations ([Chl-a], [Chl-b], [Chl-c], [PPC], [PSC]) near Bermuda. As examples, global maps of these five different pigments were also obtained from MERIS $R_{rs}(\lambda)$ imagery of 2007. The ratios of these concentrations to [Chl-a] are also presented to highlight their independence from chlorophyll *a*.

2.3. Absorption Measurements

Water samples for absorption and HPLC measurements were filtered onto a GF/F filter and stored in liquid nitrogen before laboratory measurements. Spectrophotometers were used to measure the absorbance and then to calculate the absorption coefficient of particles (a_p) and detrital matter (a_d). The phytoplankton absorption coefficient ($a_{ph}(\lambda)$) were obtained by subtracting a_d from a_p following NASA Ocean Optics Protocols, Revision 4, Volume IV protocol [72]. These $a_{ph}(\lambda)$ spectra generally cover 400–800 nm with spectral resolution around 3 nm. They were interpolated into 1 nm resolution for studies here.

Following Hoepffner and Sathyendranath [37,64] and Wang et al. [63,73], the phytoplankton absorption coefficients were decomposed into 13 Gaussian curves using the least square curve fitting technique provided in MATLAB and Statistics Toolbox (Release 2016a, MathWorks, Inc. Natick, MA, USA):

$$a_{\mathrm{ph}}(\lambda) = \sum_{i=1}^{n} a_{\mathrm{Gau}}(\lambda_i) \exp\left[-0.5\left(\frac{\lambda - \lambda_i}{\sigma_i}\right)^2\right] \quad (1)$$

where σ_i and $a_{\mathrm{Gau}}(\lambda_i)$ are the width and peak magnitude of the i-th Gaussian curve at peak center (λ_i) as shown in Table 2. The obtained $a_{\mathrm{Gau}}(\lambda)$ are used as ground truth to validate the inversion results from $R_{\mathrm{rs}}(\lambda)$.

Table 2. The 12 Gaussian curves corresponding to the phytoplankton pigment absorption coefficients, with Peak_loc as the center location of each pigment absorption peak and width as the full width at half maximum (FWHM). The relationships indicate the power-law relationships used to estimate the Gaussian peak amplitudes from the two independent variables: x_1: $a_{\mathrm{Gau}}(434)$ and x_2: $a_{\mathrm{Gau}}(492)$. Chl-a: chlorophyll *a*, Chl-b: chlorophyll *b*, Chl-c: chlorophyll *c*, PPC: photo-protective carotenoids, PSC: photosynthetic carotenoids, PE: phycoerythrin and PC: phycocyanin.

Peak	Pigments	Peak_loc (nm)	Width(FWHM) (nm)	Relationships	R^2
1	Chl-a	406	16	$1.13{x_1}^{1.01}$	0.98
2	Chl-a	434	12	x_1	–
3	Chl-c	453	12	$0.60{x_1}^{0.95}$	0.99
4	Chl-b	470	13	$0.51{x_1}^{0.97}$	0.98
5	PPC	492	16	x_2	–
6	PSC	523	14	$0.87{x_2}^{1.17}$	0.99
7	PE	550	14	$0.79{x_2}^{1.27}$	0.96
8	Chl-c	584	16	$0.40{x_2}^{1.17}$	0.96
9	PC	617	13	$0.34{x_1}^{1.14}$	0.93
10	Chl-c	638	11	$0.47{x_2}^{1.19}$	0.96
11	Chl-b	660	11	$0.30{x_2}^{1.11}$	0.94
12	Chl-a	675	10	$0.86{x_1}^{1.11}$	0.98

2.4. Pigment Concentrations

All the HPLC analyses were carried out according to the method following or adapted from Van Heukelem and Thomas [23]. The concentrations of chlorophyll *a*, *b*, *c*, photo-protective carotenoids (PPC) and photosynthetic carotenoids (PSC) were estimated from HPLC measurements as:

(A) Total chlorophyll *a* (Chl-a) = chlorophyll *a* + divinyl chlorophyll *a* + chlorophyllide *a*;
(B) Total chlorophyll *b* (Chl-b) = chlorophyll *b* + divinyl chlorophyll *b*;
(C) chlorophyll *c* (Chl-c) = chlorophyll *c*1 + chlorophyll *c*2;
(D) PPC = α-carotene + β-carotene + zeaxanthin + alloxanthin + diadinoxanthin;
(E) PSC = 19′-hexanoyloxyfucoxanthin + fucoxanthin + 19′-butanoyloxyfucoxanthin + peridinin.

MERIS [Chl-a] was estimated from the Level 3 $R_{\mathrm{rs}}(\lambda)$ following the standard algorithm OC4E provided by NASA [4]. Details about this algorithm can be found on the following webpage: https://oceancolor.gsfc.nasa.gov/atbd/chlor_a/.

2.5. Pigment Retrieval from $R_{rs}(\lambda)$

2.5.1. $a_{\mathrm{Gau}}(\lambda)$ from $R_{\mathrm{rs}}(\lambda)$

The multi-pigment inversion model (MuPI) was used to retrieve $a_{\mathrm{Gau}}(\lambda)$ from $R_{\mathrm{rs}}(\lambda)$. Wang et al. [63,73] developed this semi-analytical inversion model (MuPI) to retrieve $a_{\mathrm{Gau}}(\lambda)$ from hyper- or multi-spectral $R_{\mathrm{rs}}(\lambda)$. A brief description of MuPI is presented here. The functional relationship between $R_{rs}(\lambda)$ and inherent optical properties (IOPs) is taken from Gordon et al. [47] and Lee et al. [49]:

$$R_{rs}(\lambda) = 0.52\sum_{i=1}^{2} g_i \left[\frac{b_b(\lambda)}{a(\lambda)+b_b(\lambda)}\right]^i / \left\{1 - 1.7\sum_{i=1}^{2} g_i \left[\frac{b_b(\lambda)}{a(\lambda)+b_b(\lambda)}\right]^i\right\} \tag{2}$$

where g_1 (sr^{-1}) and g_2 (sr^{-1}) are fixed to 0.089 and 0.125 sr^{-1}. The IOP spectra, $a(\lambda)$ and $b_b(\lambda)$, are partitioned into relevant components

$$b_b(\lambda) = b_{bw}(\lambda) + b_{bp}(\lambda) \tag{3}$$

$$a(\lambda) = a_w(\lambda) + a_{ph}(\lambda) + a_{dg}(\lambda) \tag{4}$$

with $b_{bw}(\lambda)$ for seawater backscattering coefficient [74] and $a_w(\lambda)$ for seawater absorption coefficient [75,76]. Phytoplankton absorption coefficient ($a_{ph}(\lambda)$) is modeled following Equation (1); $b_{bp}(\lambda)$, particulate backscattering coefficient, is modeled following Equations (5) and (6) [49]; and the combined dissolved and detrital particulate absorption coefficient $a_{dg}(\lambda)$ is modeled using Equation (7) [52,77,78].

$$b_{bp} = b_{bp}(\lambda_0)\left(\frac{\lambda_0}{\lambda}\right)^{\eta} \tag{5}$$

$$\eta = 2\left(1 - 1.2\exp\left(-0.9\frac{R_{rs}(440)}{R_{rs}(550)}\right)\right) \tag{6}$$

$$a_{dg}(\lambda) = a_{dg}(\lambda_0)\exp(-S(\lambda - \lambda_0)) \tag{7}$$

where λ_0 is a reference wavelength (nearest to 440 nm), S is the spectral decay constant for absorption of detrital and dissolved materials and kept as an unknown within 0.007 to 0.02 nm^{-1} [52,77,78]. η is the power-law exponent for the particulate backscattering coefficient calculated from the $R_{rs}(440)$ to $R_{rs}(550)$ ratio following Lee et al. [49].

In the determination of Gaussian parameters (σ_i and λ_i) for $a_{ph}(\lambda)$ in the global scale, we also tested various combinations of parameters using data published in the literature [63–65]. The existing parameters were not successful at obtaining satisfactory results for every data range due to various reasons, including the fact that the initial datasets used to obtain the parameters had a small dynamic range unable to cover varied conditions such as coastal regions and non-bloom natural oceanic waters. Thus, a refinement of the parameters σ_i and λ_i was conducted using the $a_{ph}(\lambda)$ dataset obtained from SeaBASS to improve the overall performance of the MuPI model for global oceans. The non-linear least square fitting procedure in MATLAB was used to solve Equation (1). A set of refined Gaussian parameters for σ_i and λ_i were obtained and are presented in Table 2. For oceanic waters, as the absorption coefficient from water molecules contributes >80% of the total absorption coefficient for wavelengths >550 nm, it is difficult to obtain accurate $a_{Gau}(\lambda)$ by directly inverting $R_{rs}(\lambda)$ in the longer wavelengths. On the other hand, since the Gaussian peaks at 434 and 492 nm cover the main absorption features of the different pigments, the two $a_{Gau}(\lambda)$ at 434 and 492 nm were chosen as the independent variables in this effort.

Following Wang et al. [73], the implementation of this model used two Gaussian peak heights [$a_{Gau}(\lambda_1)$ and $a_{Gau}(\lambda_2)$] to reconstruct $a_{ph}(\lambda)$, in which empirical relationships as shown in Table 2 were used. These relationships between $a_{Gau}(\lambda)$ were obtained by regression analysis with the purpose of reducing the unknowns in the $R_{rs}(\lambda)$ inversion procedure [63,73]. With this design, there will be five unknowns [$a_{\mathrm{Gau}}(\lambda_1)$, $a_{\mathrm{Gau}}(\lambda_2)$, $b_{\mathrm{bp}}(\lambda_0)$, $a_{\mathrm{dg}}(\lambda_0)$ and S] to be retrieved from a $R_{\mathrm{rs}}(\lambda)$ spectrum, which is obtained by a minimization of the cost function (Equation (8)):

$$\delta = \frac{\sqrt{\frac{1}{N_\lambda}\sum_{i=1}^{N_\lambda}\left(\hat{R}_{\mathrm{rs}}(\lambda_i) - R_{\mathrm{rs}}(\lambda_i)\right)^2}}{\frac{1}{N_\lambda}\sum_{i=1}^{N_\lambda} R_{\mathrm{rs}}(\lambda_i)} \tag{8}$$

with N_λ as the wavelength number, $R_{\mathrm{rs}}(\lambda)$ as the measured, and $\hat{R}_{\mathrm{rs}}(\lambda)$ the modeled spectrum, respectively. Basically, δ value provides a measure of the relative difference between the input and output R_{rs} spectra. The generalized reduced gradient (GRG) nonlinear optimization procedure [79] was used to solve Equation (8).

The statistical indices used to estimate the agreement between the two values (R and R) was the unbiased absolute percentage difference (UAPD), defined as Equation (9) and root mean square error (RMSE, Equation (10)) with N as the number of samples.

$$\mathrm{UAPD} = \frac{|\hat{R} - R|}{0.5(\hat{R} + R)} \times 100\% \tag{9}$$

$$\mathrm{RMSE} = \sqrt{\frac{1}{N}(\hat{R} - R)^2} \tag{10}$$

2.5.2. $a_{\mathrm{Gau}}(\lambda)$ Versus Cpigs

Hoepffner and Sathyendranath [64] indicated that each Gaussian curve represents the absorption contributed by one or multiple pigments. However, attempts to obtain the concentration of a specific pigment from a single Gaussian curve is not always successful [39,65,66]. For a better understanding of the Gaussian curves and their relationships with Cpigs, a series of regression analyses were applied to relate $a_{\mathrm{Gau}}(\lambda)$ with Cpigs for data from SeaBASS. The t-statistics and p-value were calculated to test the significance of the parameters. Using $p < 0.05$ as the criteria, the significant contributors to each Gaussian peak and the corresponding R^2 of these parameters were obtained (see Table 3). The possible existence of other pigments that are not detectable with current HPLC techniques, such as phycoerythrin (PE) and phycocyanin (PC), likely explains the relatively lower R^2 values for Peaks 7, 8, 9 and 10.

After a series of multivariable regression analyses, it was found that Cpigs could be estimated from $a_{\mathrm{Gau}}(\lambda)$ following the function:

$$\log_{10}(\mathrm{Cpigs}) = a_0 + \sum_{i=1}^{n} a_i \log_{10}(a_{\mathrm{Gau}}(\lambda_i)) \tag{11}$$

The corresponding $a_{\mathrm{Gau}}(\lambda)$, parameters, and the R^2 value are shown in Table 4. Further, it was found that the estimated Cpigs agree with the measured values very well throughout the concentration range when the relationships were applied to the validation dataset (with data points scattered closely to the 1:1 line; Figure 2).

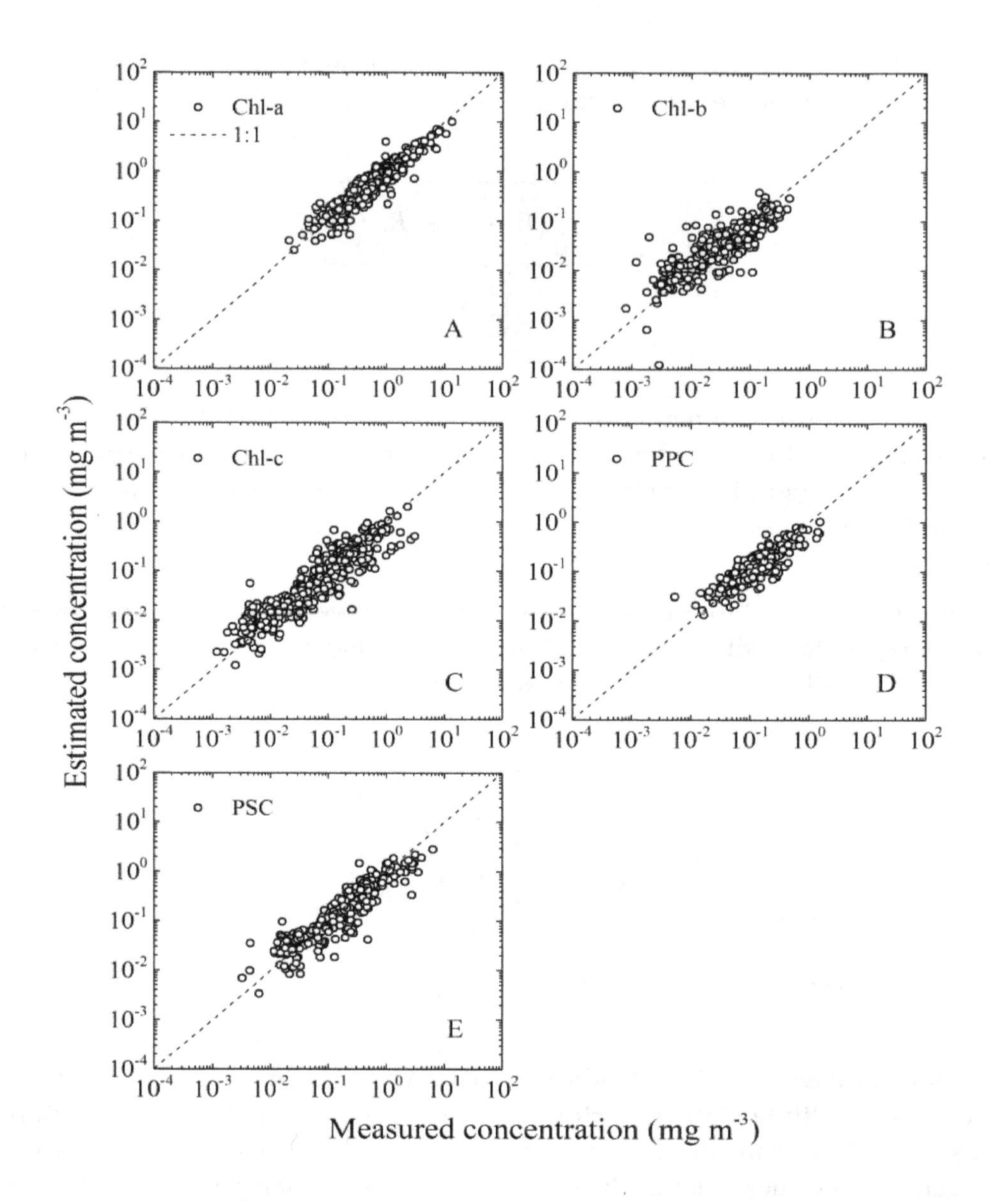

Figure 2. $a_{Gau}(\lambda)$ estimated pigment concentrations versus the measured concentrations from HPLC using the $a_{ph}(\lambda)$ and HPLC from SeaBASS: **A**: chlorophyll *a* (Chl-a), **B**: chlorophyll b (Chl-b), **C**: chlorophyll c (Chl-c), **D**: photoprotective carotenoids (PPC), and **E**: photosynthetic carotenoids (PSC).

Table 3. The *p*-value and R^2 from the t-statistics, with $a_{Gau}(\lambda)$ at 12 different wavelengths: 406, 434, 453, 470, 492, 523, 550, 584, 617, 638, 660, and 675 nm, and Chl-a: chlorophyll *a*, Chl-b: chlorophyll *b*, Chl-c: chlorophyll *c*, PPC: photo-protective carotenoids, PSC: photosynthetic carotenoids.

p-Value	Peak 406	Peak 434	Peak 453	Peak 470	Peak 492	Peak 523	Peak 550	Peak 584	Peak 617	Peak 638	Peak 660	Peak 675
Chl-a	0.01	0.00	0.00			0.01	0.03	0.00	0.00	0.00	0.00	0.00
Chl-b		0.03		0.00								
Chl-c							0.04	0.03			0.00	0.04
PPC	0.00		0.00	0.00	0.00							
PSC				0.00	0.02	0.01						
R^2	0.80	0.87	0.83	0.87	0.83	0.78	0.64	0.68	0.76	0.73	0.81	0.91

Table 4. Parameters for estimation of pigment concentrations: the pigment-specific $a_{Gau}(\lambda)$, coefficients and R^2. Chl-a: chlorophyll *a*, Chl-b: chlorophyll *b*, Chl-c: chlorophyll *c*, PPC: photo-protective carotenoids, PSC: photosynthetic carotenoids.

Pigments	$a_{Gau}(\lambda)$	Parameters ($a_0, a_1, \ldots, a_i$)	R^2
Chl-a	675	1.804, 0.975	0.89
Chl-b	434, 453, 470	−0.066, 2.470, −3.073, 1.379	0.72
Chl-c	470, 492, 523, 675	1.334, 2.022, −3.125, 0.745, 1.119	0.83
PPC	453, 470	0.734, 1.311, −0.416	0.76
PSC	470, 492, 523	1.67, 3.034, −2.670, 0.725	0.84

3. Results

3.1. Retrievals from $R_{rs}(\lambda)$

3.1.1. $a_{Gau}(\lambda)$ Validation

The MuPI model was first tested with datasets that contained different levels of chlorophyll *a* concentration from the IOCCG synthesized data and different cruises in the global ocean. The main purpose of this test was to evaluate, and validate, the implementation of the MuPI approach for a wide range of environments. A mean UAPD of 36% was obtained between $a_{Gau}(\lambda)$ from $R_{rs}(\lambda)$ inversion and $a_{Gau}(\lambda)$ from water samples throughout the data range for different datasets (see Figure 3). The differences in statistical results, noted in Table 5 for $a_{Gau}(\lambda)$ retrieval from different datasets, are strongly influenced by their different dynamic ranges and characteristics as implied in Section 2.1.

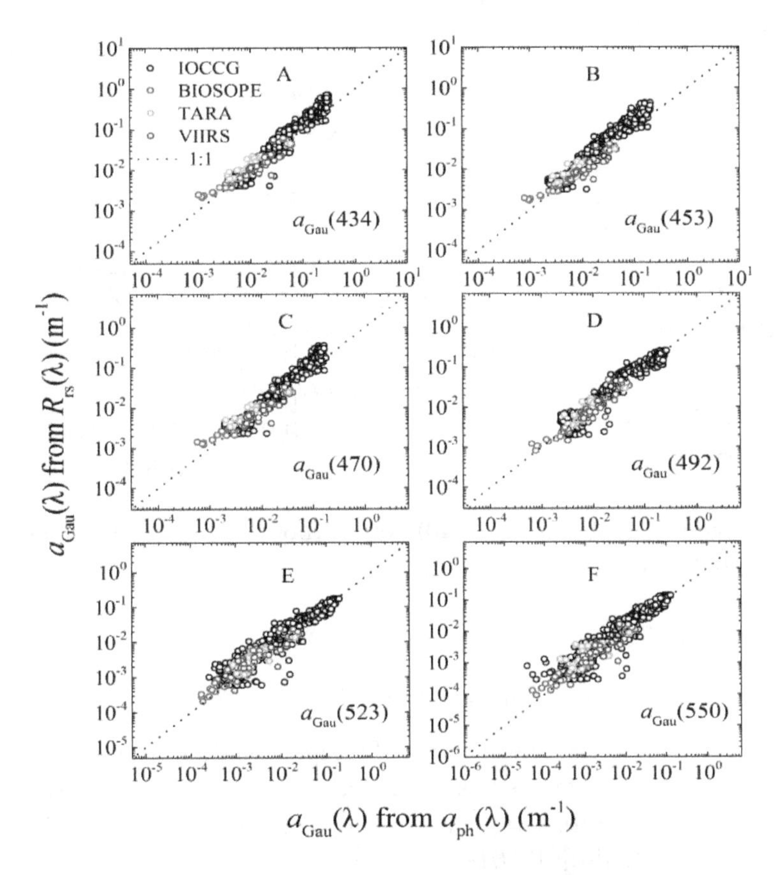

Figure 3. $a_{Gau}(\lambda)$ heights retrieved from $R_{rs}(\lambda)$ at MEdium Resolution Imaging Spectrometer (MERIS) bands versus measured Gaussian peak (decomposed from $a_{ph}(\lambda)$) for the data from different datasets (shown in different colors).

Table 5. Mean (Mea.) and median (Med.) of the unbiased percentage errors for $a_{Gau}(\lambda)$ retrieval from $R_{rs}(\lambda)$ for different datasets.

Peak Center	IOCCG		Tara Oceans		BIOSOPE		VIIRS Cruises	
	Mea.	**Med.**	**Mea.**	**Med.**	**Mea.**	**Med.**	**Mea.**	**Med.**
406	45	45	34	27	34	28	28	20
434	37	36	34	28	26	25	28	13
453	47	49	28	24	23	18	27	15
470	35	34	29	25	30	27	31	18
492	34	31	26	21	22	18	29	18
523	44	34	38	28	34	29	48	44
550	45	35	53	41	37	34	41	35
584	55	48	48	37	38	36	53	57
617	51	45	47	38	36	29	37	40
638	54	42	66	68	41	35	41	35
660	52	48	35	23	32	29	43	34
675	46	40	30	26	60	56	32	21

3.1.2. $b_{bp}(\lambda)$ and $a_{dg}(\lambda)$ Validation

The backscattering coefficients of particles and absorption coefficients of detrital and dissolved materials retrieved from $R_{rs}(\lambda)$ by MuPI have also been validated with the IOCCG dataset (with values at 440 nm as examples). The $b_{bp}(440)$ showed very high accuracy with mean UAPD of 4.8%, and all samples showed no bias in the entire data range as presented in Figure 4A. The estimated $a_{dg}(440)$ also showed very good agreement with the simulated values, with the mean UAPD of 21.3%. No inter-comparisons were made for the products $a_{dg}(\lambda)$ and $b_{bp}(\lambda)$ for other datasets because of lacking corresponding measured data.

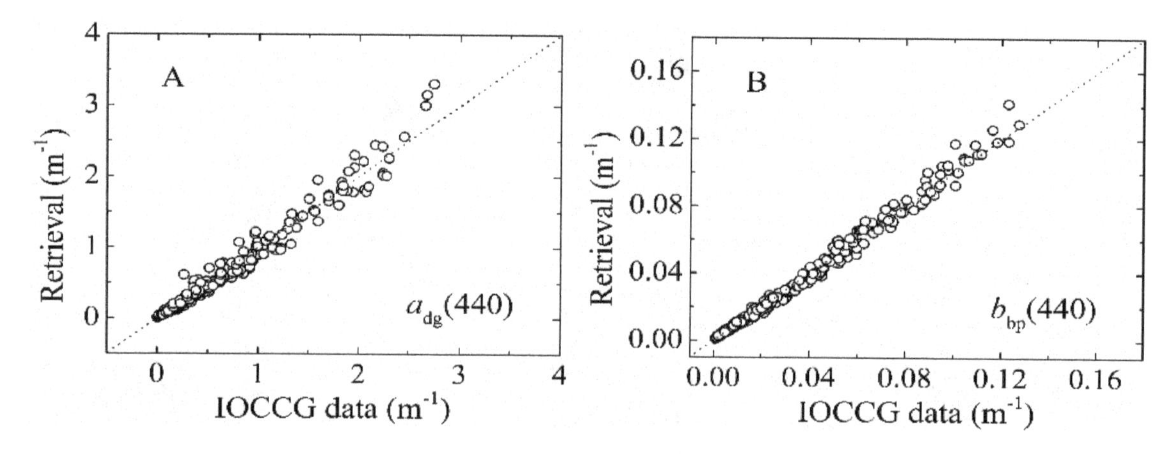

Figure 4. MuPI retrieved $a_{dg}(440)$ and $b_{bp}(440)$ versus those from the International Ocean-Colour Coordinating Group (IOCCG) dataset.

3.2. *Cpigs from Satellite Remote Sensing*

3.2.1. Cpigs Validation and Their Seasonal Variation

The ability of MuPI to capture the magnitudes of Cpigs and their seasonal variability from satellite $R_{rs}(\lambda)$ was then validated using a time series of HPLC measurements at BATS from the years 2002 to 2012. During this period of time, Cpigs varied in these ranges: [Chl-a]: 0.016–0.486 $mg{\cdot}m^{-3}$, [Chl-b]: 0.001–0.108 $mg{\cdot}m^{-3}$, [Chl-c]: 0.001–0.206 $mg{\cdot}m^{-3}$, [PPC]: 0.004–0.147 $mg{\cdot}m^{-3}$, and [PSC]: 0.003–0.106 $mg{\cdot}m^{-3}$.

To obtain Cpigs from satellite $R_{rs}(\lambda)$, $a_{Gau}(\lambda)$ were inverted first from $R_{rs}(\lambda)$ using MuPI, then Equation (11) was applied to convert the retrieved $a_{Gau}(\lambda)$ to Cpigs. As shown in Figure 5, it is found that there are good matches in the magnitudes and the seasonal cycles for the five pigments, with mean UAPD values as 38%, 78%, 65%, 36%, and 47% (and the medians are 34%, 79%, 64%, 30% and 55%) for [Chl-a], [Chl-b], [Chl-c], [PPC] and [PSC], respectively. The [Chl-a] accuracy is comparable with the NASA adopted standard [Chl-a] algorithms, for which the color index (CI) algorithm [5] showed a mean UAPD of 38.6% and OC4E of 46.7%. There are many reasons for the relatively low accuracy in the retrieval of [Chl-b] and [Chl-c], which include very low concentrations (e.g., in situ [Chl-b] and [Chl-c] were close to the HPLC detection minimum), as well as uncertainties in satellite measured $R_{rs}(\lambda)$, and the derived $a_{Gau}(\lambda)$ from $R_{rs}(\lambda)$.

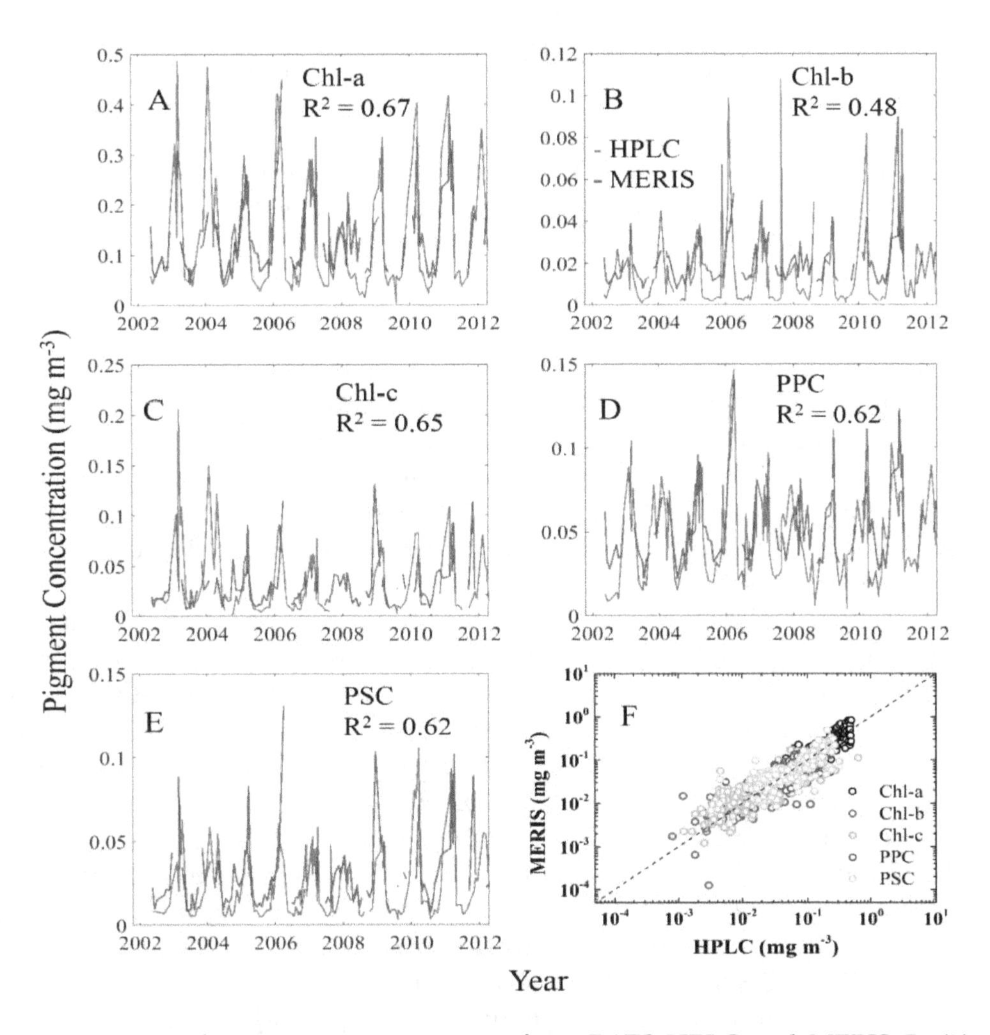

Figure 5. Time series of pigment concentrations from BATS HPLC and MERIS $R_{rs}(\lambda)$, and the determination coefficients (R^2). **A**: Chl-a: chlorophyll *a*, **B**: Chl-b: chlorophyll *b*, **C**: Chl-c: chlorophyll *c*, **D**: PPC: photoprotective carotenoids, **E**: PSC: photosynthetic carotenoids, **F**: the scatterplot of estimated versus in situ pigment concentrations.

Beyond the seasonal cycles in Cpig magnitudes, variation in pigment composition over time implied in the change of pigment ratios was also noticed in Figure 6A. On further examination, we found the ratios derived from MERIS $R_{rs}(\lambda)$ using MuPI can pick up the variation in the [Chl-b], [Chl-c], [PPC], [PSC] to [Chl-a] ratios observed from HPLC measurements very well with the mean UAPD of 50%, 47%, 25%, 37%, and median of 38%, 39%, 19% and 29% respectively (Figure 6B). Since these phytoplankton pigment ratios do not co-vary with the [Chl-a] product and cannot be empirically estimated from [Chl-a] alone, the Gaussian peaks and multiple pigments retrieved here provide a valuable glimpse into potential applications of these ratios in ocean changes that can be studied at large spatial and high temporal scales.

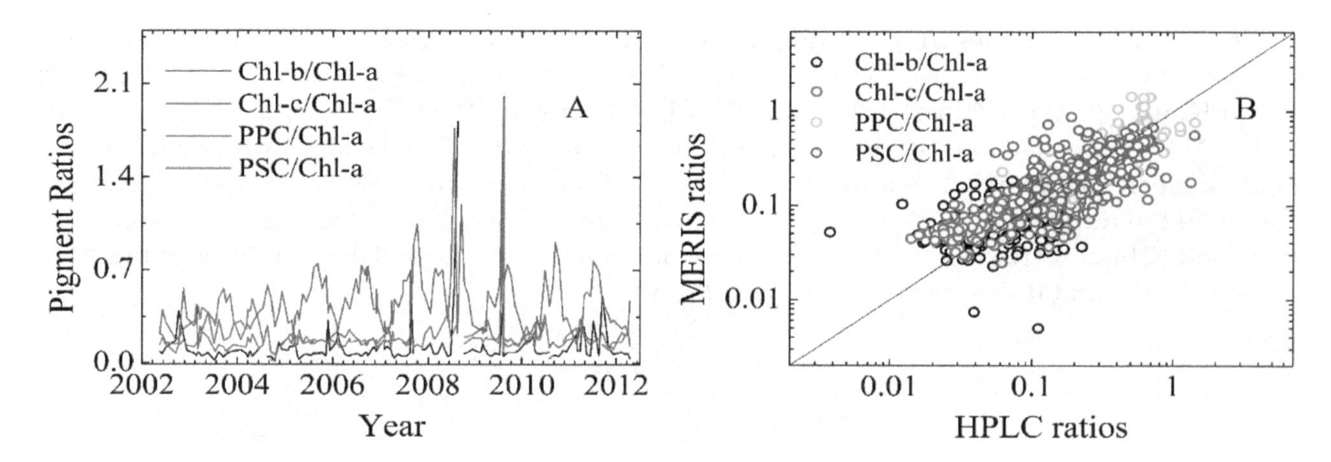

Figure 6. A: Time series of phytoplankton pigment to chlorophyll *a* (Chl-a) ratios at BATS from HPLC measurements. **B**: Chl-b, Chl-c, PPC and PSC to Chl-a ratios from HPLC versus from MERIS measured $R_{rs}(\lambda)$ using MuPI.

3.2.2. Global Distribution of Cpigs

As an example, MuPI was applied to MERIS L3 global annual (2007) average $R_{rs}(\lambda)$ to obtain the global distribution of the five different pigment concentrations (Figures 7 and 8) and their ratios to [Chl-a] (Figure 9). The global patterns of [Chl-a], [Chl-b], [Chl-c], [PPC], and [PSC] mimic the major gyre systems and other large-scale circulation features of the world ocean. High values of Cpigs are found within regions of persistent large-scale upwelling (e.g., subarctic gyres, equatorial divergences, eastern boundary currents, etc.), while low values are observed where large-scale downwelling is observed (e.g., subtropical gyres).

The basin-scale [Chl-a] distribution follows the pattern reported in the literature [4,5]. In comparing the global map of [Chl-a] derived in this study with the standard product from NASA ocean color website (Figure 7), some differences were noticed. In the previous section, when using HPLC data from BATS for validation, the retrieved [Chl-a] showed higher accuracy via MuPI than that from the standard OC4E algorithm. To better understand the differences in [Chl-a] distribution at the global scale, a validation dataset from SeaBASS was used. This dataset was obtained by searching match-ups of in situ measured [Chl-a] with those from MERIS $R_{rs}(\lambda)$. A dataset containing 608 pairs of [Chl-a] and MERIS $R_{rs}(\lambda)$ were obtained in which [Chl-a] concentrations ranged from ~0.017 to ~40.3 $mg{\cdot}m^{-3}$ (locations shown in Figure 7A). In comparing the estimated [Chl-a] from two different methods with in situ measurements (Figure 7C), MuPI and OC4E showed comparable results with mean UAPD of 48.8% and RMSE of 4.51 $mg{\cdot}m^{-3}$ for OC4E and mean UAPD of 49.3% and RMSE of 4.05 $mg{\cdot}m^{-3}$ for MuPI. As shown in Figure 7C, the [Chl-a] estimated from OC4E is biased slightly high (~10%) in the range of 1–10 $mg{\cdot}m^{-3}$ compared with results from MuPI. This range of [Chl-a] (1–10 $mg{\cdot}m^{-3}$) is mainly from coastal and inland waters for which the influences from colored dissolved and detrital matter result in lower accuracy in band-ratio estimated [Chl-a] [80]. For several samples, the [Chl-a] values from MuPI are biased low (~70%) as shown in Figure 7C. There are two possible reasons for this: (1) bad input MERIS $R_{rs}(\lambda)$, not only the values but also the spectral shape, especially at the blue bands that are susceptible to poor atmospheric correction, where negative values are often observed for coastal waters [81]; (2) the limitation of the algorithm as a result of the empirical parameters used to reduce the unknowns in MuPI, and the low contribution of $a_{ph}(\lambda)$ to the total absorption coefficients in the 400–750 nm range, which will be further discussed in Section 4.

The pigment ratios to [Chl-a] showed complicated patterns (Figure 9). In high [Chl-a] regions, [Chl-b]/[Chl-a] and [PPC]/[Chl-a] are low. In some low [Chl-a] regions, the ratios of [Chl-c]/[Chl-a] and [PSC]/[Chl-a] are relatively high, such as in the East Pacific Ocean. These results agree with previous findings about the global distribution of phytoplankton groups and pigment ratios, as lower [PPC]/[Chl-a] ratios correspond to high [Chl-a] and larger particle size [9,17,82–84]. As recorded in the literature, in different regions of the global ocean, the quality and quantity of light and nutrient, as well as temperature, is highly variable [85]. These highly unpredictable and rapid changes of the environment usually result in phytoplankton taxonomic composition variation (long-term adaptation) or physiological acclimation (short-term acclimation) [24]. The variation in pigment ratios obtained from MuPI can directly reflect these changes in phytoplankton and provide valuable information for phytoplankton studies in large spatial and high temporal scales.

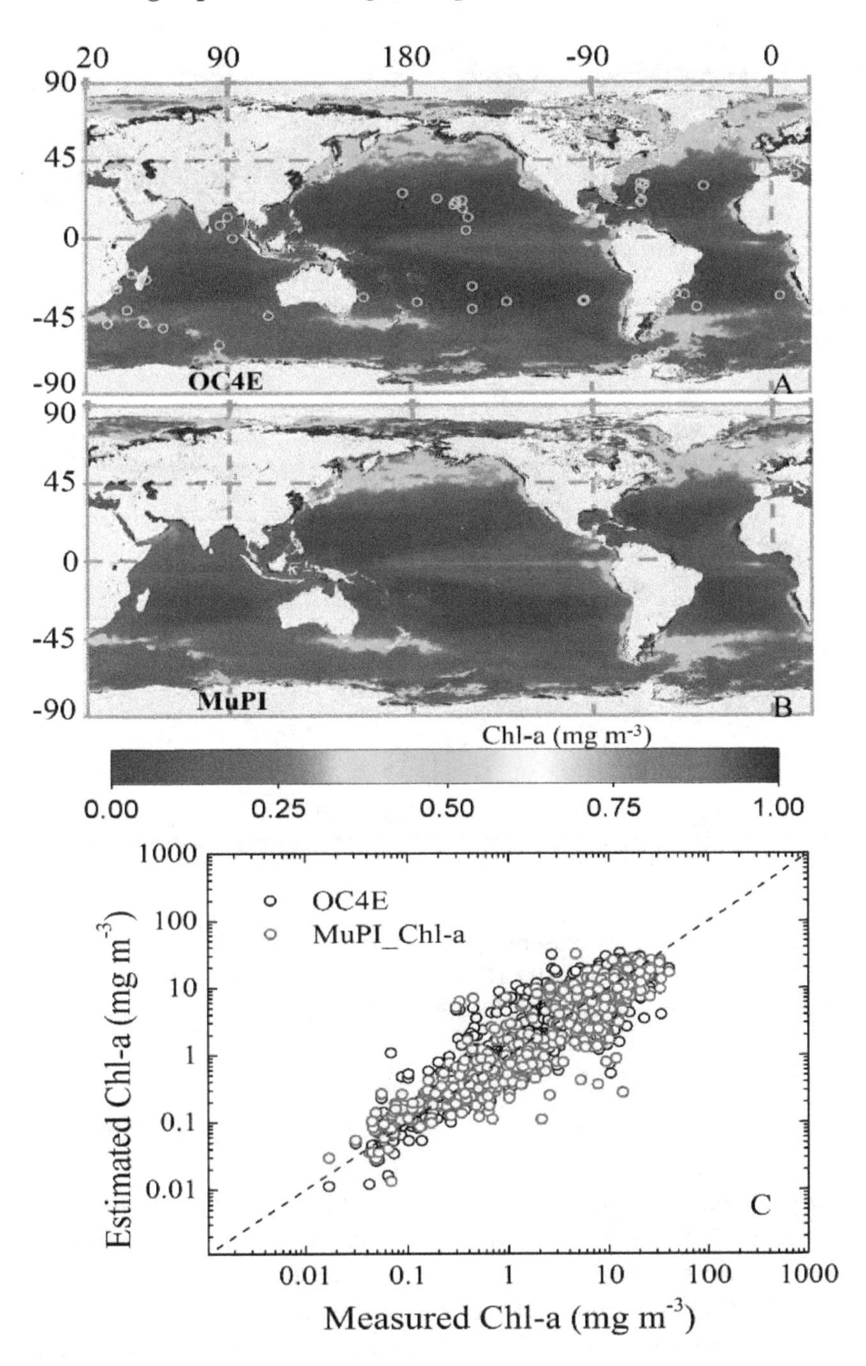

Figure 7. Global distributions of chlorophyll *a* concentration estimated from 2007 MERIS L3 $R_{rs}(\lambda)$ imagery using NASA standard algorithm OC4E (**A**) and MuPI model (**B**). The locations (**o**) of in situ Chl-a and MERIS $R_{rs}(\lambda)$ matchups for further comparison of OC4E and MuPI were plotted on the OC4E Chl-a map. **C**: Chlorophyll *a* concentration (Chl-a) from in situ measurements and from those estimated from matchups of MERIS $R_{rs}(\lambda)$ using OC4E and MuPI algorithms with mean UAPD of 48.8% and RMSE of 4.51 $mg \cdot m^{-3}$ for OC4E and mean UAPD of 49.3% and RMSE of 4.05 $mg \cdot m^{-3}$ for MuPI.

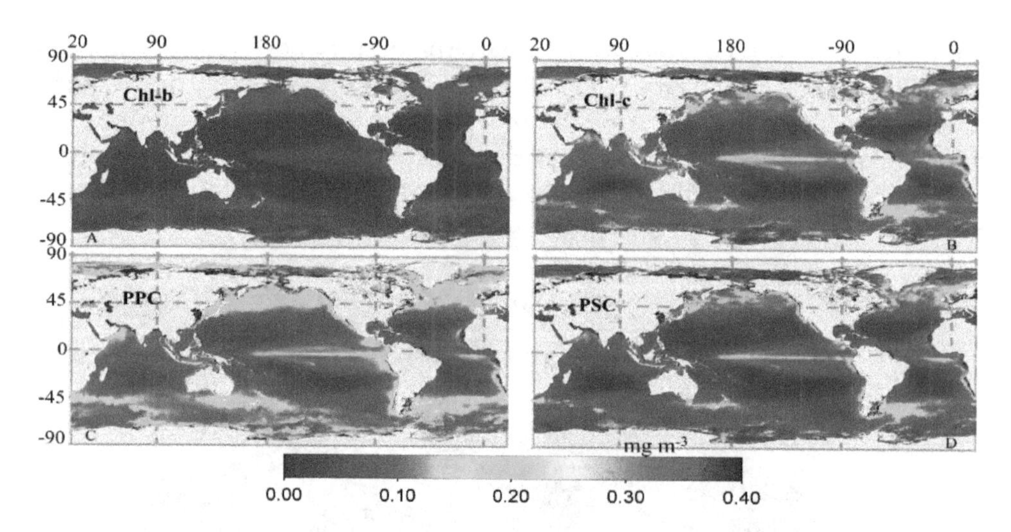

Figure 8. Global distributions of chlorophyll *b* (Chl-a), chlorophyll *c* (Chl-c), photoprotective carotenoids (PPC) and photosynthetic carotenoids (PSC) from 2007 L3 annual MERIS $R_{rs}(\lambda)$ imagery.

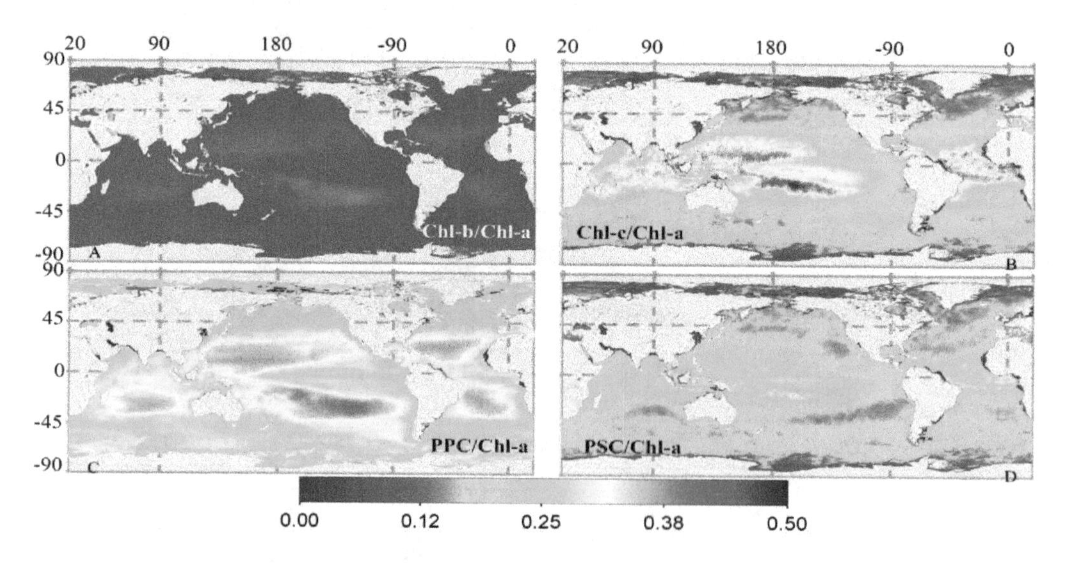

Figure 9. Global distributions of the accessory pigment to chlorophyll *a* ratios: ratio of concentrations of chlorophyll *b* (Chl-b/Chl-a), chlorophyll *c* (Chl-c/Chl-a), photoprotective carotenoids (PPC/Chl-a), and photosynthetic carotenoids to chlorophyll *a* (PSC/Chl-a).

4. Discussion

Based on the initial model of Wang et al. [63] that was developed for bloom waters, we have demonstrated that using a set of refined parameters for the Gaussian curves for global waters (Table 2), the MuPI model demonstrates consistent performance in $a_{Gau}(\lambda)$ retrievals on a global scale, as shown in Figures 3 and 4 and Table 5. Compared with HPLC measurements, the estimates of [Chl-a], [Chl-b], [Chl-c], [PPC] and [PSC] from satellite $R_{rs}(\lambda)$ showed reasonable accuracies, with the mean UAPD of 38%, 78%, 65%, 36%, and 47% respectively (Figure 5). Further, the distribution of these pigments and their ratios to [Chl-a], were obtained from MERIS measurements on the global scale (Figures 8 and 9).

The phytoplankton pigment ratios are critical indicators of the variation in phytoplankton groups and species due to their physiological adaptation to changes in nutrients, temperature, and light availability over time and space. The HPLC measured pigment concentrations and ratios have been widely used for determining the phytoplankton taxonomic composition and estimating the biomass of different groups, such as in CHEMTAX [27] and PFT analyses [10]. However, lacking effective methods, the estimation of phytoplankton pigments from satellite remote sensing has been limited to only [Chl-a] in the past decades [3–5], and the efforts made to obtain the accessory pigments have been more or less based on empirical relationships with Chl-a [45,46]. Thus, these products provided limited ability to capture the variation in the physiological status of phytoplankton. However, these unknown physiological variations in phytoplankton pigments have been one of the main uncertainties in traditional phytoplankton remote sensing models that use [Chl-a] as input to represent the phytoplankton biomass, such as the traditional [Chl-a] based primary productivity [14,20]. MuPI, as shown in this study, fills in this gap by obtaining not only the accessory pigment concentrations but also their ratios to [Chl-a] on the global scale, and reasonable accuracy has been obtained in validation with in situ data.

[Chl-a] is by far the easiest quantity to validate as it is routinely measured. An independent evaluation of the model has been conducted with the MERIS match-up dataset (Figure 7). This dataset contains nearly simultaneous in situ [Chl-a] measurements and MERIS $R_{rs}(\lambda)$ at coincident locations. To evaluate other pigments, an independent evaluation was conducted with MERIS $R_{rs}(\lambda)$ and in situ HPLC match-ups from 2002 to 2012 near Bermuda (Figure 5). The match-up dataset contains data from different seasons over a decadal scale. Use of the model with the match-up dataset from BATS confirms good overall behavior of the MuPI model for pigment concentration and pigment ratio retrievals, demonstrating the ability of the model to obtain accurate information from satellite ocean color imagery. However, because of the limitation of in situ pigment concentrations, the potential of obtaining PE and PC concentrations from satellite remote sensing data was not addressed in this study.

The main difficulty in making the model more applicable with any waterbody comes from the parameterization of the Gaussian curves, particularly the empirical relationships among $a_{Gau}(\lambda)$. Although it is reasonably straightforward to optimize the parameters with each dataset to obtain better retrievals, it would be extremely difficult (if not impossible) to do so when in situ measurements from the target location are hard to obtain. Instead, a set of globally optimized parameters were obtained using a dataset that covers a large dynamic range of the global ocean. Another challenge for $a_{Gau}(\lambda)$ retrieval in the open ocean lies in the low contribution of pigment absorption to the total absorption coefficient around 550–650 nm. Thus, it is difficult to directly invert $R_{rs}(\lambda)$ to obtain pigment absorption coefficients at the longer wavelengths (>550 nm).

This version of the MuPI model should be considered interim because the model could be further updated when more global data become available. In its present form, the model is optimized to work with $R_{rs}(\lambda)$ data from the first nine MERIS bands. As a first step, several components of the Gaussian model were deliberately formulated by use of empirical relationships to limit the number of unknowns to be solved via the spectral optimization procedure. This is particularly true for estimation of 13 $a_{Gau}(\lambda)$ from two independent Gaussian curves, which significantly reduced the unknowns. Instead of the Gaussian scheme, another potential way to obtain different pigment information from remote sensing data is through the specific absorption coefficients as adopted by many studies [46,86]. However, the specific absorption coefficients have significant limitations, such as the variation of the coefficients in different waters and the lack of routine measurements of some pigments, such as PE and PC [46,86,87].

With the information of accessory pigments obtained from MuPI, different biogeochemical studies could be conducted: 1. Remotely sensed PE and PC concentrations could be validated and applied to the estimation of cyanobacteria on a global scale. 2. The pigment ratios could be used as a direct indicator for estimation of phytoplankton functional types or functional traits, and phytoplankton physiological variation over space and time. 3. The pigment absorption coefficients (photoprotective and photosynthetic) could be estimated from satellite remote sensing data and incorporated into models for more accurate estimation of primary productivity.

5. Conclusions

The multi-pigment inversion model, namely MuPI, which semi-analytically obtains concentrations of multiple pigments from remote sensing reflectance, has been validated and applied to MERIS $R_{rs}(\lambda)$ imagery to obtain not only [Chl-a], but also [Chl-b], [Chl-c], [PPC], [PSC] (and subsequently their ratios to [Chl-a]) in the global oceans. The obtained pigment concentrations and the pigment ratios showed good agreement with in situ HPLC data, with the mean UAPD of 38%, 78%, 65%, 36%, and 47% respectively. Further, at the global scale, the MuPI obtained [Chl-a] from MERIS showed comparable results with those estimated from the widely used OC4E algorithm with mean UAPD of 48.8% and RMSE of 4.51 mg·m^{-3} for OC4E and mean UAPD of 49.3% and RMSE of 4.05 mg·m^{-3} for MuPI. However, unlike OC4E, MuPI as a semi-analytical model provided reasonable retrievals of several parameters {[Chl-a], [Chl-b], [Chl-c], [PPC], [PSC], $a_{dg}(440)$ and $b_{bp}(440)$} simultaneously from satellite obtained remote sensing reflectance. The information of these accessory pigments would extend the application of satellite ocean color data in global biogeochemical studies that was previously limited due to [Chl-a] as the only available pigment.

Author Contributions: G.W. conceived and designed the analysis, collected the data and wrote the paper; Z.L. and C.B.M. supervised the project, analyzed results and wrote the paper.

Acknowledgments: We are grateful for the two anonymous reviewers who have provided constructive suggestions to improve the quality of this manuscript. We thank Charles W. Kovach very much for the careful reading of the manuscript and all the helps he provided in improving the language. We thank Jianwei Wei and Junfang Lin in helping to take in situ measurements of remote sensing reflectance, Chuanmin Hu's group for sharing the phytoplankton absorption coefficients, and National Aeronautics and Space Administration (NASA) for sharing the HPLC measurements from 2014–2015 VIIRS Cal/Val cruises. We really appreciate those who shared the data from cruises of the Tara Oceans expedition and BIOSOPE and made it publicly accessible. We thank ESA for providing the MERIS imagery and NASA for SeaWiFS Bio-optical Archive and Storage System (SeaBASS) dataset, the Bermuda Institute of Ocean Science for the BATS dataset, and the International Ocean Color Coordinating Group for the IOCCG dataset. Support from the National Aeronautics and Space Administration (NASA) Ocean Biology and Biogeochemistry and Water and Energy Cycle Programs, the National Oceanic and Atmospheric Administration (NOAA) JPSS VIIRS Ocean Color Cal/Val Project and cruises are greatly appreciated.

References

1. Falkowski, P.G. The role of phytoplankton photosynthesis in global biogeochemical cycles. *Photosynth. Res.* **1994**, *39*, 235–258. [CrossRef] [PubMed]
2. Kiørboe, T. Turbulence, phytoplankton cell size, and the structure of pelagic food webs. In *Advances in Marine Biology*; Academic Press: Cambridge, MA, USA, 1993; Volume 29, pp. 1–72.
3. Gordon, H.R.; Clark, D.K.; Brown, J.W.; Brown, O.B.; Evans, R.H.; Broenkow, W.W. Phytoplankton pigment concentrations in the Middle Atlantic Bight: Comparison of ship determinations and CZCS estimates. *Appl. Opt.* **1983**, *22*, 20–36. [CrossRef] [PubMed]
4. O'Reilly, J.E.S.; Maritorena, B.G.; Mitchell, D.A.; Siegel, K.L.; Carder, S.A.; Garver, M.; Kahru, C.R. McClain, Ocean color chlorophyll algorithms for SeaWiFS. *J. Geophys. Res.* **1998**, *103*, 24937–24953. [CrossRef]
5. Hu, C.; Lee, Z.; Franz, B. Chlorophyll algorithms for oligotrophic oceans: A novel approach based on three-band reflectance difference. *J. Geophys. Res.* **2012**, *117*. [CrossRef]
6. Alvain, S.; Moulin, C.; Dandonneau, Y.; Bréon, F.M. Remote sensing of phytoplankton groups in case 1 waters from global SeaWiFS imagery. *Deep Sea Res. Part I Oceanogr. Res. Pap.* **2005**, *52*, 1989–2004. [CrossRef]
7. Bracher, A.; Vountas, M.; Dinter, T.; Burrows, J.P.; Röttgers, R.; Peeken, I. Quantitative observation of cyanobacteria and diatoms from space using PhytoDOAS on SCIAMACHY data. *Biogeosciences* **2009**, *6*, 751–764. [CrossRef]
8. Ciotti, A.M.; Bricaud, A. Retrievals of a size parameter for phytoplankton and spectral light absorption by colored detrital matter from water-leaving radiances at SeaWiFS channels in a continental shelf region off Brazil. *Limnol. Oceanogr. Methods* **2006**, *4*, 237–253. [CrossRef]
9. Brewin, R.J.; Sathyendranath, S.; Hirata, T.; Lavender, S.J.; Barciela, R.M.; Hardman-Mountford, N.J. A three-component model of phytoplankton size class for the Atlantic Ocean. *Ecol. Model.* **2010**, *221*, 1472–1483. [CrossRef]
10. Hirata, T.; Hardman-Mountford, N.J.; Brewin, R.J.W.; Aiken, J.; Barlow, R.; Suzuki, K.; Yamanaka, Y. Synoptic relationships between surface Chlorophyll-a and diagnostic pigments specific to phytoplankton functional types. *Biogeosciences* **2011**, *8*, 311–327. [CrossRef]
11. Mouw, C.B.; Yoder, J.A. Optical determination of phytoplankton size composition from global SeaWiFS imagery. *J. Geophys. Res. Oceans* **2010**, *115*. [CrossRef]

12. Sathyendranath, S.; Cota, G.; Stuart, V.; Maass, H.; Platt, T. Remote sensing of phytoplankton pigments: A comparison of empirical and theoretical approaches. *Int. J. Remote Sens.* **2001**, *22*, 249–273. [CrossRef]
13. Gitelson, A.A.; Schalles, J.F.; Hladik, C.M. Remote chlorophyll-a retrieval in turbid, productive estuaries: Chesapeake Bay case study. *Remote Sens. Environ.* **2007**, *109*, 464–472. [CrossRef]
14. Behrenfeld, M.J.; Falkowski, P.G. Photosynthetic rates derived from satellite-based chlorophyll concentration. *Limnol. Oceanogr.* **1997**, *42*, 1–20. [CrossRef]
15. Stumpf, R.P.; Culver, M.E.; Tester, P.A.; Tomlinson, M.; Kirkpatrick, G.J.; Pederson, B.A.; Soracco, M. Monitoring Karenia brevis blooms in the Gulf of Mexico using satellite ocean color imagery and other data. *Harmful Algae* **2003**, *2*, 147–160. [CrossRef]
16. Lehman, P.W. Comparison of chlorophyll a and carotenoid pigments as predictors of phytoplankton biomass. *Mar. Biol.* **1981**, *65*, 237–244. [CrossRef]
17. Schitüter, L.; Riemann, B.; Søndergaard, M. Nutrient limitation in relation to phytoplankton carotenoid/chlorophyll a ratios in freshwater mesocosms. *J. Plankton Res.* **1997**, *19*, 891–906. [CrossRef]
18. Breton, E.; Brunet, C.; Sautour, B.; Brylinski, J.M. Annual variations of phytoplankton biomass in the Eastern English Channel: Comparison by pigment signatures and microscopic counts. *J. Plankton Res.* **2000**, *22*, 1423–1440. [CrossRef]
19. Kruskopf, M.; Flynn, K.J. Chlorophyll content and fluorescence responses cannot be used to gauge reliably phytoplankton biomass, nutrient status or growth rate. *New Phytol.* **2006**, *169*, 525–536. [CrossRef] [PubMed]
20. Behrenfeld, M.J.; O'Malley, R.T.; Boss, E.S.; Westberry, T.K.; Graff, J.R.; Halsey, K.H.; Brown, M.B. Revaluating ocean warming impacts on global phytoplankton. *Nat. Clim. Chang.* **2016**, *6*, 323. [CrossRef]
21. Bidigare, R.R.; Morrow, J.H.; Kiefer, D.A. Derivative analysis of spectral absorption by photosynthetic pigments in the western Sargasso Sea. *J. Mar. Res.* **1989**, *47*, 323–341. [CrossRef]
22. Jeffrey, S.W.; Vesk, M. Introduction to marine phytoplankton and their pigment signature. In *Phytoplankton Pigments in Oceanography*; UNESCO Publishing: Paris, France, 1997; p. 3784.
23. Kirkpatrick, G.J.; Millie, D.F.; Moline, M.A.; Schofield, O. Optical discrimination of a phytoplankton species in natural mixed populations. *Limnol. Oceanogr.* **2000**, *45*, 467–471. [CrossRef]
24. Van Heukelem, L.; Thomas, C.S. Computer-assisted high-performance liquid chromatography method development with applications to the isolation and analysis of phytoplankton pigments. *J. Chromatogr. A* **2001**, *910*, 31–49. [CrossRef]
25. Roy, S.; Llewellyn, C.A.; Egeland, E.S.; Johnsen, G. (Eds.) *Phytoplankton Pigments: Characterization, Chemotaxonomy and Applications in Oceanography*; Cambridge University Press: Cambridge, UK, 2011.
26. Simis, S.G.; Peters, S.W.; Gons, H.J. Remote sensing of the cyanobacterial pigment phycocyanin in turbid inland water. *Limnol. Oceanogr.* **2005**, *50*, 237–245. [CrossRef]
27. Wynne, T.; Stumpf, R.; Tomlinson, M.; Warner, R.; Tester, P.; Dyble, J.; Fahnenstiel, G. Relating spectral shape to cyanobacterial blooms in the Laurentian Great Lakes. *Int. J. Remote Sens.* **2008**, *29*, 3665–3672. [CrossRef]
28. Mackey, M.D.; Mackey, D.J.; Higgins, H.W.; Wright, S.W. CHEMTAX—A program for estimating class abundances from chemical markers: Application to HPLC measurements of phytoplankton. *Mar. Ecol. Prog. Ser.* **1996**, *144*, 265–283. [CrossRef]
29. Vidussi, F.; Claustre, H.; Manca, B.B.; Luchetta, A.; Marty, J.C. Phytoplankton pigment distribution in relation to upper thermocline circulation in the eastern Mediterranean Sea during winter. *J. Geophys. Res. Oceans* **2001**, *106*, 19939–19956. [CrossRef]
30. Uitz, J.; Claustre, H.; Morel, A.; Hooker, S.B. Vertical distribution of phytoplankton communities in open ocean: An assessment based on surface chlorophyll. *J. Geophys. Res. Oceans* **2006**, *111*. [CrossRef]
31. Sathyendranath, S.; Aiken, J.; Alvain, S.; Barlow, R.; Bouman, H.; Bracher, A.; Clementson, L.A. Phytoplankton functional types from Space. In *Reports of the International Ocean-Colour Coordinating Group (IOCCG)*; International Ocean-Colour Coordinating Group: Dartmouth, Canada, 2014; pp. 1–156.
32. Uitz, J.; Stramski, D.; Reynolds, R.A.; Dubranna, J. Assessing phytoplankton community composition from hyperspectral measurements of phytoplankton absorption coefficient and remote-sensing reflectance in open-ocean environments. *Remote Sens. Environ.* **2015**, *171*, 58–74. [CrossRef]
33. Catlett, D.; Siegel, D.A. Phytoplankton pigment communities can be modeled using unique relationships with spectral absorption signatures in a dynamic coastal environment. *J. Geophys. Res. Oceans* **2018**, *123*, 246–264. [CrossRef]
34. Bracher, A.; Bouman, H.A.; Brewin, R.J.; Bricaud, A.; Brotas, V.; Ciotti, A.M.; Hardman-Mountford, N.J.

Obtaining phytoplankton diversity from ocean color: A scientific roadmap for future development. *Front. Mar. Sci.* **2017**, *4*, 55. [CrossRef]

35. Bricaud, A.; Mejia, C.; Blondeau-Patissier, D.; Claustre, H.; Crepon, M.; Thiria, S. Retrieval of pigment concentrations and size structure of algal populations from their absorption spectra using multilayered perceptrons. *Appl. Opt.* **2007**, *46*, 1251–1260. [CrossRef] [PubMed]
36. Ciotti, A.M.; Lewis, M.R.; Cullen, J.J. Assessment of the relationships between dominant cell size in natural phytoplankton communities and the spectral shape of the absorption coefficient. *Limnol. Oceanogr.* **2002**, *47*, 404–417. [CrossRef]
37. Devred, E.; Sathyendranath, S.; Stuart, V.; Platt, T. A three component classification of phytoplankton absorption spectra: Application to ocean-color data. *Remote Sens. Environ.* **2011**, *115*, 2255–2266. [CrossRef]
38. Hoepffner, N.; Sathyendranath, S. Determination of the major groups of phytoplankton pigments from the absorption spectra of total particulate matter. *J. Geophys. Res. Oceans* **1993**, *98*, 22789–22803. [CrossRef]
39. Hirata, T.; Aiken, J.; Hardman-Mountford, N.; Smyth, T.J.; Barlow, R.G. An absorption model to determine phytoplankton size classes from satellite ocean colour. *Remote Sens. Environ.* **2008**, *112*, 3153–3159. [CrossRef]
40. Lohrenz, S.E.; Weidemann, A.D.; Tuel, M. Phytoplankton spectral absorption as influenced by community size structure and pigment composition. *J. Plankton Res.* **2003**, *25*, 35–61. [CrossRef]
41. Moisan, J.R.; Moisan, T.A.; Linkswiler, M.A. An inverse modeling approach to estimating phytoplankton pigment concentrations from phytoplankton absorption spectra. *J. Geophys. Res. Oceans* **2011**, *116*. [CrossRef]
42. Organelli, E.; Bricaud, A.; Antoine, D.; Uitz, J. Multivariate approach for the retrieval of phytoplankton size structure from measured light absorption spectra in the Mediterranean Sea (BOUSSOLE site). *Appl. Opt.* **2013**, *52*, 2257–2273. [CrossRef] [PubMed]
43. Mouw, C.B.; Hardman-Mountford, N.J.; Alvain, S.; Bracher, A.; Brewin, R.J.; Bricaud, A.; Hirawake, T. A consumer's guide to satellite remote sensing of multiple phytoplankton groups in the global ocean. *Front. Mar. Sci.* **2017**, *4*, 41. [CrossRef]
44. Jensen, A.; Sakshaug, E. Studies on the phytoplankton ecology of the trondheemsfjord. II. Chloroplast pigments in relation to abundance and physiological state of the phytoplankton. *J. Exp. Mar. Biol. Ecol.* **1973**, *11*, 137–155. [CrossRef]
45. Suggett, D.J.; Moore, C.M.; Hickman, A.E.; Geider, R.J. Interpretation of fast repetition rate (FRR) fluorescence: Signatures of phytoplankton community structure versus physiological state. *Mar. Ecol. Prog. Ser.* **2009**, *376*, 1–19. [CrossRef]
46. Pan, X.; Mannino, A.; Russ, M.E.; Hooker, S.B.; Harding, L.W., Jr. Remote sensing of phytoplankton pigment distribution in the United States northeast coast. *Remote Sens. Environ.* **2010**, *114*, 2403–2416. [CrossRef]
47. Moisan, T.A.; Rufty, K.M.; Moisan, J.R.; Linkswiler, M.A. Satellite observations of phytoplankton functional type spatial distributions, phenology, diversity, and ecotones. *Front. Mar. Sci.* **2017**, *4*, 189. [CrossRef]
48. Gordon, H.R.; Brown, O.B.; Evans, R.H.; Brown, J.W.; Smith, R.C.; Baker, K.S.; Clark, D.K. A semianalytic radiance model of ocean color. *J. Geophys. Res. Atmos.* **1988**, *93*, 10909–10924. [CrossRef]
49. Lee, Z.; Carder, K.L.; Mobley, C.D.; Steward, R.G.; Patch, J.S. Hyperspectral remote sensing for shallow waters. I. A semianalytical model. *Appl. Opt.* **1998**, *37*, 6329–6338. [CrossRef] [PubMed]
50. Lee, Z.; Carder, K.L.; Arnone, R.A. Deriving inherent optical properties from water color: A multiband quasi-analytical algorithm for optically deep waters. *Appl. Opt.* **2002**, *41*, 5755–5772. [CrossRef] [PubMed]
51. Lee, Z. *Remote Sensing of Inherent Optical Properties: Fundamentals, Tests of Algorithms, and Applications*; International Ocean-Colour Coordinating Group: Dartmouth, Canada, 2006; Volume 5.
52. Maritorena, S.; Siegel, D.A.; Peterson, A.R. Optimization of a semianalytical ocean color model for global-scale applications. *Appl. Opt.* **2002**, *41*, 2705–2714. [CrossRef] [PubMed]
53. Werdell, P.J.; Franz, B.A.; Bailey, S.W.; Feldman, G.C.; Boss, E.; Brando, V.E.; Mangin, A. Generalized ocean color inversion model for retrieving marine inherent optical properties. *Appl. Opt.* **2013**, *52*, 2019–2037. [CrossRef] [PubMed]
54. Werdell, P.J.; McKinna, L.I.; Boss, E.; Ackleson, S.G.; Craig, S.E.; Gregg, W.W.; Stramski, D. An overview of approaches and challenges for retrieving marine inherent optical properties from ocean color remote sensing. *Prog. Oceanogr.* **2018**. [CrossRef]
55. Brando, V.E.; Dekker, A.G.; Park, Y.J.; Schroeder, T. Adaptive semianalytical inversion of ocean color radiometry in optically complex waters. *Appl. Opt.* **2012**, *51*, 2808–2833. [CrossRef] [PubMed]
56. Brewin, R.J.; Raitsos, D.E.; Dall'Olmo, G.; Zarokanellos, N.; Jackson, T.; Racault, M.F.; Hoteit, I. Regional ocean-colour chlorophyll algorithms for the Red Sea. *Remote Sens. Environ.* **2015** *165*, 64–85. [CrossRef]

57. Bukata, R.P.; Jerome, J.H.; Kondratyev, A.S.; Pozdnyakov, D.V. *Optical Properties and Remote Sensing of Inland and Coastal Waters*; CRC Press: Boca Raton, FL, USA, 2018.
58. Devred, E.; Sathyendranath, S.; Stuart, V.; Maass, H.; Ulloa, O.; Platt, T. A two-component model of phytoplankton absorption in the open ocean: Theory and applications. *J. Geophys. Res. Oceans* **2006**, *111*. [CrossRef]
59. Garver, S.A.; Siegel, D.A. Inherent optical property inversion of ocean color spectra and its biogeochemical interpretation: 1. Time series from the Sargasso Sea. *J. Geophys. Res. Oceans* **1997**, *102*, 18607–18625. [CrossRef]
60. Hoge, F.E.; Lyon, P.E. Satellite retrieval of inherent optical properties by linear matrix inversion of oceanic radiance models: An analysis of model and radiance measurement errors. *J. Geophys. Res. Oceans* **1996**, *101*, 16631–16648. [CrossRef]
61. Roesler, C.S.; Perry, M.J. In situ phytoplankton absorption, fluorescence emission, and particulate backscattering spectra determined from reflectance. *J. Geophys. Res. Oceans* **1995**, *100*, 13279–13294. [CrossRef]
62. Wang, P.; Boss, E.S.; Roesler, C. Uncertainties of inherent optical properties obtained from semianalytical inversions of ocean color. *Appl. Opt.* **2005**, *44*, 4074–4085. [CrossRef] [PubMed]
63. Loisel, H.; Stramski, D.; Dessailly, D.; Jamet, C.; Li, L.; Reynolds, R.A. An Inverse Model for Estimating the Optical Absorption and Backscattering Coefficients of Seawater from Remote-Sensing Reflectance over a Broad Range of Oceanic and Coastal Marine Environments. *J. Geophys. Res. Oceans* **2018**, *123*, 2141–2171. [CrossRef]
64. Wang, G.; Lee, Z.; Mishra, D.R.; Ma, R. Retrieving absorption coefficients of multiple phytoplankton pigments from hyperspectral remote sensing reflectance measured over cyanobacteria bloom waters. *Limnol. Oceanogr. Methods* **2016**, *14*, 432–447. [CrossRef]
65. Hoepffner, N.; Sathyendranath, S. Effect of pigment composition on absorption properties of phytoplankton. *Mar. Ecol. Prog. Ser.* **1991**, *73*, 1–23. [CrossRef]
66. Chase, A.P.; Boss, E.; Cetinić, I.; Slade, W. Estimation of phytoplankton accessory pigments from hyperspectral reflectance spectra: Toward a global algorithm. *J. Geophys. Res. Oceans* **2017**. [CrossRef]
67. Lutz, V.A.; Sathyendranath, S.; Head, E.J.H. Absorption coefficient of phytoplankton: Regional variations in the North Atlantic. *Mar. Ecol. Prog. Ser.* **1996**, 197–213. [CrossRef]
68. Mitchell, B.G. Algorithms for determining the absorption coefficient for aquatic particulates using the quantitative filter technique. In *Ocean Optics X*; International Society for Optics and Photonics: Bellingham, WA, USA, 1990; Volume 1302, pp. 137–149.
69. Mobley, C.D. *Light and Water: Radiative Transfer in Natural Waters*; Academic Press: Cambridge, MA, USA, 1994.
70. Lee, Z.; Pahlevan, N.; Ahn, Y.H.; Greb, S.; O'Donnell, D. Robust approach to directly measuring water-leaving radiance in the field. *Appl. Opt.* **2013**, *52*, 1693–1701. [CrossRef] [PubMed]
71. Mueller, J.L. *Ocean Optics Protocols for Satellite Ocean Color Sensor Validation, Revision 4: Radiometric Measurements and Data Analysis Protocols*; Goddard Space Flight Center: Greenbelt, MD, USA, 2003; Volume 3.
72. Bailey, S.W.; Werdell, P.J. A multi-sensor approach for the on-orbit validation of ocean color satellite data products. *Remote Sens. Environ.* **2006**, *102*, 12–23. [CrossRef]
73. Mitchell, B.G.; Kahru, M.; Wieland, J.; Stramska, M. Determination of spectral absorption coefficients of particles, dissolved material and phytoplankton for discrete water samples. In *Ocean Optics Protocols for Satellite Ocean Color Sensor Validation, Revision 4, Volume IV: Inherent Optical Properties: Instruments, Characterizations, Field Measurements and Data Analysis Protocols*; NASA/TM-2003-211621; Mueller, J.L., Fargion, G.S., McClain, C.R., Eds.; NASA Goddard Space Flight Center: Greenbelt, MD, USA, 2003; pp. 39–64.
74. Wang, G.; Lee, Z.; Mouw, C. Multi-spectral remote sensing of phytoplankton pigment absorption properties in cyanobacteria bloom waters: A regional example in the western basin of Lake Erie. *Remote Sens.* **2017**, *9*, 1309. [CrossRef]
75. Morel, A. Optical properties of pure water and pure sea water. *Opt. Asp. Oceanogr.* **1974**, *1*, 22.
76. Pope, R.M.; Fry, E.S. Absorption spectrum (380–700 nm) of pure water. II. Integrating cavity measurements. *Appl. Opt.* **1997** *36*, 8710–8723. [CrossRef] [PubMed]
77. Lee, Z.; Wei, J.; Voss, K.; Lewis, M.; Bricaud, A.; Huot, Y. Hyperspectral absorption coefficient of "pure"

seawater in the range of 350–550 nm inverted from remote sensing reflectance. *Appl. Opt.* **2015**, *54*, 546–558. [CrossRef]

78. Nelson, N.B.; Siegel, D.A.; Michaels, A.F. Seasonal dynamics of colored dissolved material in the Sargasso Sea. *Deep Sea Res. Part. I Oceanogr. Res. Pap.* **1998**, *45*, 931–957. [CrossRef]
79. Babin, M.; Stramski, D.; Ferrari, G.M.; Claustre, H.; Bricaud, A.; Obolensky, G.; Hoepffner, N. Variations in the light absorption coefficients of phytoplankton, nonalgal particles, and dissolved organic matter in coastal waters around Europe. *J. Geophys. Res. Oceans* **2003**, *108*. [CrossRef]
80. Lasdon, L.S.; Waren, A.D.; Jain, A.; Ratner, M. Design and testing of a generalized reduced gradient code for nonlinear programming. *ACM Trans. Math. Softw. (TOMS)* **1978**, *4*, 34–50. [CrossRef]
81. Gilerson, A.A.; Gitelson, A.A.; Zhou, J.; Gurlin, D.; Moses, W.; Ioannou, I.; Ahmed, S.A. Algorithms for remote estimation of chlorophyll-a in coastal and inland waters using red and near infrared bands. *Opt. Express* **2010**, *18*, 24109–24125. [CrossRef] [PubMed]
82. Ruddick, K.; Park, Y.; Astoreca, R.; Neukermans, G.; Van Mol, B. Validation of MERIS water products in the Southern North Sea. In *Proceedings of the 2nd MERIS—(A) ATSR Workshop*; ESA Publications Office Frascati: Frascati, Spain, 2008.
83. Claustre, H. The trophic status of various oceanic provinces as revealed by phytoplankton pigment signatures. *Limnol. Oceanogr.* **1994**, *39*, 1206–1210. [CrossRef]
84. Aiken, J.; Pradhan, Y.; Barlow, R.; Lavender, S.; Poulton, A.; Holligan, P.; Hardman-Mountford, N. Phytoplankton pigments and functional types in the Atlantic Ocean: A decadal assessment, 1995–2005. *Deep Sea Res. Part II Top. Stud. Oceanogr.* **2009**, *56*, 899–917. [CrossRef]
85. Descy, J.P.; Sarmento, H.; Higgins, H.W. Variability of phytoplankton pigment ratios across aquatic environments. *Eur. J. Phycol.* **2009**, *44*, 319–330. [CrossRef]
86. Behrenfeld, M.J.; O'Malley, R.T.; Siegel, D.A.; McClain, C.R.; Sarmiento, J.L.; Feldman, G.C.; Boss, E.S. Climate-driven trends in contemporary ocean productivity. *Nature* **2006**, *444*, 752. [CrossRef] [PubMed]
87. Bricaud, A.; Claustre, H.; Ras, J.; Oubelkheir, K. Natural variability of phytoplanktonic absorption in oceanic waters: Influence of the size structure of algal populations. *J. Geophys. Res. Oceans* **2004**, *109*. [CrossRef]

Assessing the Impact of a Two-Layered Spherical Geometry of Phytoplankton Cells on the Bulk Backscattering Ratio of Marine Particulate Matter

Lucile Duforêt-Gaurier [1,*], David Dessailly [1], William Moutier [2] and Hubert Loisel [1]

1 Univ. Littoral Cote d'Opale, Univ. Lille, CNRS, UMR 8187, LOG, Laboratoire d'Océanologie et de Géosciences, F 62930 Wimereux, France; david.dessailly@univ-littoral.fr (D.D.); hubert.loisel@univ-littoral.fr (H.L.)
2 Royal Meteorological Institute of Belgium, 1180 Brussels, Belgium; william.moutier@gmail.com
* Correspondence: lucile.duforet@univ-littoral.fr

Abstract: The bulk backscattering ratio ($\tilde{b_{bp}}$) is commonly used as a descriptor of the bulk real refractive index of the particulate assemblage in natural waters. Based on numerical simulations, we analyze the impact of modeled structural heterogeneity of phytoplankton cells on $\tilde{b_{bp}}$. $\tilde{b_{bp}}$ is modeled considering viruses, heterotrophic bacteria, phytoplankton, organic detritus, and minerals. Three case studies are defined according to the relative abundance of the components. Two case studies represent typical situations in open ocean, oligotrophic waters, and phytoplankton bloom. The third case study is typical of coastal waters with the presence of minerals. Phytoplankton cells are modeled by a two-layered spherical geometry representing a chloroplast surrounding the cytoplasm. The $\tilde{b_{bp}}$ values are higher when structural heterogeneity is considered because the contribution of coated spheres to light backscattering is higher than homogeneous spheres. The impact of heterogeneity is; however, strongly conditioned by the hyperbolic slope ξ of the particle size distribution. Even if the relative abundance of phytoplankton is small (<1%), $\tilde{b_{bp}}$ increases by about 58% (for $\xi = 4$ and for oligotrophic waters), when the heterogeneity is taken into account, in comparison with a particulate population composed only of homogeneous spheres. As expected, heterogeneity has a much smaller impact (about 12% for $\xi = 4$) on $\tilde{b_{bp}}$ in the presence of suspended minerals, whose increased light scattering overwhelms that of phytoplankton.

Keywords: ocean optics; backscattering ratio; phytoplankton; coated-sphere model; bulk refractive index; seawater component

---

1. Introduction

Seawater constituents (water molecules, suspended particles, dissolved substances, and air bubbles) impact the propagation of light through absorption and scattering processes. In natural waters, suspended particulate matter is mostly composed of phytoplankton, heterotrophic organisms, viruses, biogenic detritus, and mineral particles. Absorbing and scattering characteristics of water constituents are described by the inherent optical properties (IOP) [1] which do not depend on the radiance distribution but solely on the concentration and chemical composition of dissolved organic matter, and the concentration, size distribution and chemical composition of particulate matter. All IOPs can be defined from the absorption coefficient, a, and the volume scattering function, β. For instance, the scattering, b, and backscattering, b_b, coefficients are obtained from the integration of β over all scattering angles, and only backward scattering angles, respectively.

Owing to the availability of commercial optical backscattering sensors and flow-through attenuation and absorption meters, in situ measurements of bulk IOP have now been routinely performed for more than two decades. While these measurements allow a better description of the IOP variability in natural waters, they can also be used as proxies for the estimation of the bulk particulate matter. For instance, the spectral slope of the particulate beam attenuation coefficient, c_p, is tightly linked to the slope of the particle size distribution (PSD), ξ, assuming a Junge-type distribution of PSD [2–4]. The particulate backscattering ratio b_{bp}/b_p is used to obtain information about the particle composition. Indeed, based on the Lorentz-Mie scattering calculations that assume marine particles as homogeneous spheres, an analytical relationship between b_{bp}/b_p, ξ and $\tilde{n}_r$ was generated [5]. This latter equation is used in conjunction with in situ measurements of b_{bp}, b_p, and c_p to describe the variability of the physical nature (i.e., refractive index) of the bulk particulate matter in oceanic and coastal environments [6–10].

In the past, many theoretical and experimental studies, mainly dedicated to phytoplankton, showed that while the absorption, attenuation and total scattering of algal cells are correctly described using the homogeneous sphere model, such model is less appropriate for simulating backscatter. Indeed, the structural heterogeneity and inner complexity of phytoplankton cells (gas vacuoles, chloroplast, silica wall, etc.) explain why the measured backscattering signal is higher than predicted by the Lorentz-Mie theory for homogeneous spheres [11–20]. The underestimation of b_{bp} by homogeneous spheres may explain the fact that in situ observations of backscattering are significantly higher than theoretical simulations [21–23].

In this paper, we examine the impact of particle structural heterogeneity on the bulk backscattering ratio for realistic combinations of optically significant constituents. The purpose of our study is not to provide a new analytical relationship for b_{bp}/b_p as a function of $\tilde{n}_r$ and ξ but rather to assess the sensitivity of b_{bp}/b_p to the modeled structural heterogeneity of phytoplankton cells for some realistic water bodies. Typical phytoplankton bloom and no bloom conditions, as defined in Stramski and Kiefer [24], will first be examined. Then, the last case study will account for the presence of mineral particles, which have a great effect on the scattering properties.

Because the bulk scattering (b) and backscattering (b_b) coefficients of a water body result from additive contributions of all individual constituents that scatter light, we will consider various sub-populations of marine particles, namely organic detritus, minerals, heterotrophic bacteria, viruses, and phytoplankton. Robertson Lain et al. [23] showed that the two-layered sphere model is appropriate for modeling of remote-sensing reflectance and IOPs in high biomass Case 1 waters. The real refractive index of the chloroplast and the relative volume of the chloroplast are key parameters impacting the backscattering efficiency of phytoplankton cells. This was recently confirmed by two studies where measurements of light scattering by phytoplankton cultures were well reproduced by the two-sphere model [15,16]. For these reasons, in this study, phytoplankton optical properties have been simulated considering a two-layered sphere model. The size range of the different considered particles (viruses, bacteria, phytoplankton, and organic detritus), as well as their real and imaginary refractive index values are defined from literature [21,25].

To establish the foundations of the present study, the different theoretical considerations as well as the two different numerical codes used for the calculations are first presented. Then, we describe the different sub-populations of particles and their associated size distribution, refractive index, and internal structure used to simulate their optical properties. The impact of the modeled structural heterogeneity of phytoplankton cells is then discussed for the three realistic water bodies as mentioned previously.

2. Theoretical Considerations

2.1. Backscattering Cross Section for Polydisperse Particle Assemblages

Light scattering is produced by the presence of an object (such as a particle) with a refractive index different from that of the surrounding medium. The refractive index is expressed in complex form as $n(\lambda) = n_r(\lambda) + i\, n_i(\lambda)$, where λ is the wavelength of the radiation in vacuum in units of nm. The real part determines the phase velocity of the propagating wave and the imaginary part accounts for the absorption. Please note that the refractive index is a relative value dependent upon the surrounding medium, i.e., relative to the refractive index of the medium. The single scattering process by a particle is described by the scattering cross section $C_{sca}(D,\lambda)$ (units m^2) and the scattering phase function $\tilde{F}(D,\theta,\lambda)$ (dimensionless) as defined by Mishchenko et al. [26] (Equations (4.51)–(4.53), pp. 100–101):

$$\frac{1}{2}\int_0^{\pi} \tilde{F}(D,\lambda,\theta)\, sin\theta\, d\theta = 1 \tag{1}$$

As particles are here assumed to be spherical, $\tilde{F}$ depends only on the particle diameter D, the scattering angle θ within the arbitrary azimuthal plane of scattering, and the wavelength λ. In the following, λ is omitted for brevity. To account for polydisperse particulate assemblages, the particle size distribution (PSD) is defined. For the present study, we adopt a power-law PSD (also named the Junge-like PSD) which is commonly used to represent the size distribution of marine particles in natural waters [5,24,27,28]. The ensemble-average normalized phase function is:

$$\tilde{F}(\theta) = \int_{D_{min}}^{D_{max}} \tilde{F}(D,\theta) \times A\, D^{-\xi} dD \tag{2}$$

where D_{min} and D_{max} define the particle diameter range, ξ is the hyperbolic slope of PSD, and $A\, D^{-\xi}$ (units, μm^{-1}) is the relative differential particle size distribution. As in many theoretical studies, the relative PSD is normalized such that the integral over the entire size range is unity. It follows that $\tilde{F}(\theta)$ represents the average normalized phase function per particle. Equation (2) can be written for the scattering cross section replacing $\tilde{F}(D,\theta)$ with $C_{sca}(D)$ and $\tilde{F}(\theta)$ with C_{sca}. The backscattering cross section of the polydisperse assemblage is defined as:

$$C_{sca}^{bb} = \frac{C_{sca}}{2} \int_{\pi/2}^{\pi} \tilde{F}(\theta)\, sin\theta\, d\theta \tag{3}$$

It can be easily seen from Equations (1)–(3) that the integration of $\tilde{F}(\theta)$ between 0 and π gives C_{sca}, the scattering cross section of the polydisperse population. Many numerical codes (including those described in Section 3) use the normalized phase function $\tilde{F}(\theta)$ to describe the angular distribution of the scattered radiation. However, in hydrologic optics, the volume scattering function (VSF), $\tilde{\beta}(\theta)$ ($m^{-1}\, sr^{-1}$), is more commonly used instead of $\tilde{F}(\theta)$ [29]. The relationship between $\tilde{\beta}(\theta)$ and $\tilde{F}(\theta)$ is:

$$\tilde{\beta}(\theta) = \frac{N\, C_{sca}}{4\pi} \tilde{F}(\theta) \tag{4}$$

with N the number of particles per cubic meter.

2.2. The Bulk Backscattering Ratio

Marine particles are divided into five different categories: viruses (VIR), heterotrophic bacteria (BAC), phytoplankton (PHY), organic detritus (DET), and minerals (MIN). Table 1 displays the size ranges and the refractive indices of the different components as defined by previous studies [21,24,25]. The ensemble-average values of $\tilde{F}_j(\theta)$, $C_{sca,j}$, and $C_{sca,j}^{bb}$ are computed from Equations (1)–(3) for each particulate component j.

Table 1. Summary of the seawater constituents.

Component (j)	Sphere Model	D_{min}-D_{max} (μm)	n_r	n_i
Viruses	homogeneous	0.03–0.2	1.05	0
Heterotrophic bacteria	homogeneous	0.2–2	1.05	1.0×10^{-4}
Phytoplankton cells	two or three-layered	0.3–40	1.044 *	1.5×10^{-3} *
Organic detritus	homogeneous	0.05–500	1.04	2.3×10^{-5}
Minerals	homogeneous	0.05–500	1.18	1.0×10^{-4}

* the values represent the equivalent refractive indices (Equation (12)). The refractive indices of the spheres representing the chloroplast and cytoplast are described in Table 2. λ = 532 nm.

The total normalized phase function and total scattering cross section of the water body are obtained as in Mishchenko et al., 2002 (Equations (4.74) p. 102 and (3.13) p. 71):

$$\tilde{F}^{tot}(\theta) = \frac{\sum_{j=1}^{5} N_j\, C_{sca,j}\, \tilde{F}_j(\theta)}{\sum_{j=1}^{5} N_j\, C_{sca,j}} \tag{5}$$

$$C_{sca}^{bb,\,tot} = \sum_{j=1}^{5} N_j\, C_{sca,j}^{bb} \tag{6}$$

where N_j is the relative concentration (i.e., the relative number of particles per unit volume of water) of the considered component. C_{sca}^{tot} is defined by replacing $C_{sca,j}^{bb}$ with $C_{sca,j}$ in Equation (6).

The total (i.e., bulk) backscattering coefficient (b_{bp}) (units m^{-1}) of the water body is the sum of the relevant $b_{bp,j}$ associated with each jth group. $b_{bp,j}$ is equal to the polydisperse $C_{sca,j}^{bb}$ weighted by the particle concentration of the jth group:

$$b_{bp} = \sum_{j=1}^{5} b_{bp,j} = N_{TOT} \times C_{sca}^{bb,\,tot} \tag{7}$$

with N_{TOT} the total particle concentration (particles per m^3) in the water body. Similarly, b_p is defined from Equation (7) by replacing b_{bp} with b_p and $C_{sca}^{bb,\,tot}$ with C_{sca}^{tot}. The bulk backscattering ratio $\tilde{b_{bp}}$ is the dimensionless ratio:

$$\tilde{b_{bp}} = \frac{b_{bp}}{b_p} \tag{8}$$

In this study, we will use the bulk particulate real refractive index ($\tilde{n}_r$), which reproduces the bulk scattering properties of a water body. It represents the mean refractive index weighted by the scattering cross sections of all the particles:

$$\tilde{n}_r = \frac{\sum_{j=1}^{5} n_{r,j} \times N_j\, C_{sca,j}}{\sum_{j=1}^{5} N_j\, C_{sca,j}} \tag{9}$$

Similarly, the bulk imaginary refractive index ($\tilde{n}_i$) is defined as follows:

$$\tilde{n}_i = \frac{\sum_{j=1}^{5} n_{i,j} \times N_j\, C_{abs,j}}{\sum_{j=1}^{5} N_j\, C_{abs,j}} \tag{10}$$

where $C_{abs,j}$ is the absorption cross section of particles.

Table 2. Refractive index (n_r(chlp) + $i\,n_i$(chlp)) of the sphere representing the chloroplast for two morphological models. The refractive index of the sphere representing the cytoplast is constant ($1.02 + i\,1.336 \times 10^{-4}$). The equivalent refractive index of the cell is $1.044 + i\,1.5 \times 10^{-3}$.

Model * (%cyt-%chlp)	80%–20%	70%–30%	80%–18.5%–1.5%
n_r	1.140	1.100	1.144
n_i	6.966×10^{-3}	4.688×10^{-3}	7.531×10^{-3}

* The percentages represent the relative volume of the model cytoplasm and chloroplast.

2.3. The Scattering Coefficient as Measured by In Situ Transmissometers

In field measurements, b_p is derived from the total absorption and beam attenuation coefficients (a and c, respectively) as measured by instruments such as WETLabs ac9 and its later variants. Any detector has a finite field of view (FOV), therefore beam transmissometers are defined by their acceptance angle $\theta_{acceptance}$, which differs from 0°. If we want to compare, in a future study, our theoretical results to available in situ measurements, b_p must be derived from the scattering cross section, rebuilt from the normalized phase function integrated between $\theta_{acceptance}$ and π instead of 0 and π [30]. To make a distinction, when C_{sca} is calculated by integrating the scattering function between $\theta_{acceptance}$ and π, the symbols $C_{sca}^{\theta_a}$, $b_p^{\theta_a}$ and $\widetilde{b_{bp}^{\theta_a}}$ ($= b_{bp}/b_p^{\theta_a}$) will be used. As in Twardowski et al. [5], we set the acceptance angle to 1°, which is consistent with acceptance angles of commercially available beam transmissometers such as the WETLabs C-Star (1.2^o) or WETLabs ac9 (0.93^o) ([30] and references therein).

3. Numerical Modeling of the Marine Particle Scattering

The Meerhoff Mie program version 3.0 [31] and the ScattnLay code [32,33] are used to simulate the scattering and absorbing properties of homogeneous and multilayered spheres, respectively. Radiative transfer computations are carried out given the wavelength of the incident radiation equal to 532 nm and the refractive index of sea water equal to 1.34. The Meerhoff Mie program allows simulations of a polydisperse ensemble of spheres with a large choice of PSD. The outputs are the ensemble-average quantities per particle $\tilde{F}(\theta)$, C_{sca} and C_{sca}^{bb} (Equations (2) and (3)). The ScattnLay code performs computations only for monodisperse particles. To obtain the normalized phase function and cross sections for a polydisperse population, a numerical integration over the size range must be done separately (Figure 1, N°S2). Particular attention must be paid to the integration step to guarantee the accuracy of the numerical integration.

The Meerhoff Mie program is used to generate a first dataset named DS1 based on computations for homogeneous spheres for the same case studies as in Twardowski et al. [5]. n_r ranges from 1.02 to 1.2 (with a 0.2 increment), n_i is set to 0.005, $D_{min} = 0.012$ µm, $D_{max} = 152$ µm, and ξ is between 2.5 and 5. Please note that Twardowski et al. [5] did not mix different particle components with different refractive indices, as they studied $\widetilde{b_{bp}^{\theta_a}}$ for a polydisperse population of particles having the same refractive index. In this case, Equations (5)–(7) are not useful as $\widetilde{b_{bp}^{\theta_a}}$ is directly related to $C_{sca}^{bb}/C_{sca}^{\theta_a}$.

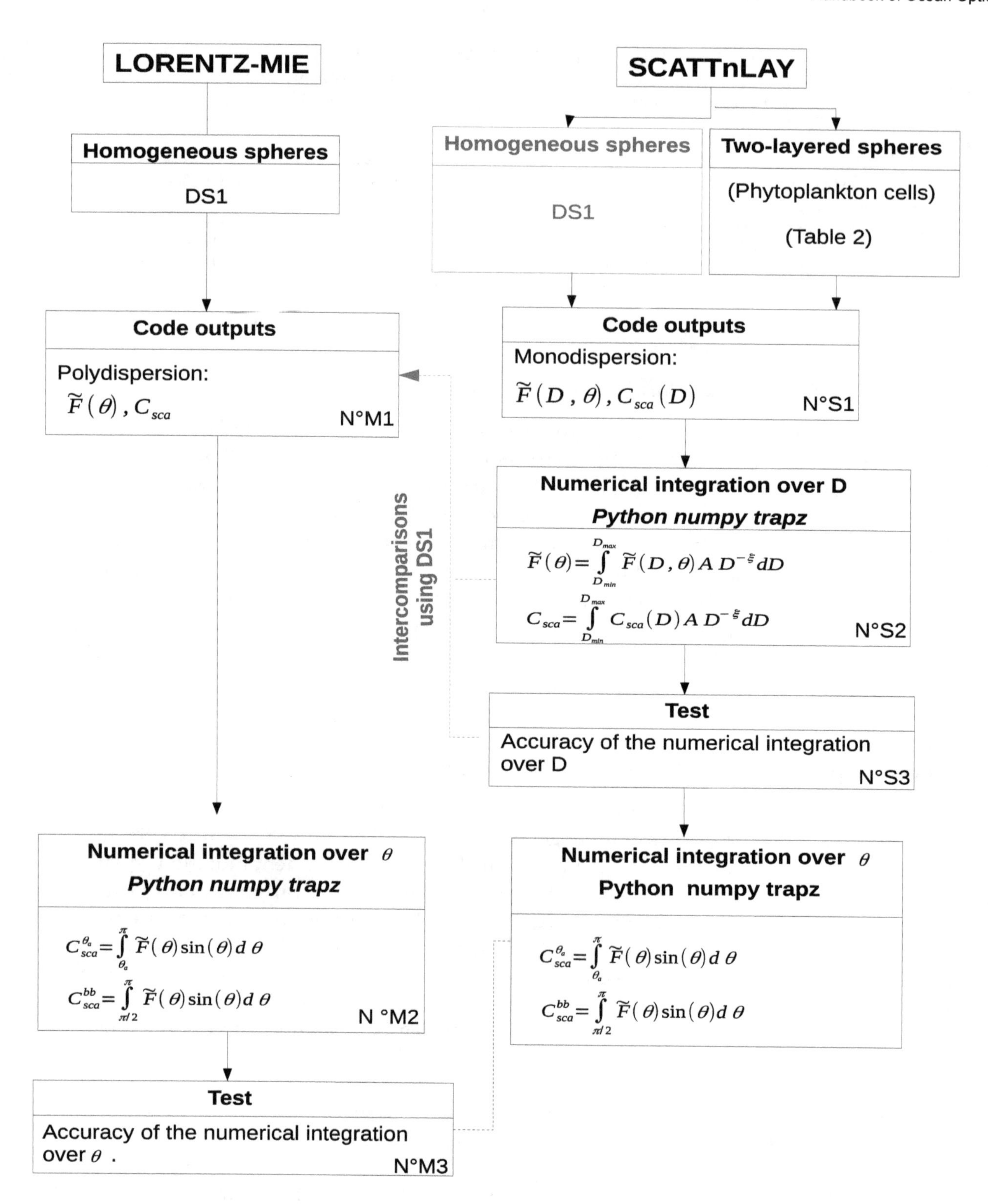

Figure 1. Flow chart of the integration procedure applied to the MIE and ScattnLay outputs.

In the second dataset (named DS2), a distinction is made between VIR, BAC, PHY, DET, and MIN in terms of internal structure, refractive index, and size range. The scattering properties of phytoplankton cells are modeled using the two-layered sphere model as in Robertson Lain et al. [23]. These investigators showed that a chloroplast layer (chlp) surrounding the cytoplasm (cyt) is an optimal morphology to simulate optical properties of algal cells. Based on their study, the value of the real part of the refractive index of the sphere representing the cytoplasm is fixed to 1.02, and the value of the imaginary part at 532 nm is 1.336×10^{-4} [23]:

$$n_i(\text{cyt}, 532\,\text{nm}) = n_i(\text{cyt}, 400\,\text{nm}) \times \exp[-0.01 \times (532 - 400)] \tag{11}$$

with n_i(cyt, 400 nm) = 0.0005. Concerning the sphere representing the chloroplast, n(chlp) is calculated according to the Gladstone and Dale formula [34]:

$$\sum_k n_k \times \vartheta_k = n_{equ}, \tag{12}$$

where n_k and ϑ_k are the complex refractive index and the relative volume of the k-th layer, and n_{equ} is the complex equivalent refractive index of the whole particle. The knowledge of the complex equivalent refractive index is useful to compare the simulations of heterogeneous spheres among themselves, regardless of the number of layers and the relative proportion of each layer. The complex equivalent refractive index is kept constant ($n_{equ} = 1.044 + i\,1.5 \times 10^{-3}$). The refractive index of the sphere representing the chloroplast is described in Table 2 according to the relative volume of the modeled chloroplast (20% or 30%). We also tested a three-layered sphere model. The outer layer represents the cell membrane. We assumed that the cell membrane is non-absorbing and have a n_r = 1.09 [13]. The second layer represents the chloroplast and the third layer the cytoplasm. The values of n_r and n_i for the cytoplasm are identical to n_r and n_i for the two-layered sphere. The values of n_r and n_i for the chloroplast are adjusted according to Equation (12) to keep the complex equivalent refractive index of the cell constant (Table 2). The relative volumes are 1.5%, 18.5% and 80% for the modeled cell membrane, chloroplast, and cytoplasm, respectively.

In DS2, multilayered sphere models are not implemented for viruses, heterotrophic bacteria, organic detritus, and minerals because of the paucity of relevant information about their optical and morphometrical properties. As we cannot gather enough accurate information about the internal structure of such particles, the homogeneous sphere model is used. The suitable n_r and n_i values for viruses, heterotrophic bacteria, organic detritus, and minerals are obtained from [25] (Table 1).

4. Abundance of the Various Particulate Components

The relative concentrations N_j associated with each particle group are chosen to realistically represent the mix of marine components and to ensure that the overall size distribution matches the Junge power law (Tables 3–5 and Figure 2).

In situ laser diffraction measurements of the PSD in different oceanic regions showed that the size distribution of marine particles can be approximated by the Junge-like power law [35–37]. As discussed by Reynolds et al. [36], the power law with a single slope is a convenient empirical descriptor of the PSD, but we have to keep in mind that, in some cases (e.g., in nearshore waters and in the presence of specific populations of phytoplankton) the particle size distribution deviates from the Junge-like power law [36–38]. Relatively steep hyperbolic slopes (around 4) are encountered in open ocean waters, whereas less steep slopes (around 3.3) are characteristic of phytoplankton bloom and/or production of particle aggregates. In the present study, results are discussed for ξ = 3, 3.5, and 4 as the vast majority of hyperbolic slopes are in this range (Figure 11 in [36,37]). Results are shown also for ξ = 2.5 and 5 but hyperbolic slopes greater than 4 are much less likely to occur. Likewise, $\xi < 3$ are rare excepted when there is biological growth in the relatively large size classes and/or aggregation. To compare with typical particulate abundances estimated in natural waters, a total abundance (N_{TOT}) of 1.1262×10^{14} particles per m^3 is considered to be in Stramski et al. [25]. Three case studies are defined. The first one represents oligotrophic-like waters with no phytoplankton bloom and no-mineral particles: the phytoplankton abundance (N_{PHY}) spans from 1.1×10^9 (for ξ = 4.9) to 4.6×10^{11} (for ξ = 2.5) particles per m^3 (0.001%–0.41% of N_{TOT}). The second one represents waters with a phytoplankton bloom and no minerals, where N_{PHY} is higher as compared to the oligotrophic-like case: N_{PHY} ranges between 8.3 × 10^9 (for ξ = 4.9) and 2.3 × 10^{12} (for ξ = 2.5) particles per m^3 (0.007%–2% of N_{TOT}). The third one represents coastal-like waters with minerals and no bloom conditions: minerals are added proportionally to obtain a bulk real refractive index $\tilde{n}_r$ around 1.1. The mineral abundance (N_{MIN}) spans from 4.8 × 10^{12} (for ξ = 4.9) to 1.3 × 10^{13} (for ξ = 2.5) particles per m^3 (4.2%–11.7% of N_{TOT}).

The abundances of the different particle components can be directly compared to the abundances provided in Stramski et al. [25] as N_{TOT} is identical. In Stramski et al. [25], ξ = 4, so comparisons are possible only for this value (Table 6). We note that N_{PHY} is of the same order of magnitude. Stramski et al. [25] have higher concentrations of DET and MIN and lower concentrations of VIR and BAC. In their paper, the authors explained that the concentrations of DET and MIN were chosen to obtain realistic contributions of detrital and mineral absorption. However, they cautioned against attaching particular significance to their selected DET and MIN concentrations in the context of how well these values can represent realistic concentrations in specific oceanic water bodies. The abundances of viruses and bacteria (N_{VIR} and N_{BAC}), used in this study, agree with the Stramski and Kiefer values [24]. Stramski and Kiefer [24] (Table 1 in their paper) used N_{VIR} between 3.0 × 10^9 and 4.6 × 10^{14} particles per m^3, N_{BAC} between 3.0 × 10^{11} and 1.5 × 10^{12} particles per m^3. Middleboe and Brussard, 2017 [39] confirmed that viral abundance can reach up to 10^{14} particles per m^3. For phytoplankton, Stramski and Kiefer made a distinction between prochlorophytes, cyanobacteria, ultrananoplankton, larger nanoplankton, and microplankton. Over these different phytoplankton groups, N_{PHY} ranges between 1.0 × 10^{11} for picoplankton to 3.0 × 10^5 for microplankton. They used $N_{PHY} \geq 5 \times 10^{11}$ particles per m^3 when there is a bloom of phototrophic picoplankton.

Table 3. Relative abundance of viruses (VIR), bacteria (BAC), phytoplankton (PHY), and organic detritus (DET) with the corresponding bulk refractive index (Equations (9) and (10)) for the water body with no bloom conditions and no minerals (oligotrophic-like).

			Relative Abundance N_j (%)			
ξ	$\tilde{n}_r$	$\tilde{n}_i$	VIR	BAC	PHY	DET
2.5	1.040	4.280×10^{-4}	78.85	5.349	0.4059	15.39
3	1.042	7.570×10^{-4}	84.74	2.120	0.1002	13.04
3.5	1.043	1.034×10^{-3}	88.50	0.8244	0.0281	10.64
4	1.045	9.931×10^{-4}	91.15	0.3178	0.0084	8.528
4.9	1.047	6.718×10^{-4}	94.35	5.651×10^{-2}	0.0010	5.588

Table 4. Relative abundance of viruses (VIR), bacteria (BAC), phytoplankton (PHY), and organic detritus (DET) with the corresponding bulk refractive index (Equations (9) and (10)) for the water body with phytoplankton bloom conditions and no minerals (phytoplankton bloom).

			Relative Abundance N_j (%)			
ξ	$\tilde{n}_r$	$\tilde{n}_i$	VIR	BAC	PHY	DET
2.5	1.041	6.195×10^{-4}	51.96	3.760	1.995	42.29
3	1.041	1.048×10^{-3}	61.91	1.599	0.6165	35.88
3.5	1.042	1.313×10^{-3}	69.84	0.6575	0.1922	29.31
4	1.043	1.362×10^{-3}	76.18	0.2650	0.0600	23.49
4.9	1.044	1.194×10^{-3}	84.55	0.0499	7.367×10^{-3}	15.40

Table 5. Relative abundance of viruses (VIR), bacteria (BAC), phytoplankton (PHY), and organic detritus (DET) with the corresponding bulk refractive index (Equations (9) and (10)) for waters with minerals and no bloom conditions (coastal-like).

			Relative Abundance N_j (%)				
ξ	$\tilde{n}_r$	$\tilde{n}_i$	VIR	BAC	PHY	DET	MIN
2.5	1.103	7.322×10^{-4}	70.96	5.311	3.650×10^{-1}	11.68	11.68
3	1.108	9.361×10^{-4}	78.04	2.105	8.801×10^{-2}	9.882	9.882
3.5	1.119	6.253×10^{-4}	83.03	0.819	2.391×10^{-2}	8.066	8.066
4	1.131	1.376×10^{-4}	86.75	0.3155	6.902×10^{-3}	6.462	6.462
4.9	1.145	9.794×10^{-6}	91.47	5.607×10^{-2}	7.782×10^{-4}	4.23	4.23

Table 6. Comparisons between abundances defined in the present study and abundances defined by Stramski et al. [25]. The hyperbolic slope ξ is 4 and N_{TOT} is 1.1262×10^{14} particles per m^3.

	Abundance (Particles per m^3)				
Case Study	VIR	BAC	PHY	DET	MIN
Oligotrophic-like	1.0265×10^{14}	3.5796×10^{11}	9.4680×10^{9}	9.6046×10^{12}	0
Phytoplankton bloom	8.5799×10^{13}	2.9846×10^{11}	6.7587×10^{10}	2.6455×10^{13}	0
Coastal-like	9.7702×10^{13}	3.5536×10^{11}	7.7733×10^{9}	7.2774×10^{12}	7.2774×10^{12}
Stramski et al. [25]	2.5000×10^{12}	1.0000×10^{11}	2.4759×10^{10}	8.2500×10^{13}	2.7500×10^{13}

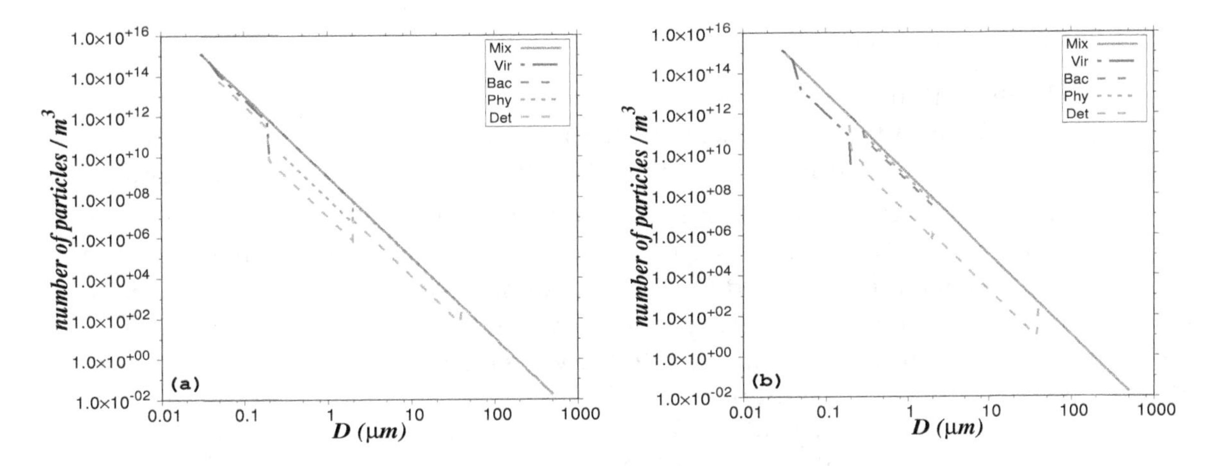

Figure 2. Composite PSD as derived from individual PSDs of the five considered particle groups for (**a**) the oligotrophic-like water body and (**b**) the phytoplankton bloom water body. $N_{TOT} = 1.1262 \times 10^{14}$ particles per m^3 and $\xi = 4$.

An indication of the total chlorophyll-a concentration is given for the oligotrophic-like, phytoplankton bloom, and coastal-like case studies. For that purpose, we considered the median intracellular chlorophyll-a values given in Brotas et al. [40]. These authors used Brewin et al. model [41] to calculate the fractional contributions of pico, nano, and microplankton to total phytoplankton chlorophyll biomass. Then, they derived the intracellular chlorophyll-a per cell for each size class from the results of cell enumeration (microscope counts and flow cytometry) and the chlorophyll-a concentration for that size class given by the Brewin et al. model. The computed median intracellular chlorophyll-a values were 0.004, 0.224, and 26.78 pg Chla cell^{-1} for pico, nano, and microplankton, respectively. In our study, we multiplied the corresponding intracellular chlorophyll-a content by the numerical abundance of pico-, nano-, and micro-plankton as derived from the PSD and we summed the chlorophyll-a concentration per class to obtain the total chlorophyll-a concentration ([Chla], units mg m^{-3}) (Table 7).

Table 7. Total chlorophyll-a concentration for the three case studies.

	Oligotrophic-Like	Phytoplankton Bloom	Coastal-Like
ξ	[Chla]	[Chla]	[Chla]
3	8.35	11.51	7.497
3.5	0.773	1.580	0.6889
4	0.102	0.341	0.0884

The chlorophyll-a concentration (mg m^{-3}) is estimated, as an indication, using the relative abundance of phytoplankton cells described in Tables 3–5 and considering N_{TOT} is 1.1262×10^{14} particles per m^3.

We emphasize that total chlorophyll-a concentrations are given as an indication as they depend on the abundance of phytoplankton, which in turn depends on N_{TOT} and ξ. For the oligotrophic-like case study, [Chla] ranges from 0.10 for $\xi = 4$ to 8.4 mg m^{-3} for $\xi = 3$. However, in oligotrophic waters, in situ measurements of PSD showed that ξ values are around 4. For ξ between 3.5 and 4, [Chla] is less than 1 mg m^{-3}, which is typical [Chla] in oligotrophic waters. In bloom conditions, the hyperbolic slope can be less than 4. For example, Buonassissi and Dierssen [35] found ξ around 3.3 in bloom conditions. For $\xi = 3.3$, we found [Chla] of 1.92 mg m^{-3}. For the coastal case study, [Chla] is low as compared to in situ [Chla] values in coastal areas. This is because we considered a high load of minerals as compared to phytoplankton abundance.

5. Results

5.1. Accuracy of Numerical Computations

A numerical integration over θ is required to derive $b_p^{\theta_a}$ and b_{bp} from the normalized phase function (Section 2). Due to the sharp increase of the normalized phase function in the forward scattering directions (Figure 3), the selection of the relevant angular step for the numerical integration is crucial. For that purpose, the impact of angular step ($\Delta\theta$) on the calculation of $\widetilde{b_p^{\theta_a}}$ is studied using Lorentz-Mie simulations in DS1 (Figure 1, N°M2, M3). The normalized phase function of polydisperse particles $\tilde{F}(\theta)$ exhibits a maximum around $\theta = 0^o$ [26]. For small ξ value, that is when the proportion of large-sized particles compared to smaller particles increases, the forward peak is sharper. Indeed, for particles with a large diameter as compared to the wavelength, $\tilde{F}(D,\theta)$ displays a sharp forward peak [26] due the concentration of light near $\theta = 0^o$ caused by diffraction. The presence of the peak in $\tilde{F}(\theta)$ requires several integration points large enough to provide the desired numerical accuracy. The numerical integration over θ (Figure 1, N°M2) is performed using the "Trapz" function from the Numpy package with Python. The "Trapz" function performs an integration along the given axis using the composite trapezoidal rule. To test the accuracy of the integration and to find the correct integration step, $\Delta\theta$, we compare the result of the numerical integration of $\tilde{F}(\theta)$ between 0 and π to its theoretical value (=2) (Figure 1, N°M3). When $\Delta\theta = 0.05^o$, corresponding to a total number of integration steps (N_θ) of 3600, the numerical integration value of $\tilde{F}(\theta)$ is in the range [1.999–2.000] for small ξ. For larger ξ, it is in the range [1.800–1.999]. When the value of the numerical integration is in the range [1.800–1.999], a renormalization factor is applied to $\tilde{F}(\theta)$ to ensure that the result of the numerical integration is exactly 2. We could also increase the number of integration points, but it would increase the computation time. Using a renormalization factor for large ξ is a good compromise to guarantee the accuracy and save computation time.

For two-layered spheres (i.e., phytoplankton cells), the ScattnLay code provides only normalized phase functions for monodisperse particles (Figure 1, N°S1), so the numerical integration over the particle diameter range (Equation (2)) is realized as a separate calculation with the Python "Trapz" function (Figure 1, N°S2). For monodisperse particles, the normalized phase function displays a forward peak as explained above but can also display a sequence of maxima and minima due to interference and resonance features [26,42]. The frequency of the maxima and minima over the range of θ increases with both increasing n_r and size parameter ($=\pi D/\lambda$). To test the accuracy of the numerical integration over the particle diameter range (Figure 1, N^{o}S3), we ran the ScattnLay code for DS1 case studies and compared $\tilde{F}(\theta)$ and C_{sca} rebuilt from Equation (2) with Lorentz-Mie computations as the Lorentz-Mie code provides the polydisperse phase functions and cross section as outputs (Figure 1, N^{o}M1). Note that even a narrow polydispersion washes out the interference and resonance features, which explains why most natural particulate assemblages do not exhibit such patterns [26,42] (Figure 3). A perfect match is obtained between the ScattnLay-rebuilt-polydisperse and Lorentz-Mie-polydisperse $\tilde{F}(\theta)$ and C_{sca} values when the integration step (ΔD) is set to 0.01 μm for D in the range [0.03, 2 μm]; 0.1 μm for D in the range [2, 20 μm]; 2.0 μm for D in the range [20, 200 μm]; and 10.0 μm for D in the range [200, 500 μm].

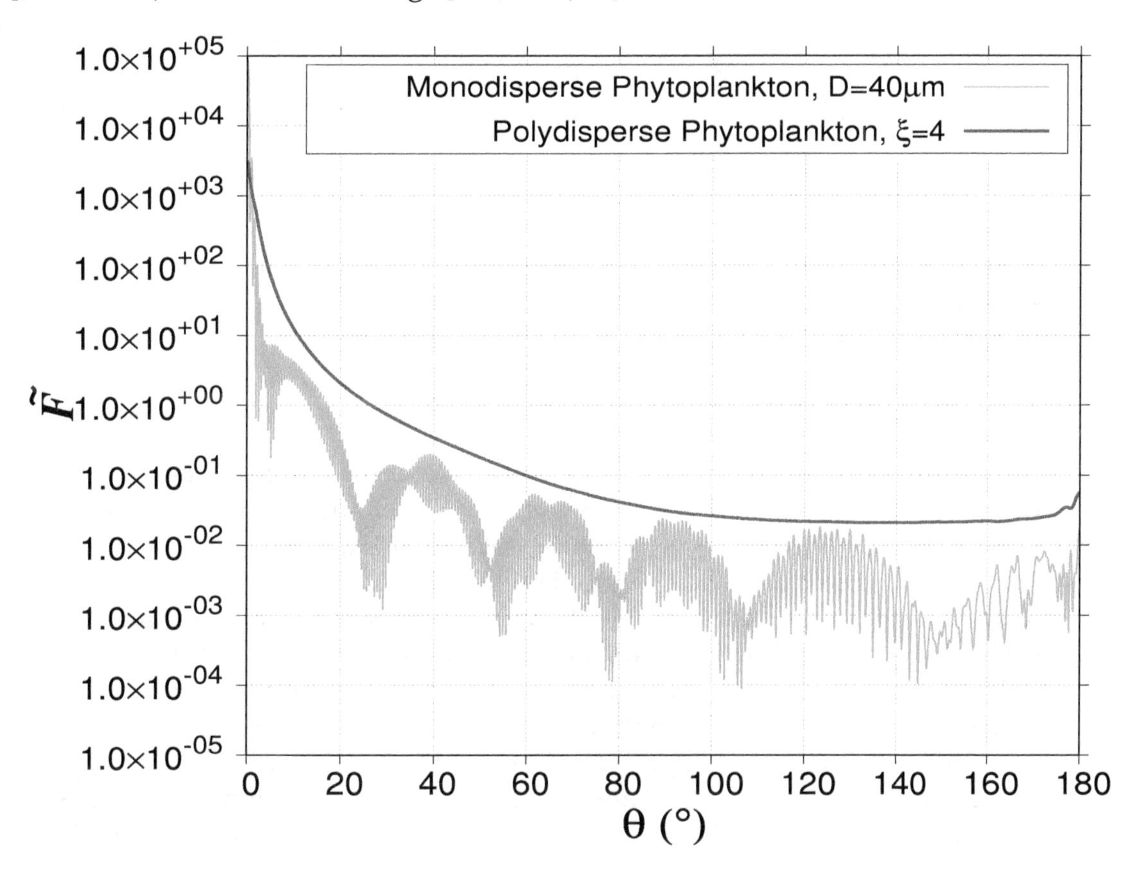

Figure 3. Interference and resonance features observed for the scattering phase function of monodisperse particles (light green). The major low-frequency maxima and minima are called the "interference structure". The high-frequency ripples are resonance features. The interference and resonance feature are washed out for a polydisperse assemblage of particles (dark green).

Using the DS1 data set, the impact of the angular integration on the backscattering ratio $\widetilde{b_{bp}^{\theta_a}}$ is examined as a function of the hyperbolic slope ξ for different values of the real refractive index and two values of total angular steps (i.e., N_θ = 750 and 3600) (Figure 4). The impact of the integration is noticeable only for ξ values lower than about 3 and relatively high n_r values. When the number of angular steps increases, the curves become flatter at low ξ values. Differences in the curve shape are reduced if we increase the angular step. For $\Delta\theta = 0.24^o$ ($N_\theta = 750$), the present results of the Lorentz-Mie calculations (solid lines in Figure 4) perfectly match those previously obtained by Twardowski et al. [5] (not shown). However, in this case (N_θ = 750), the numerical integration is not accurate enough as the integration of Equation (1) gives values between 1.999 (ξ = 4.9) and 1.04 (ξ = 2.5). In the following, $\Delta\theta$ is set to 0.05^o (N_θ = 3600) and Figure 4 (dashed lines) will be the reference figure for homogeneous spheres.

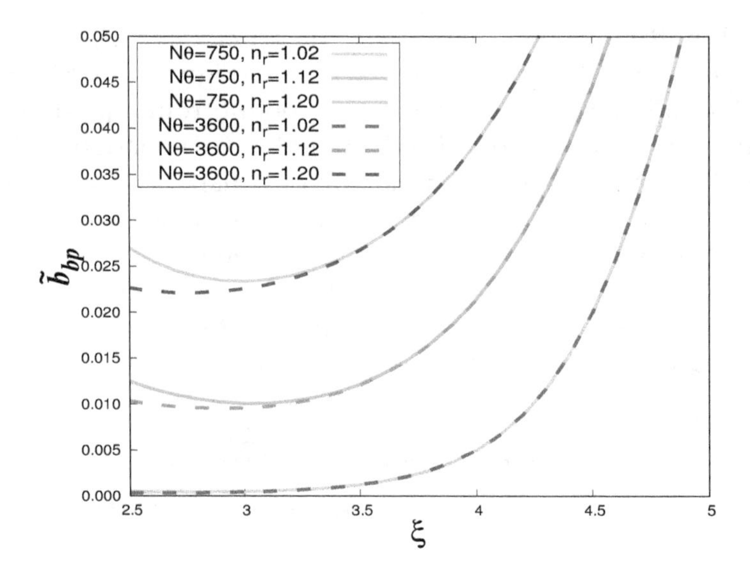

Figure 4. Results of Lorentz-Mie calculations (DS1) of the particulate backscattering ratio $\widetilde{b_{bp}^{\theta_a}}$ as a function of the hyperbolic slope, ξ, and different values of n_r and N_θ. The imaginary part of the refractive index = 0.005 as in Twardowski et al. [5]. This figure can be compared to Figure 1 in Twardowski et al. [5].

5.2. *Impact of the Structural Heterogeneity of Phytoplankton Cells on the Bulk Particulate Backscattering Ratio*

The impact of phytoplankton cell structural heterogeneity on $\widetilde{b_{bp}^{\theta_a}}$ is examined as a function of ξ for the three previously described water bodies (oligotrophic-like, phytoplankton bloom, coastal-like) considering the 80%–20% phytoplankton morphological model (Figure 5a). The real and imaginary bulk refractive indices for oligotrophic-like, phytoplankton bloom, and coastal-like case studies, vary with ξ as the relative proportions of the different particle components, having different n_r and n_i, vary with ξ (Tables 1–5 and Figure 5b,c). For the no-mineral water bodies (oligotrophic-like and phytoplankton bloom), $\tilde{n}_r$ stays around 1.04 $\pm$ 0.007 (Figure 5b). In contrast, $\tilde{n}_i$ shows large variation with ξ for both oligotrophic-like and phytoplankton bloom water bodies (Figure 5c). In bloom conditions, $\tilde{n}_i$ increases as the relative proportion of phytoplankton increases as compared to the no bloom conditions. In agreement with typical values of the oceanic bulk imaginary refractive index [43], the $\tilde{n}_i$ values for the particulate populations considered here are always lower than 0.002. In the presence of mineral particles (coastal-like), $\tilde{n}_r$ increases as MIN have a higher n_r than VIR, BAC, PHY and DET. Its values are between 1.103 ($\xi = 2.5$) and 1.145 ($\xi = 4.9$). Values of $\tilde{n}_i$ vary between 9.79×10^{-6} and 9.44×10^{-4}.

The impact of the structural heterogeneity of phytoplankton cells is evaluated by comparison with Lorentz-Mie calculations (particulate assemblages composed of homogeneous spheres only, regardless of the particle group) performed for low and high bulk refractive index. These case studies with homogeneous spheres only are named "Homogeneous reference cases". The real and imaginary values of the bulk refractive indices are 1.045 and 9.93×10^{-4} for the "Homogeneous reference case 1", 1.043 and 1.36×10^{-3} for the "Homogeneous reference case 2", and 1.131 and 1.37×10^{-4} for the "Homogeneous reference case 3", respectively (Figure 5b,c). These values of $\tilde{n}_r$ and $\tilde{n}_i$ were chosen to be equal to values of $\tilde{n}_r$ and $\tilde{n}_i$ obtained for the oligotrophic-like, phytoplankton bloom, and the coastal-like case study when ξ = 4. "Homogeneous reference cases 1 and 2" with low $\tilde{n}_r$ represent phytoplankton-dominated Case 1 waters and are compared with the oligotrophic-like and phytoplankton bloom water body, respectively. "Homogeneous reference case 3" with high $\tilde{n}_r$ represents mineral-dominated Case 2 waters and is compared with the coastal-like water body. The variation of $\widetilde{b_{bp}^{\theta_a}}$ due to structural heterogeneity of phytoplankton cells is evaluated using the relative absolute difference calculated between the homogeneous reference cases (named

x in Equation (13)) and oligotrophic, phytoplankton bloom or coastal-like water bodies (named y in Equation (13)):

$$\Delta\epsilon = \frac{|x-y|}{(x+y)} \times 200\ (\%) \tag{13}$$

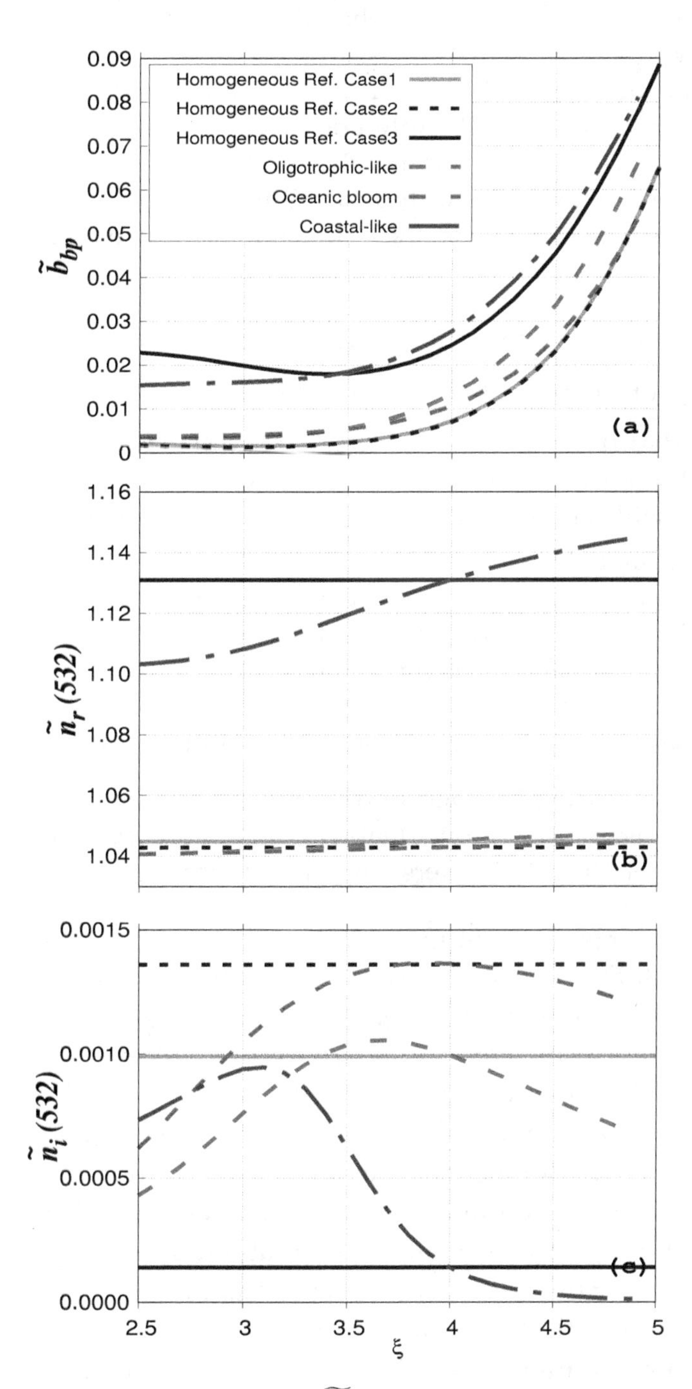

Figure 5. (**a**) Particulate backscattering ratio $\widetilde{b_{bp}^{\theta_a}}$ as a function of the hyperbolic slope for the oligotrophic-like (red dashed line), phytoplankton bloom (green dashed line), and coastal-like (brown dashed line) water bodies as described in Section 4. Black and gray lines are for homogeneous reference cases. The gray solid line corresponds to $n_r = 1.045$, $n_i = 9.93 \times 10^{-4}$, the black dashed line to $n_r = 1.1043$, $n_i = 1.36 \times 10^{-3}$, and the black solid line to $n_r = 1.131$, $n_i = 1.37 \times 10^{-4}$, respectively. Phytoplankton cells are modeled as two-layered spheres with a relative volume of the cytoplasm of 20% (%cyt-%chl = 80–20). (**b**) as in panel (**a**) but for the real refractive index. (**c**) as in panel (**a**) but for the imaginary part of the refractive index.

Even if the numerical relative abundance of phytoplankton is very small for the oligotrophic-like water body (=8.4 × 10^{-3}%), the structural heterogeneity increases the $\widetilde{b_{bp}^{\theta_a}}$ value by 58% compared to the homogeneous case ("Homogeneous reference case 1"). This is consistent with previous studies showing the large contribution of coated spheres to the backscattering signal [15,16,18,19,22,44]. The value of $\Delta\epsilon$ calculated between the oligotrophic-like and phytoplankton bloom water bodies is smaller (=22% at $\xi = 4$) even if $\tilde{n}_i$ is different (9.93 × 10^{-4} for oligotrophic-like against 1.36 × 10^{-3} for phytoplankton bloom). This latter pattern provides evidence that the structural heterogeneity (coated-sphere model) has a greater impact on the particulate backscattering ratio than the tested increase in the bulk imaginary refractive index. The relative absolute differences between the phytoplankton bloom and "Homogeneous reference case 2" is 41%. When mineral particles are taken into account, $\Delta\epsilon$ is 12% between the "Homogeneous reference case 3" and the coastal-like water body. This smaller difference is because phytoplankton have a smaller impact on the bulk scattering when highly scattering particles such as minerals are added.

The impact of the relative volume of the cytoplasm on $\widetilde{b_{bp}^{\theta_a}}$ is now evaluated by comparing the change of $\widetilde{b_{bp}^{\theta_a}}$ as a function of ξ for the 80%–20% and 70%–30% models for the oligotrophic-like and phytoplankton bloom water bodies (Figure 6). The mean relative difference in $\widetilde{b_{bp}^{\theta_a}}$ is about 5.41% with a maximum value of 11.5 % ($\xi = 3$) for oligotrophic-like case study. In bloom conditions, the mean relative difference reaches 13.0% with a maximum value of 23.5% ($\xi = 3.2$). Figure 7 compares simulated $\widetilde{b_{bp}^{\theta_a}}$ when phytoplankton cells are modeled as two-layered spheres (80%–20%) or three-layered spheres (80%–18.5%–1.5%). For the oligotrophic-like waters, relative absolute differences are small. They range between 0.0174% and 1.81% with a mean value of 0.444%. For phytoplankton bloom case study, they are between 9.84 × 10^{-3} and 2.86% with a mean value of 0.894%.

Regardless of the morphological model used to optically simulate phytoplankton cells, the $\widetilde{b_{bp}^{\theta_a}}$ reaches an asymptote when ξ decreases for phytoplankton bloom water bodies ($\widetilde{b_{bp}^{\theta_a}} = 0.005$ for $\xi = 3.5$ and 0.004 for $\xi = 2.5$). The value of the asymptote is consistent with previous observations [5], which showed the lowest backscattering ratio (about 0.005) in waters with high chlorophyll-a concentration.

The contribution of the different particle groups to the backscattering ratio is presented in Figure 8 for the 80%–20% model and $\xi = 4$. For coastal-like waters, the minerals contribute more than 80% of the total $\widetilde{b_{bp}^{\theta_a}}$, whereas they contribute only 6.5% to the total particulate abundance. This percentage agrees with the results of Stramski et al. [25] (Figure 12 in their paper). Such high contribution to backscattering is due to the high real refractive index of minerals. As in Stramski et al. [25], these results show the important role of minerals even when they are less abundant than organic living and non-living particles. In oligotrophic-like waters, the contribution of heterotrophic bacteria ranges from a few to about 30% with a maximum for ξ between 3.5 and 4, which agrees with Stramski and Kiefer [24]. The contribution of viruses is quite high, about 40–60% for ξ between 4 and 5. This high contribution is explained by the extreme value of viral abundance (around 1 × 10^{14} particles per cubic meter) used in this study [24]. As for bacteria, the contribution of phytoplankton ranges between a few and 30% with a maximum around $\xi = 3$. For a ξ value of 4, typical of oligotrophic waters, the contribution is around 10%. For the phytoplankton bloom study case, the contribution of phytoplankton cells is between 10% and 60%; maximum values are reached for a PSD slope between 3 and 3.5. Such high percentages are due to the higher backscattering cross section of phytoplankton as compared to the other particles (Figure 9). The low phytoplankton abundance is offset by the high $C_{sca,PHY}^{bb}$ so that the backscattering coefficient of phytoplankton represents a significant contribution to the total backscattering.

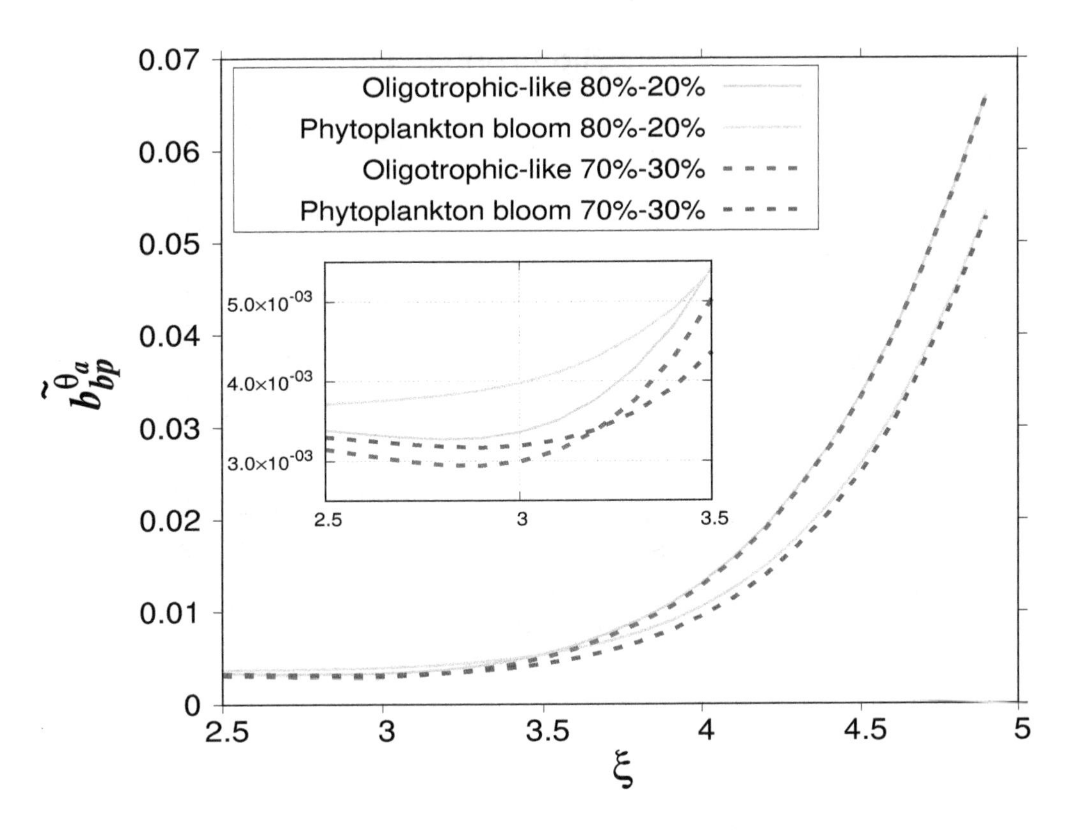

Figure 6. Particulate backscattering ratio as a function of the hyperbolic slope for oligotrophic-like and phytoplankton bloom water bodies. Phytoplankton cells are modeled as two-layered spheres with a relative volume of the chloroplast of 20 % and 30 %, as indicated.

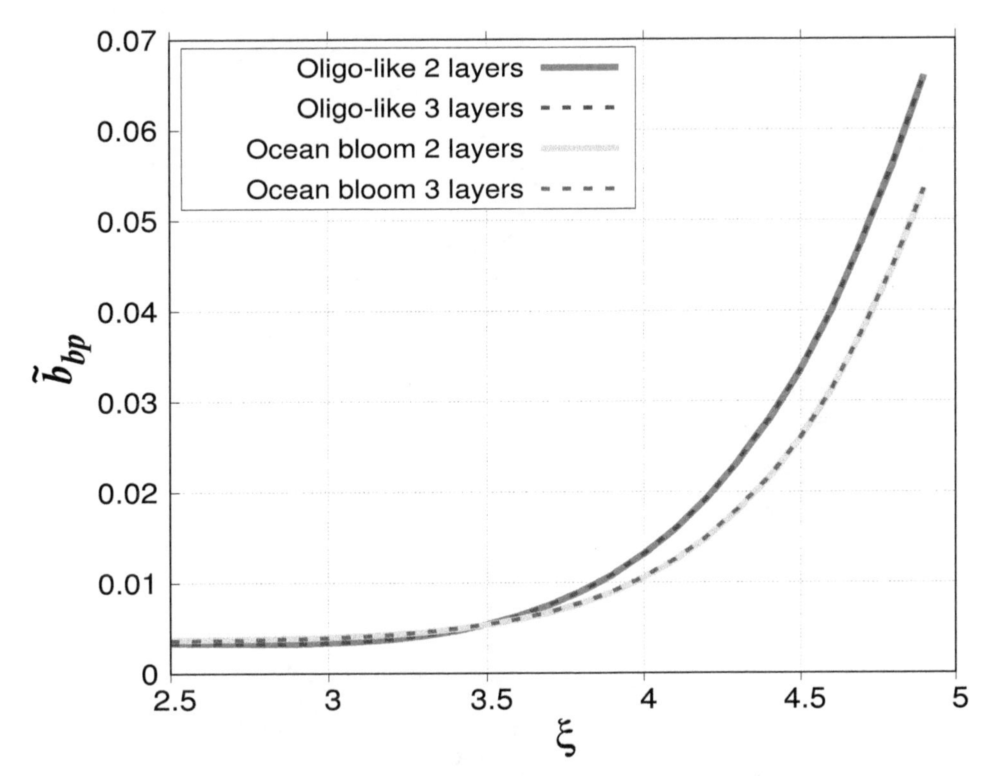

Figure 7. Particulate backscattering ratio as a function of the hyperbolic slope for oligotrophic-like and phytoplankton bloom water bodies. Phytoplankton cells are modeled as two-layered spheres (80%–20%) or three-layered spheres (80%–18.5%–1.5%), as indicated.

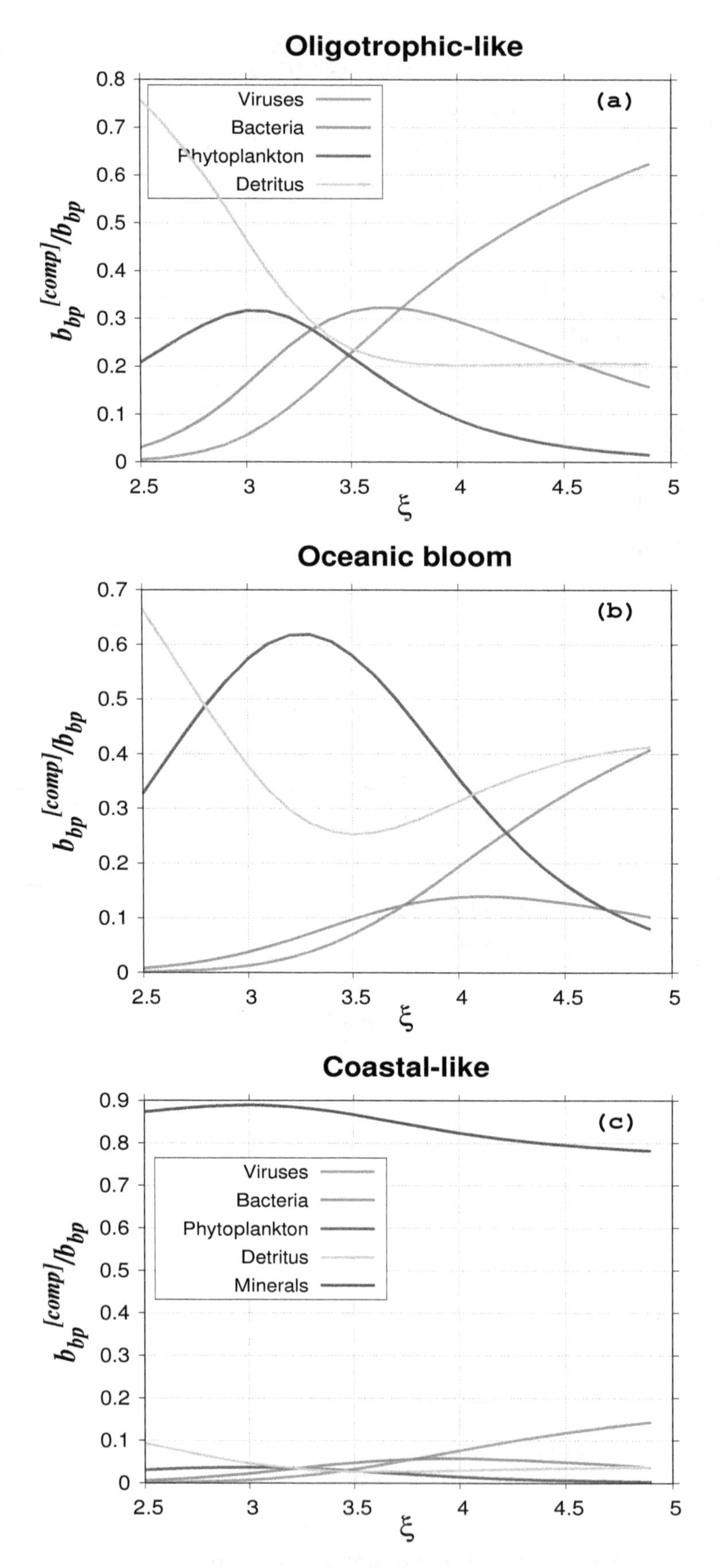

Figure 8. Contribution of the different particle groups the total bulk backscattering ratio for (**a**) oligotrophic-like, (**b**) phytoplankton bloom, and (**c**) coastal-like water bodies. The phytoplankton cells are modeled as a two-layered sphere (80%–20%).

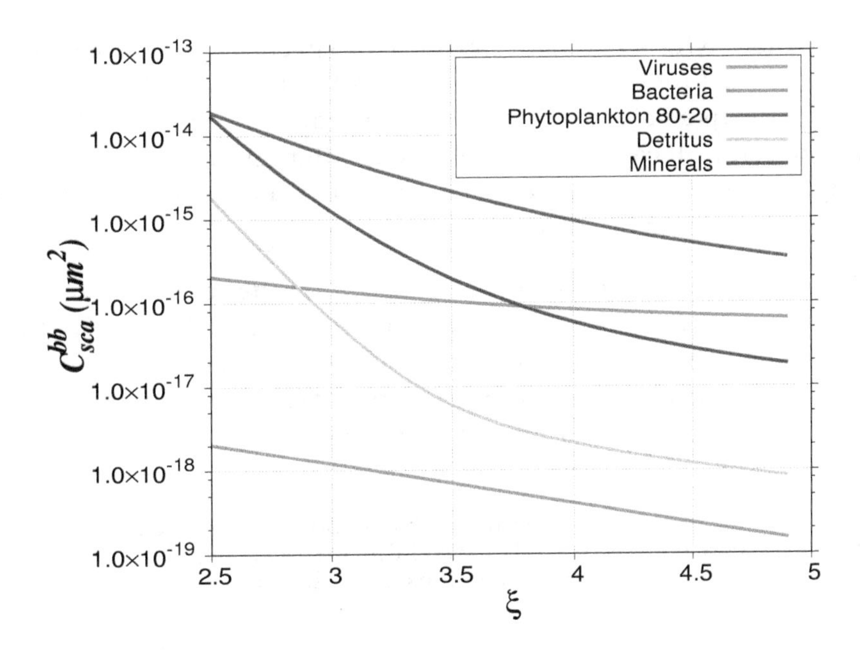

Figure 9. Backscattering cross sections, C_{sca}^{bb}, of the different particle groups. The phytoplankton cells are modeled as a two-layered sphere (80%–20%).

6. Concluding Remarks

Modeling phytoplankton cells as two-layered spheres tends to increase the bulk backscattering ratio because heterogeneous spheres are more efficient backscatterers. Even if the phytoplankton component has the smallest relative abundance, its impact on $\widetilde{b_{bp}^{\theta_a}}$ can be important depending on the hyperbolic slope of the Junge distribution. This is because phytoplankton cells have the highest backscattering cross section. For $\xi = 4$, the relative difference (in absolute value), ($\Delta\epsilon$), between $\widetilde{b_{bp}^{\theta_a}}$ for the oligotrophic-like and the reference case, having the same bulk refractive index but composed exclusively of homogeneous spheres, can reach about 58%. When minerals are added, the impact of phytoplankton decreases as the scattering by minerals dominates.

Considering different sub-populations of particles with different refractive index implies that the bulk refractive index varies with the value of the hyperbolic slope of the PSD, as the contribution of each scattering component varies. Consequently, the development of models, similar to the one proposed by Twardowski et al. [5] to derive $\tilde{n}_r$ from $\widetilde{b_{bp}^{\theta_a}}$ and ξ, but accounting for phytoplankton heterogeneity, is not straightforward. Other aspects of the problem and other parameters such as the adoption of a 2D or 3D model, the proportion between cytoplasm and chloroplast, or the variation of phytoplankton refractive index according to the considered species, would have to be taken into account. One option would be to develop a look-up table approach based on the main parameters driving the $\widetilde{b_{bp}^{\theta_a}}$ variability. For that purpose, other simulations would have to be performed to be able to identify the pertinent parameters.

In this study, a large set of numerical simulations, as well as a proper methodology have been developed to simulate the particulate scattering properties of a water body in its complexity. We show that a special care should be taken in the integration step size when the particulate scattering coefficients are calculated from the particulate scattering function, especially for relatively small values of the PSD slope. We show that an integration angular step of 0.05° ($N_\theta = 3600$) is required to obtain the required accuracy considering the inputs (refractive indices and size range) used in this study.

The method is adapted to be used repeatedly to model a very large variety of particulate assemblages. While the present study has been limited to three case studies, additional calculations can be conducted to better represent the variability encountered in oceanic environments in terms of particulate community and its complexity in terms of mixing, morphology, size, and chemical nature.

For example, in waters with significant presence of specific populations of phytoplankton or under bloom conditions, the Junge-like approximation of PSD is usually unsatisfactory [36–38,45]. We also showed that the relative volume of the model chloroplast to the cytoplasm or the presence of third layer to model the cell wall can all affect the backscattering ratio (about 5–25%). It means that changes in the phytoplankton structural heterogeneity can explain partly the natural variability of the backscattering ratio, particularly in bloom conditions.

Further work is required, mainly experimental studies, to better characterize the internal structure and chemical composition of viruses, heterotrophic bacteria, biogenic detritus, and particle aggregates. This will allow, as Robertson Lain et al. [23] did for phytoplankton, suitable models to be developed to describe properly, in numerical code, the morphological properties of various types of particles to provide more realistic simulations of their optical properties.

Author Contributions: The research is completed through collaboration of all authors. L.D.-G. and H.L. were the team leaders of this work who were responsible for coordination. D.D. was in charge of program coding and assisted in data analysis. W.M. was involved in discussions. L.D.-G. analyzed the data and wrote the paper.

Acknowledgments: The authors thank Michael Twardowski, (Harbor Branch Oceanographic Institute) for the valuable advice. We appreciate thoughtful comments from the anonymous reviewers and Dariusz Stramski, who provided time and effort to improve this manuscript. Experiments presented in this paper were carried out using the CALCULCO computing platform, supported by SCoSI/ULCO (Service COmmun du Système d'Information de l'Université du Littoral Côte d'Opale).

References

1. Preisendorfer, R.W. *Hydrologic Optics, Volume 1: Introduction*; Springfield National Technical Information Service; Office of Naval Research: Arlington, VA, USA, 1976.
2. Morel, A. *The Scattering of Light by Seawater: Experimental Results and Theoretical Approach*; Translation by George Halikas of the paper published in French in AGARD Lecture Series; N°61; North Atlantic Treaty Organization: Neuilly-sur-Seine, France, 1973.
3. Boss, E.; Pegau, W.S.; Gardner, W.D.; Zaneveld, J.R.V.; Barnard, A.H.; Twardowski, M.S.; Chang, G.C.; Dickey, T.D. The spectral particulate attenuation and particle size distribution in the bottom boundary layer of a continental shelf. *J. Geophys. Res.* **2001**, *106*, 9509–9516. [CrossRef]
4. Boss, E.; Twardowski, M.S.; Herring, S. Shape of the particulate beam attenuation spectrum and its relation to the size distribution of oceanic particles. *Appl. Opt.* **2001**, *40*, 4885–4893. [CrossRef] [PubMed]
5. Twardowski, M.; Boss, E.; Macdonald, J.; Pegau, W.; Barnard, A.; Zaneveld, J. A model for estimating bulk refractive index from optical backscattering ratio and the implications for understanding particle composition in case I and case II waters. *J. Geophys. Res.* **2001**, *106*, 14129–14142. [CrossRef]
6. Boss, E.; Pegau, W.S.; Lee, M.; Twardowski, M.; Shybanov, E.; Korotaev, G.; Baratange, F. Particulate backscattering ratio at LEO 15 and its use to study particle composition and distribution. *J. Geophys. Res.* **2004**, *109*, C01014. [CrossRef]
7. Loisel, H.; Mériaux, X.; Berthon, J.F.; Poteau, A. Investigation of the optical backscattering to scattering ratio of marine particles in relation to their biogeochemical composition in the eastern English Channel and southern North Sea. *Limnol. Oceanogr.* **2007**, *52*, 739–752. [CrossRef]
8. Nasiha, H.J.; Shanmugam, P. Estimating the Bulk Refractive Index and Related Particulate Properties of Natural Waters from Remote-Sensing Data. *IEEE J. Sel. Top. Appl. Earth Obs. Remote Sens.* **2015**, *8*, 5324–5335. [CrossRef]
9. Snyder, W.A.; Arnone, R.A.; Davis, C.O.; Goode, W.; Gould, R.W.; Ladner, S.; Lamela, G.; Rhea, W.J.; Stavn, R.; Sydor, M.; et al. Optical scattering and backscattering by organic and inorganic particulates in U.S. coastal waters. *Appl. Opt.* **2008**, *47*, 666–677. [CrossRef]
10. Sullivan, J.; Twardowski, M.; Donaghay, P.; Freeman, S. Use of optical scattering to discriminate particle types in coastal waters. *Appl. Opt.* **2005**, *44*, 1667–1680. [CrossRef]

11. Meyer, R.A. Light scattering from biological cells: Dependence of backscatter radiation on membrane thickness and refractive index. *Appl. Opt.* **1979**, *18*, 585–588. [CrossRef]
12. Bricaud, A.; Zaneveld, J.R.V.; Kitchen, J.C. Backscattering efficiency of coccolithophorids: Use of a three-layered sphere model. *Proc. SPIE* **1992**, *1750*, 27–33.
13. Kitchen, J.C.; Zaneveld, J.R.V. A three-layered sphere model of the optical properties of phytoplankton. *Limnol. Oceanogr.* **1992**, *37*, 1680–1690. [CrossRef]
14. Stramski, D.; Piskozub, J. Estimation of scattering error in spectrophotometric measurements of light absorption by aquatic particles from three-dimensional radiative transfer simulations. *Appl. Opt.* **2003**, *42*, 3634–3646. [CrossRef]
15. Moutier, W.; Duforêt-Gaurier, L.; Thyssen, M.; Loisel, H.; Mériaux, X.; Courcot, L.; Dessailly, D.; Rêve, A.H.; Grégori, G.; Alvain, S.; et al. Evolution of the scattering properties of phytoplankton cells from flow cytometry measurements. *PLoS ONE* **2017**, *12*. [CrossRef] [PubMed]
16. Poulin, C.; Zhang, X.; Yang, P.; Huot, Y. Diel variations of the attenuation, backscattering and absorption coefficients of four phytoplankton species and comparison with spherical, coated spherical and hexahedral particle optical models. *J. Quant. Spectrosc. Radiat. Transf.* **2018**, *217*, 288–304. [CrossRef]
17. Quirantes, A.; Bernard, S. Light scattering by marine algae: Two-layer spherical and nonspherical models. *J. Quant. Spectrosc. Radiat. Transf.* **2004**, *89*, 311–321. [CrossRef]
18. Vaillancourt, R.D.; Brown, C.W.; Guillard, R.L.; Balch, W.M. Light backscattering properties of marine phytoplankton: Relationships to cell size, chemical composition and taxonomy. *J. Plankton Res.* **2004**, *26*, 191–212. [CrossRef]
19. Volten, H.; Haan, J.F.; Hovenier, J.W.; Schreurs, R.; Vassen, W.; Dekker, A.G.; Hoogenboom, H.J.; Charlton, F.; Wouts, R. Laboratory measurements of angular distributions of light scattered by phytoplankton and silt. *Limnol. Oceanogr.* **1998**, *43*, 1180–1197. [CrossRef]
20. Witkowski, K.; Król, T.; Zielinski, A.; Kuten, E. A light-scattering matrix for unicellular marine phytoplankton. *Limnol. Oceanogr.* **1998**, *43*, 859–869. [CrossRef]
21. Stramski, D.; Boss, E.; Bogucki, D.; Voss, K.J. The role of seawater constituents in light backscattering in the ocean. *Prog. Oceanogr.* **2004**, *61*, 27–56. [CrossRef]
22. Whitmire, A.L.; Pegau, W.S.; Karp-Boss, L.; Boss, E.; Cowles, T.J. Spectral backscattering properties of marine phytoplakton cultures. *Opt. Express* **2010**, *18*, 15073–15093. [CrossRef]
23. Robertson Lain, L.; Bernard, S.; Evers-King, H. Biophysical modelling of phytoplankton communities from first principles using two-layered spheres: Equivalent Algal Populations (EAP) model. *Opt. Express* **2014**, *22*, 16745–16758. [CrossRef] [PubMed]
24. Stramski, D.; Kiefer, D.A. Light scattering by microorganisms in the open ocean. *Prog. Oceanogr.* **1991**, *28*, 343–383. [CrossRef]
25. Stramski, D.; Bricaud, A.; Morel, A. Modeling the inherent optical properties of the ocean based on the detailed composition of the planktonic community. *Appl. Opt.* **2001**, *40*, 2929–2945. [CrossRef] [PubMed]
26. Mishchenko, M.I.; Travis, L.D.; Lacis, A.A. *Scattering, Absorption and Emission of Light of Small Particles*; Cambridge University Press: Cambridge, UK, 2002; ISBN 9780521782524.
27. Jonasz, M. Particle size distribution in the Baltic. *Tellus* **1983**, *B35*, 346–358. [CrossRef]
28. Loisel, H.; Nicolas, J.M.; Sciandra, A.; Stramski, D.; Poteau, A. Spectral dependency of optical backscattering by marine particles from satellite remote sensing of the global ocean. *J. Geophys. Res.* **2006**, *111*, C09024. [CrossRef]
29. Morel, A.; Bricaud, A. Inherent optical properties of algal cells, including picoplankton. Theoretical and experimental results. *Can. Bull. Fish. Aquat. Sci.* **1986**, *214*, 521–559.
30. Boss, E.; Slade, W.H.; Behrenfeld, M.; Dall'Olmo, G. Acceptance angle effects on the beam attenuation in the ocean. *Opt. Express* **2009**, *17*. [CrossRef]
31. Dolman, V.L. *Meerhoff Mie Program User Guide*; Internal Report Astronomy Department, Free University: Amsterdam, The Netherlands, 1989.
32. Peña, O.; Pal, U. Scattering of electromagnetic radiation by a multilayered sphere. *Comput. Phys. Commun.* **2009**, *180*, 2348–2354. [CrossRef]
33. Yang, W. Improved recursive algorithm for light scattering by a multilayered sphere. *Appl. Opt.* **2003**, *42*, 1710–1720. [CrossRef]

34. Aas, E. Refractive index of phytoplankton derived from its metabolite composition. *J. Plankton Res.* **1996**, *18*, 2223–2249. [CrossRef]
35. Buonassissi, C.J.; Dierssen, H.M. A regional comparison of particle size distributions and the power law approximation in oceanic and estuarine surface waters. *J. Geophys. Res.* **2010**, *115*, C10028. [CrossRef]
36. Reynolds, R.A.; Stramski, D.; Wright, V.M.; Woźniak, S.B. Measurements and characterization of particle size distributions in coastal waters. *J. Geophys. Res.* **2010**, *115*, C08024. [CrossRef]
37. Reynolds, R.A.; Stramski, D.; Neukermans, G. Optical backscattering by particles in Arctic seawater and relationships to particle mass concentration, size distribution, and bulk composition. *Limnol. Oceanogr.* **2016**, *61*, 1869–1890. [CrossRef]
38. Woźniak, S.B.; Stramski, D.; Stramska, M.; Reynolds, R.A.; Wright, V.M.; Miksic, E.Y.; Cichocka, M.; Cieplak, A.M. Optical variability of seawater in relation to particle concentration, composition, and size distribution in the nearshore marine environment at Imperial Beach, California. *J. Geophys. Res.* **2010**, *115*, C08027. [CrossRef]
39. Middleboe, M.; Brussaard, C.P.D. Marine Viruses: Key Players in Marine Ecosystems. *Viruses* **2017**, *9*, 302. [CrossRef] [PubMed]
40. Brotas, V.; Brewin, R.; Sá, C.; Brito, A.C.; Silva, A.; Mendes, C.R; Diniz, T.; Kaufmann, M.; Tarran, G.; Groom, S.B.; et al. Deriving phytoplankton size classes from satellite data: Validation along a trophic gradient in the eastern Atlantic Ocean. *Remote Sens. Environ.* **2013**, *134*, 66–77. [CrossRef]
41. Brewin, R.J.W.; Sathyendranath, S.; Hirata, T.; Lavender, S.; Barciela, R.M.; Hardman-Mountford, N.J. A three-component model of phytoplankton size class for the Atlantic Ocean. *Ecol. Model.* **2010**, *221*, 1472–1483. [CrossRef]
42. Mishchenko, M.; Lacis, A. Manifestations of morphology-dependent resonances in Mie scattering matrices. *Appl. Math. Comput.* **2000**, *116*, 167–179. [CrossRef]
43. Bricaud, A.; Roesler, C.; Zaneveld, J.R.V. In situ methods for measuring the inherent optical properties of ocean waters. *Limnol. Oceanogr.* **1995**, *40*, 393–410. [CrossRef]
44. Zaneveld, J.R.V.; Kitchen, J.C. The variation in the inherent optical properties of phytoplankton near an absorption peak as determined by various models of cell structure. *J. Geophys. Res.* **1995**, *100*, 309–313. [CrossRef]
45. Reynolds, R.A.; Stramski, D.; Mitchell, B.G. A chlorophyll-dependent semianalytical reflectance model derived from field measurements of absorption and backscattering coefficients within the Southern Ocean. *J. Geophys. Res.* **2001**, *106*, 7125–7138. [CrossRef]

10

Modeling Sea Bottom Hyperspectral Reflectance

Georges Fournier *, Jean-Pierre Ardouin and Martin Levesque

DRDC Valcartier Research Centre, Québec, QC G3J1X5, Canada;
jean-pierre.ardouin@drdc-rddc.gc.ca (J.-P.A.); martin.levesque@drdc-rddc.gc.ca (M.L.)
* Correspondence: grfournier1@gmail.com

Featured Application: Hyperspectral Bathymetry and near-shore bottom mapping. Retrieving both depth and bottom types from hyperspectral remote-sensing reflectance requires inverting the remote-sensing reflectance profile to fit both the inherent optical properties of the water column and the bottom spectral reflectance profile. In order to obtain a robust fit, the number of parameters required to characterize the bottom reflectance spectrum must be kept to a minimum. The model which we have developed allows one to model a good approximation to bottom spectra by using at most three parameters.

Abstract: Over the near-ultraviolet (UV) and visible spectrum the reflectance from mineral compounds and vegetation is predominantly due to absorption and scattering in the bulk material. Except for a factor of scale, the radiative transfer mechanism is similar to that seen in murky optically complex waters. We therefore adapted a semi-empirical algebraic irradiance model developed by Albert and Mobley to calculate the irradiance reflectance from both mineral compounds and vegetation commonly found on the sea bottom. This approach can be used to accurately predict the immersed reflectance spectra given the reflectance measured in air. When applied to mineral-based compounds or various types of marine vegetation, we obtain a simple two-parameter fit that accurately describes the key features of the reflectance spectra. The non-linear spectral combination effect as a function of the thickness of vegetation growing on a mineral substrate is then accounted for by a third parameter.

Keywords: remote-sensing reflectance; bathymetry; hyperspectral; bottom mapping; radiative transfer

---

1. Introduction

The application that supplied the primary impetus for the present work was bathymetry and near shore bottom mapping. Both problems require inverting the remote-sensing reflectance profile to simultaneously fit both the inherent optical properties of the water column and the bottom spectral reflectance profile. In order to obtain robust and reliable results, the number of fitted parameters must be kept to a minimum. The parameters required to model the water column are already well known from numerous and extensive remote-sensing reflectance studies and detailed in the semi-empirical algebraic irradiance model developed by Albert and Mobley [1] that we use as a basis for our work. The model we are proposing here produces a good approximation to bottom reflectance spectra by using only three parameters.

In the near-ultraviolet (UV) and visible, the reflectance from mineral compounds and vegetation is predominantly due to absorption and backscattering in the bulk material. For most inorganic liquids or solids such as minerals the absorption comes from the broadened far wing of electronic transitions in the deep UV [2–4] and the backscattering is dominated by reflections at the interface between the crystalline grains of the material. For vegetation the absorption is primarily due to the chlorophyll-a and accessory pigments contained in the plant chloroplasts while the backscattering is due to reflections at the membranes of the cells and their inner components. The radiative transfer processes in both minerals and vegetation, even though occurring on a much smaller scale, are very similar to that occurring in murky waters. This prompted us to adapt a semi-empirical model for murky type II waters due to Albert and Mobley [1] and generalize its results with another model due to Aas [5].

Except for the obvious size scale factor, the key difference between the radiative transfer that occurs in minerals and vegetation against that found in murky waters is due to the physics of the backscattering term. We assume the backscattering term comes from the reflection of the interfaces between the structural elements of the solid. The surfaces of the interfaces are modeled to be rough and randomly oriented. The formulas for this type of backscattering are identical to those derived for randomly oriented particles with rough surfaces [6,7]. The formulas scale as a function of the relative index of refraction of the solid grains and the material of the gap. If the original reflectance was measured for dry samples, the gaps contain air. If the sample is immersed the gaps are water filled and the relative index is smaller. This occurs at or just below the surface of solid rocks and depends on the porosity and on the state and time of immersion. The same effect occurs to an even greater depth when the mineral is in powdered form such as sand We have used this effect to predict the immersed reflectance spectra given the reflectance measured in air. This new model allows one to use the vast library of spectral reflectance signatures measured in air to the underwater environment. We have also used the model in our bathymetric work by measuring the hyperspectral signature of the coastline and modifying it to use as bottom reflectance. We have found this approach to be particularly effective with sand beaches. The only parameter that needs to be estimated is the mean index of refraction of the sand grains which is very close to either silica or in some cases calcite.

To properly model vegetation absorption several effects must be accounted for. The absorption spectrum of chlorophyll-a and accessory pigments at low concentrations is modified by saturation of the absorption through the chloroplasts as the concentration increases. This is known as the package effect and has been extensively studied for spherical chloroplast by Morel and Bricaud [8]. We extend this work to include disk-shaped chloroplast. We then use the resulting formulas to fit with a single parameter the measured phytoplankton absorption spectra as a function of concentration [9,10]. The backscattering cellular interfaces are assumed to be composed of cellulose and the reflectance spectra are computed for several types of algae and underwater vegetation.

The spectra show that, as is well known, vegetation is actually translucent which means that when it grows over a mineral substrate the reflectance spectra changes significantly as a non-linear function of the thickness. We use a normalized version of the Albert and Mobley model for finite depth [1] to evaluate this effect. The reflectance from the mineral substrate replaces the bottom reflectivity in the model and the water column absorption and backscatter properties are replaced by those of the vegetation. If the vegetation cover is complete over one pixel, the complete vegetation model depends on three parameters: the concentration of chlorophyll-a, the chloroplast absorption saturation parameters and the thickness.

The aim of the present work is to help limit the number of fit parameters in order to better constrain the water depth value. This is particularly significant in conditions were there is little or no a priori knowledge of the bottom type. For convenience, Table 1 lists the symbols we use and their definitions and units.

Table 1. List of abbreviations, symbols, definitions and units.

Symbol or Abbreviation	Definition, Units
$a(\lambda)$	Absorption coefficient, m^{-1}
$a_{cl}(\lambda)$	Cellulose absorption coefficient, m^{-1}
$a_w(\lambda)$	Pure water absorption coefficient, m^{-1}
$a^*(\lambda)$	Extended Bricaud specific absorption coefficient, m^2/mg

Table 1. *Cont.*

Symbol or Abbreviation	Definition, Units
$a_o^*(\lambda)$	Specific absorption coefficient at low concentration, m^2/mg
$a_r^*(\lambda)$	Specific absorption coefficient at the reference concentration
$a_v^*(\lambda)$	Specific absorption coefficient at any concentration, m^2/mg
A_1, A_2	Coefficients for the finite thickness translucent model
α_o	Amplitude coefficient for the mineral fit, units $\lambda^{-\nu}$
$b_b(\lambda)$	Backscattering coefficient in air, m^{-1}
$b_{bw}(\lambda)$	Backscattering coefficient in water, m^{-1}
<cos>	Mean scattering cosine
$<d>$	Mean diameter of the scattering structures, m^{-1}
$\delta(\lambda)$	Bottom reflectance attenuation coefficient, m^{-1}
f	Fitting parameter for alternate $R_{spc-\infty}$ formula
f_{cl}	Mass fraction of cellulose in a vegetation cell
f_{vp}	Fraction of vegetation cover per pixel
$\gamma(\lambda)$	Translucent substance irradiance attenuation coefficient, m^{-1}
$\kappa_0, \kappa_{1W}, \kappa_{2W}, \kappa_{1b}, \kappa_{2b}$	Coefficients for the finite thickness translucent model
λ	Wavelength in air, microns
λ_0	Wavelength coefficient for the mineral fit, microns
ν	Power coefficient for the mineral fit, dimensionless
$\mu_{(d)}$	Mean value of the cell size, microns
n	Real Index of refraction in air
n_{cw}	Index of refraction of cell walls
n_w	Real Index of refraction in water
n_c	Number of cells per unit volume, m^{-3}
n_{cp}	Number of chloroplasts per cell
N	Number of scattering elements per unit volume, m^3
$p_1, p_2, p_3, p_4, p_5,$	Coefficients of the irradiance reflectance model
$p(\theta, \lambda)$	Total scattering phase function.
Q_a	Absorption efficiency, dimensionless
R_∞	Irradiance reflectance with no bottom contribution
$R_{spc-\infty}$	Spectralon reference normalized irradiance reflectance
R_b	Bottom irradiance reflectance
R_m	Mixed pixel irradiance reflectance
R_t	Irradiance reflectance for translucent materials
ρ_{chl}	Chlorophyll-a mass density, mg/m^3
ρ_{cp}	Chlorophyll-a mass density inside the chloroplasts, mg/m^3
ρ_r	Chlorophyll-a mass density concentration reference, mg/m^3
σ_g	Geometric cross-section, m^2
$\sigma_b(\lambda)$	Backscattering cross section, m^2
$\sigma_{(d)}$	Standard deviation of the cell size, microns
$\sigma_r(\%)$	Standard deviation of the relative error, units %
θ_s	Sun zenith angle in water
τ_{cp}	Thickness of the disk shaped chloroplasts, m
u_b	Backscattering coefficient times z_b, dimensionless
u_{cp}	$\rho_{cp}\tau_{cp}$, units, mg/m^2
u_r	u_{cp} at the reference chlorophyll-a concentration ρ_r
V_c	Volume of vegetation cell, m^3
V_{cp}	Volume of chloroplast, m^3
V_m	Volume of vegetation filled by cells, m^3
$x(\lambda)$	Backscattering albedo in air, dimensionless, range 0 to 1
$x_{ba}(\lambda)$	Backscattering albedo in air, dimensionless, range 0 to 1
$x_{bw}(\lambda)$	Backscattering albedo in water, dimensionless, range 0 to 1
z_b	Translucent material layer thickness, m^{-1}
ω_b	Backscattering reflection coefficient for random orientation
ω_t	Reflection coefficient for random orientation, range 0 to 1

2. Materials and Methods

2.1. Basic Model

The key parameter in any radiative transfer model of reflectance is a parameter we will refer to in this paper as the backscattering albedo $x(\lambda)$. This defined as the ratio of the total scattering in the back hemisphere to the sum of the absorption and total backscattering.

$$x(\lambda) = \frac{b_b(\lambda)}{a(\lambda) + b_b(\lambda)}, \quad (1)$$

In the above expression $a(\lambda)$ is the absorption coefficient while $b_b(\lambda)$ is the backscattering coefficient. The backscattering coefficient is defined in standard form by the following expression.

$$b_b(\lambda) = 2\pi \int_{\pi/2}^{\pi} p(\theta, \lambda) \sin\theta \, d\theta, \quad (2)$$

In the expression above $p(\theta, \lambda)$ is the scattering phase function. The main aim of our work from now on is to obtain expressions for the various contributions to both $b_b(\lambda)$ and $a(\lambda)$. Figure 1 shows graphically the various mechanisms discussed above and will serve as a guide in this task.

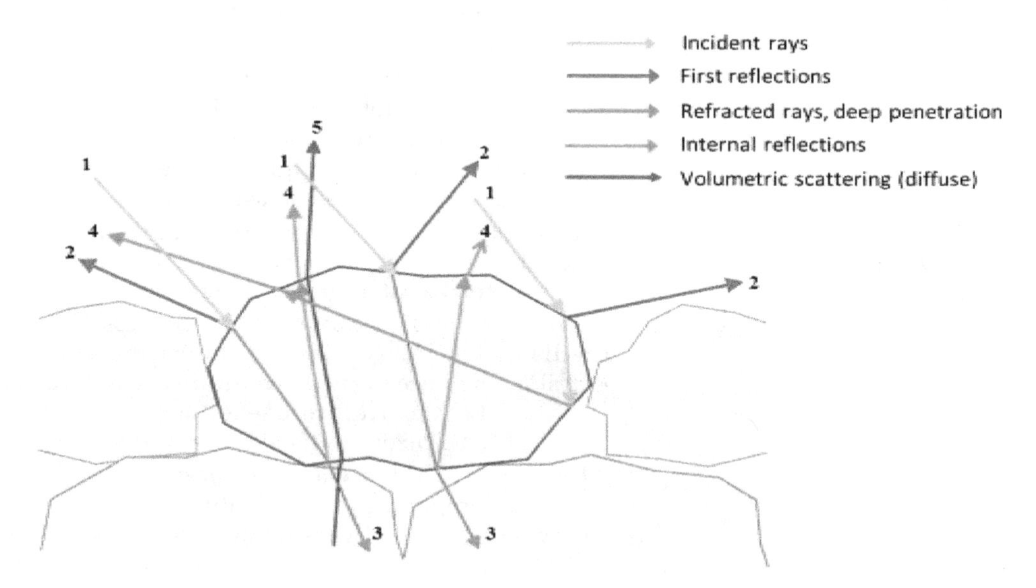

Figure 1. This figure is a schematic of the microstructure elements relevant to scattering and absorption for both minerals (grains) and vegetation (cells). The incident light rays (1) are reflected (2) and transmitted at the first surface (3). The rays transmitted through the first surface are subsequently both reflected back from the inner surfaces of the grains (4) and absorbed. The rays that penetrate deeper (5) are multiply scattered before coming back to the surface and have a near Lambertian (uniform) scattering distribution.

The materials of interest to us, minerals and vegetation, absorb little per grain or cell and the light ray will encounter many inner surfaces before being scattered back out. In our evaluation of $b_b(\lambda)$ we will treat the reflection from the first surface boundary as similar to that of the inner deeper boundaries. The formulas we will use for the backscattering from the inner surfaces of minerals are those which describe reflection from rough surface elements with random orientation. For vegetation we will use the formulas for backscattering from smooth surfaces which is a more appropriate representation. This model has been used recently to describe the backscattering from complex naturally occurring structures such as coccoliths [6,7]. The backscattering cross-section function for the reflection from a randomly oriented set of uniform (Lambertian) diffusers is given by:

$$\sigma_b(\lambda) = \sigma_g\, \omega_t(n) 2\pi \int_{\pi/2}^{\pi} \frac{4}{3\pi} (\sin\theta - \theta\cos\theta) \sin\theta d\theta = \frac{5}{6}\, \sigma_g\, \omega_t(n)\,, \tag{3}$$

σ_g is the geometric cross-section of an individual scattering structure and $\omega_t(n)$ is the Fresnel reflectance integrated over a set of randomly oriented surfaces of relative index of refraction n that together compose the surface of those structures. The total backscattering coefficient of the ensemble of the scattering elements is by definition:

$$b_b(\lambda) = \frac{5}{6}\, \omega_t(n)\, N\, \sigma_g\,, \tag{4}$$

Assuming that the number density of the scattering structures N is such that the sum of their geometric cross-sections is equal to the area of the material normal to the impinging light, we obtain the following formula for the backscattering coefficient of the material.

$$b_b(\lambda) = \frac{5}{6} \frac{\omega_t(n)}{d}\,, \tag{5}$$

where <d> is the mean diameter of the scattering structures. $\omega(n)$ for unpolarised light is given by the following formulas [3,4].

$$\omega_t = \left(\frac{\omega_\perp + \omega_\parallel}{2} \right), \tag{6}$$

$$\omega_\perp = \frac{(3n+1)(n-1)}{3(n+1)^2}\,, \tag{7}$$

$$\begin{aligned} \omega_\parallel = & \tfrac{1}{(n^2+1)^3 (n^2-1)^2} \{ (n^4-1)(n^6-4n^5-7n^4+4n^3-n^2-1) \\ & +2n^2 \left[(n^2-1)^4 \ln\left(\tfrac{n-1}{n+1}\right) + 8n^2(n^4+1)\ln(n) \right] \}, \end{aligned} \tag{8}$$

Corresponding formulas for smooth surfaces are:

$$\omega_{b\perp} = \frac{3n^4 - 16n^3 + 12n^2 - 1 + 2(2n^2-1)^{3/2}}{6(n^2-1)^2}\,, \tag{9}$$

$$\omega_{b\parallel} = \omega_{b\perp} \left[(3 - \ln 16) + \frac{37}{40} \left(\frac{n-1}{n+1} \right) \right], \tag{10}$$

$$\omega_b = \left(\frac{\omega_{b\perp} + \omega_{b\parallel}}{2} \right) \tag{11}$$

$$b_b(\lambda) = \frac{\omega_b(n)}{<d>}\,, \tag{12}$$

The formulas above were derived assuming the same Fresnel coefficients for both the entrance and exit faces of the scattering structures. We do this because for most randomly oriented convex objects the outgoing light ray has a nearly symmetrical angular relationship with the incoming light ray which implies close to identical surface reflectivity. This symmetrical relationship is strictly true for the extreme cases of spherical, cylindrical and flat plate shapes. Given the near universality of the relationship we expect that in almost all cases of interest to us any deviation from it will be small and to first order can be neglected.

2.2. Dry to Wet Reflectance Ratio

Note that the wavelength dependence of the backscattering coefficient is a direct consequence of the wavelength dependence of the relative index of refraction. One important consequence of this dependence on the relative index of refraction is the reduction in $b_b(\lambda)$ when the interfaces between the grains are filled with water instead of air. This effect is the source of the lowering of the irradiance reflectance of materials and vegetation immersed in water. Because the interstitial gaps are small the grain structure and spacing <d> is the same in both cases, we can estimate the water to air ratio directly.

$$\frac{b_{bw}(\lambda)}{b_b(\lambda)} = \frac{\omega_t(n/n_w)}{\omega_t(n)}, \tag{13}$$

Since the absorption does not change, we can directly estimate the ratio of backscattering albedo.

$$\frac{x_{bw}(\lambda)}{x_{ba}(\lambda)} = \frac{\frac{a(\lambda)}{b_b(\lambda)} + 1}{\left[\frac{\omega_t(n)}{\omega_t(n/n_w)}\right]\frac{a(\lambda)}{b_b(\lambda)} + 1}, \tag{14}$$

$x_{bw}(\lambda)$ is the backscattering albedo in water while $x_{ba}(\lambda)$ is the corresponding backscattering albedo in air.

We can at this time estimate the wet to dry reflectivity factors for three of the most important and frequently found components of materials and vegetation, crystalline quartz, calcite and cellulose.

These are shown in Figure 2. The detailed formulas as a function of wavelength for these important indices are given in Appendix A. These indices can be found in references [11–15].

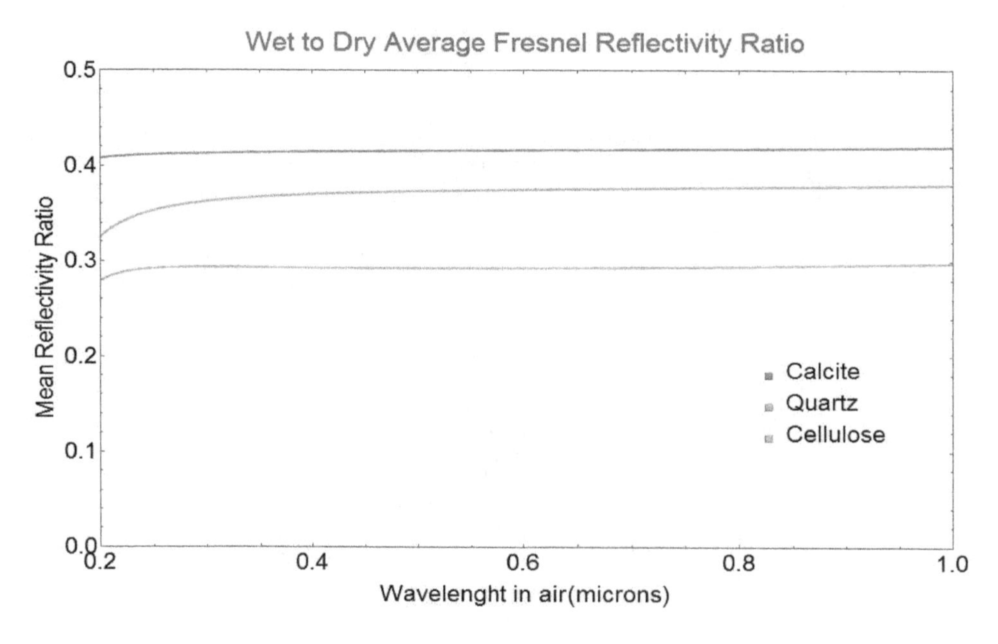

Figure 2. Wet to dry angularly averaged Fresnel reflectivity factors as a function of wavelength for important components of materials and vegetation: crystalline quartz, calcite and cellulose.

2.3. Basic Irradiance Reflectance Model

To estimate the irradiance reflectance from the backscattering albedo we use the Albert and Mobley algebraic radiance model valid for the infinitely deep medium case. The model is based on a careful analysis of solutions of the radiative transfer equation by the Hydrolight code for over 177,000 cases that encompass the full range of parameters for optically complex waters [1].

$$R_\infty = p_1 x\left(1 + p_2 x + p_3 x^2 + p_4 x^3\right)\left(1 + p_5 \frac{1}{\cos\theta_s}\right), \tag{15}$$

In this formula R_∞ is the irradiance reflectance over waters deep enough that there is no contribution from the bottom reflectance. θ_s is the sun angle just below the surface of the scattering medium. Standard irradiance reflectance measurements are carried out by comparing the signal from a high-quality diffuse reflecting surface (Spectralon) that fills the field of view of the portable spectrometer to the signal from the substance to be measured under the same illumination conditions. Therefore, we must normalize the original Albert and Mobley expression to have $R_\infty = 1$ when $x = 1$.

$$R_{spc-\infty} = \frac{p_1 x(1 + p_2 x + p_3 x^2 + p_4 x^3)}{p_1(1 + p_2 + p_3 + p_4)}, \tag{16}$$

We have used the notation $R_{spc-\infty}$ in Equation (16) to clearly note that we are referring to the calibrated Spectralon normalized irradiance reflectance but from now on we will simply assume that all reflectances have been properly normalized. Table 2 gives the coefficients of Equations (15) and (16).

Table 2. Coefficients of the Albert-Mobley model for the infinite medium depth case.

Coefficient	R_∞
p_1	0.1034
p_2	3.3586
p_3	−6.5358
p_4	4.6638
p_5	2.4121

The expression that Albert and Mobley use is based on an extensive empirical survey carried out with an exact radiative code. The results of this survey are fitted as a fourth order polynomial which is an inconvenient form to use if we need in some cases to reverse the process and, for instance, estimate x from $R_{spc-\infty}$. Aas [5] developed a two-stream radiative model and obtained approximate solutions for the irradiance reflectance from an infinite depth medium. We found that we could closely match the result of Albert and Mobley by parametrizing the formulas given by Aas. This approach yields simpler more general formulas that can easily be inverted as desired.

$$R_{spc-\infty} = \frac{(1 + f^2) - \sqrt{(1 + f^2)^2 - 4f^2x^2}}{2\, f^2 x}, \tag{17}$$

f is an empirical parameter that varies from 0 to 1. A very close fit to the results of Albert and Mobley is obtained with $= 0.79$. We have used the notation $R_{spc-\infty}$ in Equations (16) and (17) to clearly note that we are referring to the calibrated Spectralon normalized irradiance reflectance but from now on we will simply assume that all reflectance have been properly normalized to unity. Formula (17) is easily inverted to obtain x as a function of R if required.

$$x = \left(\frac{1 + f^2}{f}\right)\left(\frac{fR}{1 + fR^2}\right), \tag{18}$$

The value of f is correlated with but not equal to the mean cosine of the total scattering function <cos> which is defined as follows.

$$< \cos > = 2\pi \int_0^{\pi} p(\theta, \lambda) \cos\theta \sin\theta \, d\theta\,, \tag{19}$$

when $f = 0$, the single scattering is nearly isotropic and the irradiance reflectance is equal to the backscattering albedo. When f approaches 1 the single scattering becomes highly forward peaked. Note that $p(\theta, \lambda)$ is the total scattering function and it includes both the reflected and the transmitted part of the radiation. The transmitted part is controlled by refraction and diffraction which dominate scattering in the forward hemisphere for grains or cells much larger that the wavelength. In the cases that concern us in this work the grains or cells are large enough that the transmitted part controls the value of the mean cosine.

2.4. Irradiance Reflectance Model for Translucent Subtances

There is one more common case we have to concern ourselves with: translucent organic materials growing on a mineral substrate. In order to model this situation we use the irradiance reflection model for finite bottom depth of Albert and Mobley [1]. The irradiance reflectance of the underlying material is used as a bottom irradiance reflectance R_b in this case. The irradiance reflectance of the combination of translucent overlay of reflectance R_∞ and thickness z_b with a substrate of reflectance R_b is modeled by the following equations.

$$R_t = R_\infty \left[1 - A_1 e^{-\gamma(\lambda) z_b}\right] + R_b A_2 e^{-\delta(\lambda) z_b}\,, \tag{20}$$

with:

$$\delta(\lambda) = \left[\kappa_0 + (1 + x(\lambda))^{\kappa_{1w}}(1 + \kappa_{2w})\right]\left(\frac{b_b(\lambda)}{x(\lambda)}\right), \tag{21}$$

$$\gamma(\lambda) = \left[\kappa_0 + (1 + x(\lambda))^{\kappa_{1b}}(1 + \kappa_{2b})\right]\left(\frac{b_b(\lambda)}{x(\lambda)}\right), \tag{22}$$

Table 3 gives the coefficients of Equations (20)–(22).

Table 3. Coefficients of the Albert-Mobley model for the finite medium depth case.

Coefficient	R_∞
A_1	1.0000
κ_0	1.0546
κ_{1W}	1.9991
κ_{2W}	0.2995
A_2	1.0000
κ_{1b}	1.2441
κ_{2b}	0.5182

The new parameter that controls the behavior of the combined solution is $u_b = b_b(\lambda) z_b$. Assuming the overlaying vegetation completely covers the substrate then, as the thickness of the overlay z_b increases, the combined reflectance R_t goes to the reflectance of the overlay R_∞ while when z_b becomes small the combined reflectance approaches R_b as expected. What the model above shows is that the usual approach of linearly combining the separate reflectance signatures of the mineral substrate and the organic cover according to the weights of their relative areas only works in the limit where the cover is thick enough. For thin organic covers there is an exponential transfer of signature from substrate to cover which is a strong function of wavelength through the backscattering albedo of the translucent overlay $x(\lambda)$.

3. Results

3.1. Specific Properties of Minerals

As mentioned briefly in the introduction, in the near-UV and visible the reflectance from mineral compounds is predominantly due to absorption and backscattering in the bulk material. For minerals this absorption comes from the broadened far wing of the lowest energy electronic transitions in the deep UV [2–4] and the backscattering is dominated by reflections at the interface between the crystalline grains of the material. The usual method for obtaining the absorption spectrum of mineral compounds is to measure the transmission loss through a sample of known thickness made from mineral powder that has been pressed and sintered. This is a time-consuming process that requires great care to obtain sufficiently low backscatter. Using our model opens up the possibility of obtaining the relative absorption spectra in the visible near-infrared (IR) region by simply measuring their irradiance reflectance.

Given the irradiance reflectance R we first obtain the backscattering albedo x from Equation (18). Using the definition of the backscattering albedo (1) and the Formula (5) we derived for $b_b(\lambda)$ we obtain the following expression for the absorption spectrum:

$$<d>\, a(\lambda) = \frac{5}{6}\, \omega_t(n) \frac{(1-x)}{x}\,, \tag{23}$$

Except for the scale factor of the mean crystalline grain size <d> we can now directly obtain the absorption spectrum for any substance for which we have measured an irradiance reflectance. We will use a simple approximate empirical functional form for the far wing absorption spectrum of an electronic transition which includes the cases of broadening due to internal collision and Van der Wall like interactions in the bulk of the material.

$$<d>\, a(\lambda) = \frac{\alpha_o}{(\lambda - \lambda_o)^{\nu}}\,, \tag{24}$$

Note that there is still a considerable amount of physical meaning to the parameters in the formula above. λ_o is an estimate of the central wavelength of the lowest energy electronic transition. The value of ν is a function of the shape of the interaction potential of the molecular components of the crystalline grain. In the limiting case of an abrupt delta function like interaction potential $\nu = 2$ and Equation (24) becomes a far wing Lorentzian profile [4] which is the standard abrupt collison lineshape. In the case of a sample of Trenton Limestone measured on the shore of Lake Ontario we obtain a very good fit of <d> $a(\lambda)$ using the $\nu = 1$ solution. This solution is indicative of a smoothly varying interaction potential similar to that of a linear spring.

$$<d>\, a(\lambda) = \frac{0.081}{(\lambda - 0.183)}\,, \tag{25}$$

Many other values of ν are obviously possible and depend on the form of the interaction potential. Using expression (25) we can reconstruct the reflectance spectrum. Figure 3 shows a graph of the fit between the original reflectance and the one computed using our formulas.

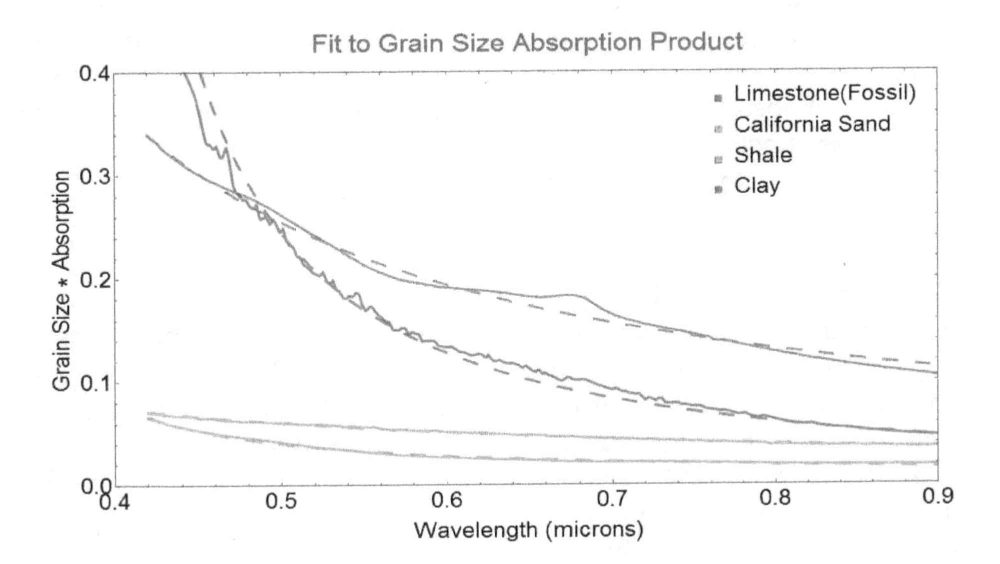

Figure 3. Comparison of the dimensionless parameter grain size times absorption coefficient estimated using Formula (23) for various mineral compounds (solid lines) with the fit (dashed lines) obtained using Equation (25). The fit parameters are given in the corresponding entries of Table 4. The spectral features seen in the experimental reflectance of the Trenton limestone sample are due to an interstitial chlorophyll-a residue lying on top of the limestone.

The fits are quite accurate over the visible spectrum from 0.35 to 0.90 micron and we expect similar accuracy for materials whose absorption is not dominated by inclusions containing color centers. The results presented in Table 4 demonstrate that this is indeed the case.

Table 4 is the result of the fit for a set of materials of interest that could possibly be found on the bottom of the water column. The fit was constrained to a region from 0.42 to 0.90 microns. This wavelength zone was chosen to avoid the reflectance measurement accuracy problems that notoriously plague the near-UV and deep blue region of the spectrum. The standard deviation of the relative error in percent between the model and the data $\sigma_r(\%)$ is given in the last column.

Table 4. This is a table of the functional fits to various minerals according to the formula.

Substance	a_o	ν	λ_o	$\sigma_r(\%)$
California Sand	0.01055	0.708	0.346	5.0
Hawaii Sand	0.04631	1.438	0.102	4.8
Greenland Sand	0.03323	0.453	0.350	4.9
Limestone (Trenton)	0.08401	1.217	0.102	3.8
Limestone (Fossil)	0.01662	0.688	0.348	6.6
Clay	0.02180	1.273	0.350	5.9
Sandy Loam	0.01087	1.882	0.195	4.1
Gray Silty Loam	0.00850	1.636	0.298	3.3
Brown Loam	0.01292	1.595	0.286	3.4
Dark Loam	0.02345	1.395	0.350	4.5
Granite	0.02102	0.737	0.304	8.6
Schist	0.11558	0.134	0.338	0.9
Shale	0.02988	0.738	0.113	1.5
Shale	0.04426	1.165	0.101	6.5
Shale	0.02364	0.219	0.344	2.8
Shale	0.11558	0.368	0.350	3.3
Siltstone	0.02669	0.229	0.345	3.9
Siltstone	0.01803	0.377	0.343	4.2

The low standard deviation of the relative error shows that the fits are very close and are in several cases within the instrumental reflectance measurement variation. Formula (24) can, therefore, serve to fit experimentally measured reflectances. We originally hoped that in the limit, the values of the parameters λ_o, ν and α_o could even be used as markers to identify an unknown material. The results given in Table 3 are not encouraging in this respect as there is a great deal of variability even for similar materials. The situation is, however, not hopeless as we have noted that several of the signatures are affected by the presence of absorbtion by organic compounds and by the colour centers of mineral inclusions. Whether these effects can be properly adressed will require further studies. We begin to address the problem of the presence vegetation in the following sections.

3.2. *Specific Properties of Vegetation*

Absorption in vegetation is controlled by the absorption of the chlorophyll-a filled chloroplasts in the cell. As the concentration of chlorophyll-a and/or the size of the chloroplasts increases the absorption through the cell increases until the chloroplast absorbs more of the light at a given wavelength until in the limit of large concentrations and/or size it becomes a dark spot masking all the light its surface intercepts at this wavelength. This absorption saturation effect was first extensively studied by Morel and Bricaud [8] who called it the package effect. This is the factor that dominantes the variability in the absorption spectrum for different types of vegetation.

To compute this effect first we need formulas for the absorption efficiency Q_a of the chloroplasts. These are derived in Appendix B for both the original model that asssumed a spherical shape for the chloroplasts and for a new model that assumes disk-like chloroplasts.

As mentioned in Appendix B, the exact formulas can be approximated to a sufficent accuracy by a simpler exponential model. From now on we will use the more realistic disk-like shape to model the absorption saturation effect.

$$a_v^*(\lambda) = \left(\frac{1}{2\,u_{cp}}\right)\left(1 - e^{-a_o^*(\lambda)2\,u_{cp}}\right), \tag{26}$$

With:

$$u_{cp} = \rho_{cp}\tau_{cp}\,, \tag{27}$$

$a_o^*(\lambda)$ is the specific mass absorption coefficient of chlorophyll-a at low concentration in units of $m^2\ gr^{-1}$. ρ_{cp} is the chlorophyll-a mass density inside the chloroplast in gr m^{-3} and τ_{cp} is the thickness of the chloroplast disk in meters. The mean thickness of a randomly oriented set of disks is 2 τ_{cp} which explains the factor of 2 seen in Equation (26).

We verified the validity of this model by first comparing the theory for disks given by Equation (26) with the Bricaud et al. [9] empirical formula for chlorophyll-a absorption in type I waters which is based on in-depth analysis of a compilation of most of the available datasets. To do this we rewrite Equation (26) as a specific absorption gain function:

$$\frac{a_v^*(\lambda)}{a_o^*(\lambda)} = \left(\frac{1}{a_o^*(\lambda)\,2\,u_{cp}}\right)\left(1 - e^{-a_o^*(\lambda)2\,u_{cp}}\right), \tag{28}$$

As can be seen in Figure 4, the overall behavior of the absorption is captured by the gain formula and this over three orders of magnitude in chlorophyll-a density. We note that the hysteresis seen in the empirical curves is due to an additional wavelength shift as a function of chlorophyll-a density. This effect was in fact observed by Gitelson [16]. The results shown in Figure 4 are a strong indication that the dominant effect in the spectral variation as a function of chlorophyll-a density is the absorption saturation effect. There was a large amount of variability in the original experimental data sets on which the empirical formulas are based so the discrepancies are not surprising. However, in the case that concerns us, which is the absorption in vegetation itself, the number density of phytoplankton which is the main uncontrolled empirical variable becomes severely constrained. The bulk chlorophyll-a mass density ρ_{chl} is given by:

$$\rho_{chl} = n_c\,n_{cp}\,\rho_{cp}V_{cp}\,, \tag{29}$$

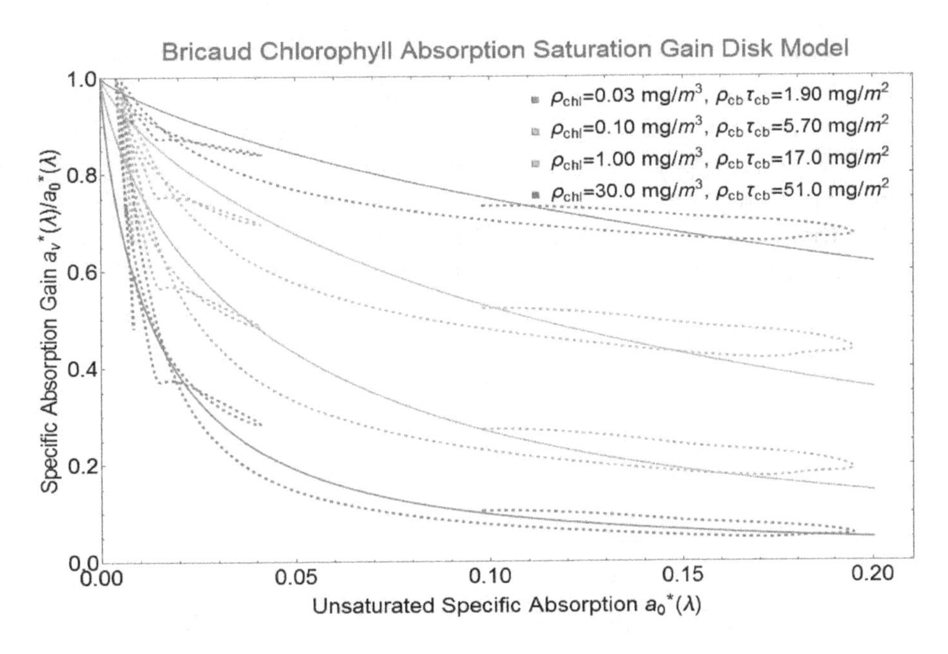

Figure 4. Graph of the ratio of the specific absorption gain to the unsaturated absorption gain. The dotted lines are from the empirical formula of Bricaud et al. [9] for chlorophyll-a. The solid lines are from Equation (21) with the parameters noted.

The number density of cells is n_c , the number of chloroplasts per cell is n_{cp} and the volume of each chloroplast is V_{cp}. In the open ocean, n_c may be weakly correlated with the other parameters while in a continuous block of cells as is the case in vegetation there is a very strongly constrained relationship. We can see this as follows. For a volume of vegetation V_m filled by cells with a volume V_c we have:

$$\rho_{chl} V_m = \left(\frac{V_m}{V_c}\right) \rho_{cp}\, n_{cp}\, V_{cp} = V_m\, \rho_{cp} \left(\frac{n_{cp}\, V_{cp}}{V_c}\right), \tag{30}$$

$$\rho_{chl} = \rho_{cp} \left(\frac{n_{cp}\, V_{cp}}{V_c}\right), \tag{31}$$

We expect the ratio of the total volume of chloroplasts n_{cp} V_{cp} to the cell volume V_c to be almost constant. The variability induced by n_c for open water has disappeared and the bulk chlorophyll-a concentration is now as expected simply proportional to the chlorophyll-a concentration inside the chloroplasts that we use to estimate the absorption saturation.

The main implication of the discussion above is that we expect to be able to model the spectral shape of the absorption spectrum of chlorophyll-a with a single fitting parameter u_{cp}. Before this becomes feasible, there are, however, several significant hurdles which have to be overcome. First, the Bricaud formula for chlorophyll-a absorption can be scaled to concentration no matter how small even for ranges that lie well outside the zone of the data used for the original fit. This creates a problem when trying to determine $a_o^*(\lambda)$ as a limiting value for low chlorophyll-a concentrations as we can extrapolate back to unphysically small values of concentration. Ciotti et al. [10] used a different approach to model the chlorophyll-a absorption from naturally occurring populations of organisms. They determine the absorption spectra for two limiting populations of organisms, the nano population and micro population. For low concentrations of chlorophyll-a the absorption spectrum of the nano population applies while for high concentrations the spectrum of the micro population is the appropriate one to use. As the concentration of chlorophyll-a increases, the spectrum evolves as a linear combination of both these extreme cases. We first attempted to use the nano population spectrum from Ciotti et al. [10] as the limiting case $a_o^*(\lambda)$ for low chlorophyll-a concentrations. Unsurprisingly, we found that the difference between our model and the Bricaud form diverged significantly at the higher concentrations.

This is problematic since for vegetation, which is the case of interest for us, the chlorophyll-a concentrations are expected to be large. In fact, they exceed the range of validity of the Bricaud formulas. To handle these extreme cases with a reasonable expectation of accuracy we decided to take a different approach. The technique is based on using a Bricaud spectrum at a given reference value with sufficiently high chlorophyll-a concentration but in a zone where the fit is still valid and extending the range from that point using the gain saturation equations. The rationale to do this is based on the fact that the concentration exceeds the measurement range and until data is available there is no other valid approach. This extension method proceeds as follows. Defining u_r as the value of u_{cp} at a reference bulk concentration ρ_r we have:

$$a_r^*(\lambda) 2u_r = \left(1 - e^{-a_o(\lambda)2u_r}\right), \tag{32}$$

$$a_o(\lambda) 2u_r = \ln[1/(1 - a_r^*(\lambda)2u_r)], \tag{33}$$

$$a_o(\lambda) = \left(\frac{1}{2u_r}\right) \ln[1/(1 - a_r^*(\lambda)2u_r)], \tag{34}$$

This $a_o(\lambda)$ is completely determined by the reference spectrum $a_r^*(\lambda)$ and the value we choose for u_r. Note that for $a_o(\lambda)$ to stay finite at all wavelengths there is a maximum value that u_r can take:

$$\max u_r = \left(\frac{1}{\max[2\, a_r^*(\lambda)]}\right), \tag{35}$$

Using (34) we can write that:

$$a_o(\lambda)2\,u_{cp} = \left(\frac{u_{cp}}{u_r}\right)\ln[1/(1-a_r^*(\lambda)2u_r)]\,, \tag{36}$$

Finally, we obtain the following general expression for an extended Bricaud absorption spectrum that can be used at chlorophyll-a densities appropriate for vegetation:

$$a^*(\lambda) = \left(\frac{1}{2\,u_{cp}}\right)\left(1-e^{-a_o(\lambda)2u_{cp}}\right)\,, \tag{37}$$

As a final practical step we need to determine what value of ρ_r we will use a reference spectrum and what value of u_r leads to the most reliable extrapolation. To do this we first choose the Bricaud spectrum for 5.0 mg/m^3 which is a value at the high end of the bulk concentration range but still well below the 20.0 mg/m^3 extreme limit of the data on which the formula was based. To determine the best value of u_r we varied that parameter until we obtained the best fit to the Bricaud spectra at 1.0, 3.0 and 10.0 mg/m^3. In all these cases we found that the optimum reference u_r asymptotically approached $\max u_r(\lambda)$. In practice, therefore, we recommend using a value of 0.99 $\max u_r(\lambda)$.

To completely model the absorption due to vegetation we need to include the absorption of water and of the cellulose that makes up the walls and internal structures of the cell. The absorption spectrum of water is taken from the data of Pope and Fry [17] for the zone from 0.38 to 0.70 microns and from the data of Kou [18] normalized to the data of Pope and Fry in their wavelength overlap zone for the 0.65 to 2.5 microns range.

The specific absorption spectrum of naturally occurring lignin cellulose from 0.4 to 2.5 microns is given in [19]. The spectrum of crystalline cellulose from 0.2 to 0.5 micron in arbitrary units can be found in reference [20]. We used the overlap zone from 0.4 to 0.5 microns with the calibrated spectrum in [16] to transform the UV-visible spectrum given in [20] to specific absorption in units of $m^2{\cdot}gr^{-1}$. We will use the pure water, chlorophyll-a and cellulose absorption spectra to model the irradiance reflectance spectra of algae and other marine vegetation.

3.3. Modeling Algae

We are now in a position to analyse the spectral signature of algae and other marine vegetation. The spectra we will be using were collected on the shores of Janvrin Island in Nova Scotia. These calibrated reflectance spectra range from 0.35 to 2.5 microns. This range extends beyond the wavelength band over which chlorophyll-a absorption has a significant amplitude. This is fortunate in as much as we can use the reflectance measured in the wavelength range over 0.90 micron to obtain a measurement of $<d>$. This is because the components which dominate absorption in that wavelength range are water and cellulose and their absolute values and relative abundances are well known. In that spectral band, therefore, we have:

$$<d> = \left(\frac{1-x(\lambda)}{x(\lambda)}\right)\frac{\omega_b(n_{cw}/n_w)}{[a_w(\lambda)(1-f_{cl})+f_{cl}\,a_{cl}(\lambda)]}\,, \tag{38}$$

f_{cl} is the mass fraction of cellulose in the cell and n_{cw} is the index of refaction of the walls of the cell and its subcomponents and n_w is the index of refraction of water. This cell wall index has been estimated for both the mesophyll and antidermal cell walls by Baranoski [12]. Since both are quite close to one another we use their average value as an estimate for n_{cw}.

We can use the fact that the mean spacing between backscattering layers $<d>$ should be independent of wavelength to estimate f_{cl}. To do this, we vary f_{cl} to minimize the variance in the estimate of $<d>$ as a function of wavelength computed with Equation (39). We use the wavelength range from 0.90 to 1.35 micron. We need to be above 0.90 microns to ensure that we are completely out of the zone where there could be remaining absorption by chlorophyll-a and other pigments in the algae. We also must stay below 1.35 micron to remain below the large water absorption band which reduces the irradiance signal to levels where instrument noise totally dominates. Figure 5 and Table 5 show the result of such a fit to the reflectance spectra of wet *Fucus* sp. and a drying mixture of *Fucus* sp. and *F. serratus* from Janvrin Island in Nova Scotia. These samples were chosen because they represent the extreme values of the reflectance spectra we measured. The results show the potential of this approach to estimate the status of the vegetation. In this case the *Fucus* sp. is much more saturated with water than the drying mixture sample while the mean backscattering feature size of the mixture is larger. The ratio of the standard deviation to the mean value of <*d*> is of the order of 5% in both cases which shows that a constant mean value <*d*> is a good model for the data.

Table 5. Functional fits to the mean spacing of backscatter layers <*d*>.

Substance	f_{cl}	$<d>(microns)$	$\sigma_{<d>}$	$\sigma_{<d>}/\mu_{<d>}$
Fucus sp.	0.043	3.29	0.15	0.048
Fucus sp. & *F. serratus*	0.164	4.07	0.21	0.054

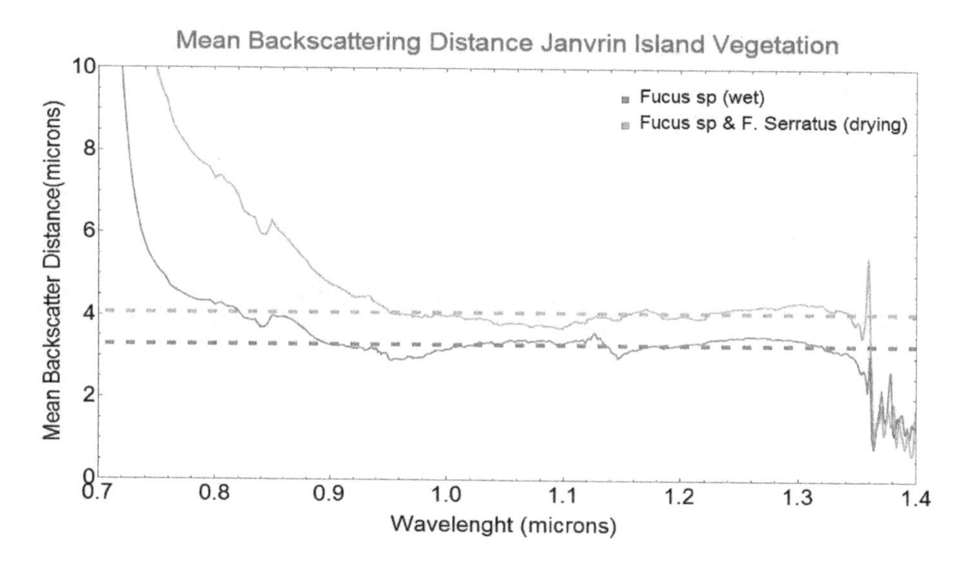

Figure 5. Fit of the mean backscatter distance derived from the irradiance reflectance of wet *Fucus* sp. and a drying mixture of *Fucus* sp. and *F. serratus* from Janvrin Island in Nova Scotia. In the zone below 0.90 microns the absorption of the chlorophyll-a and various pigments starts to dominate while in the zone above 1.35 microns the absorption of water becomes large enough that the resulting irradiance reflectance signal is dominated by noise. The full parameters of the fit are given in Table 5.

Once we have obtained the value of <*d*> we can use it in the wavelength range where the absorption of cholorophyll is dominant.

$$a_v(\lambda) = \rho_{chl} a_v^*(\lambda) = \left(\frac{1}{<d>}\right)\left(\frac{1-x(\lambda)}{x(\lambda)}\right) \omega_b(n_{cw}/n_w) - \left[a_w(\lambda)(1-f_{cl}) + f_{cl}\, a_{cl}(\lambda)\right], \quad (39)$$

As a final step we can now estimate the absorption saturation parameter u_{cp} and the bulk cholorophyll concentration ρ_{chl} by using Equation (39) and performing a non-linear least squares fit on the ratio of the experimental absorption obtained with the procedure described above to the low chlorophyll-a concentration limit $a_o^*(\lambda)$.

$$\left(\frac{\rho_{chl}}{2\,u_{cp}}\right)\left(1 - e^{-a_o^*(\lambda) 2\, u_{cp}}\right) = \left(\frac{1}{<d>}\right)\left(\frac{1-x(\lambda)}{x(\lambda)}\right) \omega_b(n_{cl}/n_w) - \left[a_w(\lambda)(1-f_{cl}) + f_{cl}\, a_{cl}(\lambda)\right], \quad (40)$$

Figure 6 shows the result of such a fit for a reflectance spectrum of wet *Fucus* sp. and a drying mixture of *Fucus* sp. and *F. serratus*. Note the significant noise increase in the short wavelength region.

This is due to the signal to noise of the reflectance measuring instrument in the blue and near UV. This significant spectral variation of the signal to noise forces us to use of an appropriate weighing function in performing the fit.

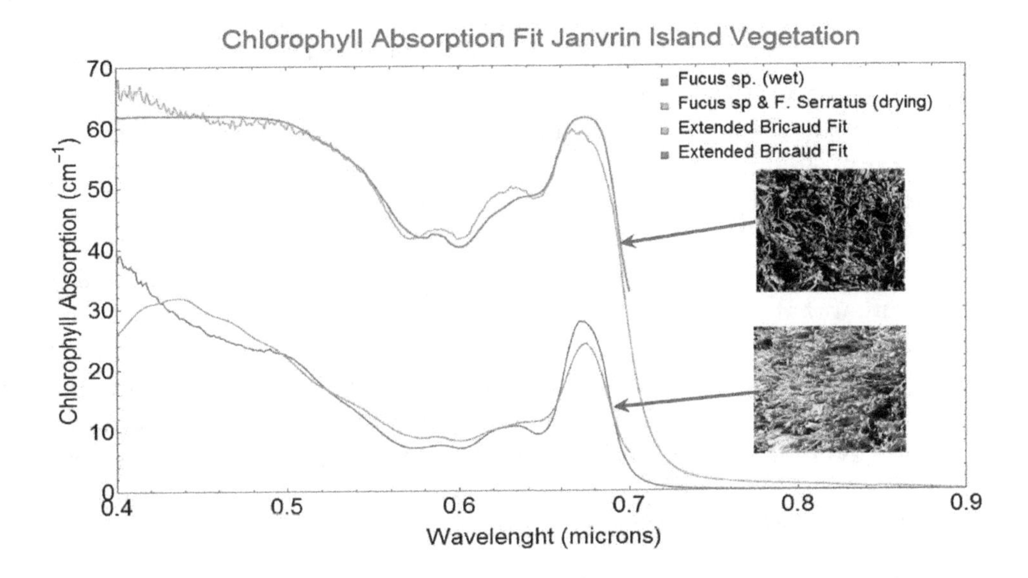

Figure 6. Fit of the absorption spectrum of wet *Fucus* sp. and a drying mixture of *Fucus* sp. and *F. serratus* with the extended Bricaud model. Note the noise due to the instrumental signal to noise degradation in the blue wavelength range. This effect was compensated by weighing the fit function inversely proportional to the S/N. The blue and yellow curves are the derived spectrum from the reflectance measurements and Equation (29). The green and red curves are the fit using Equation (37).

Once this weighing is applied we can see that the resulting modeled absorption spectrum approaches the experimental results.

In order to further verify the accuracy of the predictions of the model we have used the parameters of the fit to compute directly the predicted irradiance reflectance spectra and compare them to the original measured spectra. The results are shown in Figure 7 below.

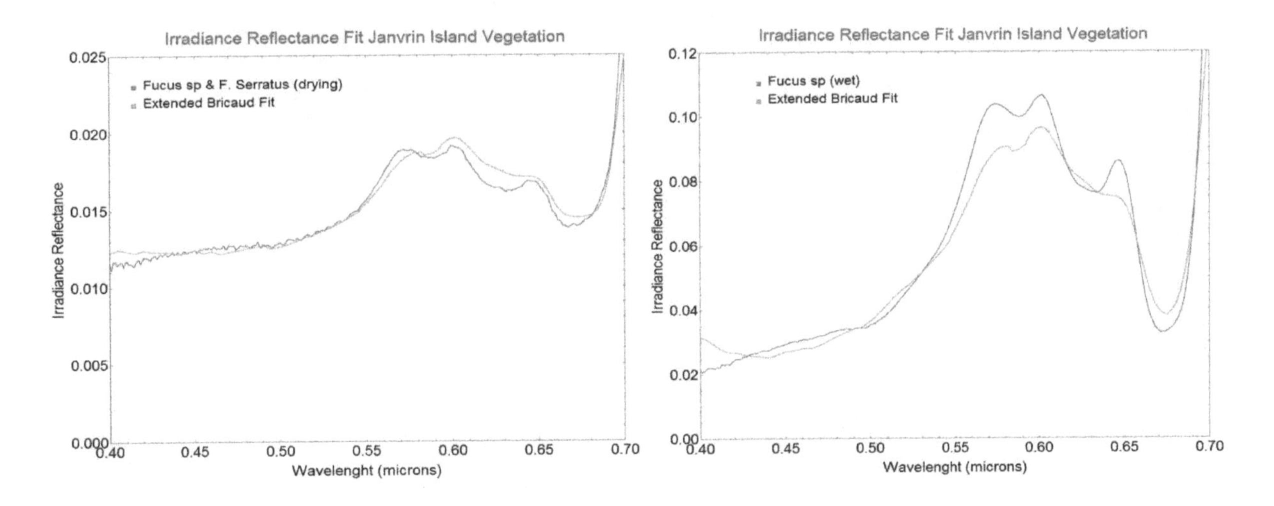

Figure 7. Comparison of modeled irradiance reflectance spectrum of wet *Fucus* sp. and a drying mixture of *Fucus* sp. and *F. serratus* (yellow curves) with the experimental measurements (blue curves). Note the noise degradation of the instrumental signal in the blue wavelength range. This reduced sensitivity may explain part of the incipient discrepancy in that spectral region.

The error between the modeled and measured reflectance signatures is of the same relative magnitude as the corresponding error in the absorption fit shown in Figure 6. Given the simplicity and generality of the model the overall precision of the fit is sufficient to satisfy our original purpose of hyperspectral bathymetry in uncharted waters.

One should note that the present model does not explicitely involve accessory pigments such as fucoxanthin which is known to be present in *Fucus*. It is based on the transformation due to absorption saturation in the chloroplasts (package effect) of the spectrum of phytoplankton. This phytoplankton spectrum is taken here as an archetype of a naturally occuring assemblage of various pigments dominated by chlorophyll-a. The absorption saturation effect shifts the resulting reflectance spectrum to the yellow and red which accounts for the relative closeness of the fit even without specific contributions from the accessory pigments. In the bathymetry application which most concerns us, the overall spectral shift and absolute level of the absorption are the key parameters needed to obtain reliable estimates of the depth. Given the exact pigment composition one could obviously improve the fit to the reflectance spectrum. However, this would defeat the purpose of obtaining the depth and the bottom spectrum without any a priori information other than the reflectance spectrum of the shoreline. This approach can, however, be improved by an iterative technique as we shall see in the discussion.

3.4. Non-Linear Effects of Vegetation Cover

We can now compute the effect of translucent vegetation growing over a mineral substrate. Because we have already obtained the mean cell size of the algae <*d*> we can directly compute its $b_b(\lambda)$ from Equation (5). This allows us to evaluate all the terms in Equation (16) and solve for the reflectance spectrum as a function of the actual thickness of the vegetation layer.

Figure 8 shows the variation in the spectral reflectance signature as a function of thickness for fucus over Trenton limestone. In the near IR, the spectra evolve from a high reflectance translucent signature for the pure *Fucus* to the low reflectivity of the wet limestone while in the visible zone that trend is reversed and the spectra go from the low reflectance of pure *Fucus* to the higher reflectance of wet limestone.

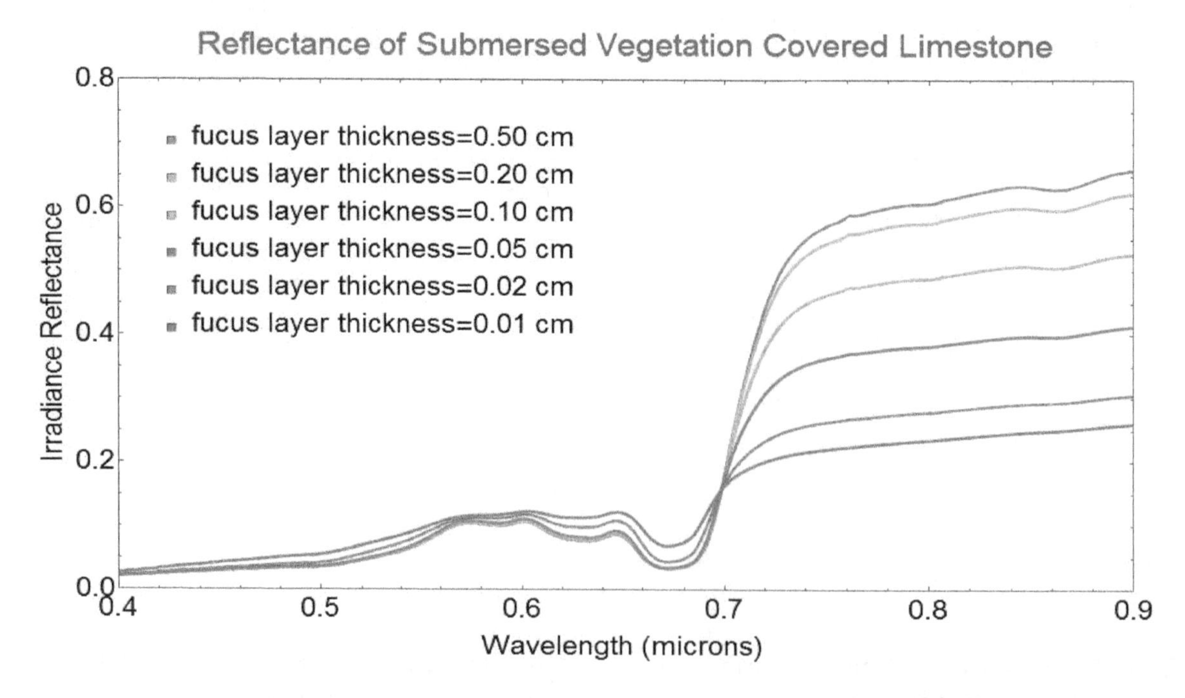

Figure 8. Computed variation of the spectral signature of translucent fucus vegetation over Trenton limestone as a function of the thickness of the layer. The mean spacing between scattering surfaces <*d*> is 3.3 microns.

Note that in cases where we don't have a separate estimate of $b_b(\lambda)$ we simply need to use directly the parameter $u_b = b_b(\lambda)\, z_b$. Given that we know the limiting spectra for the pure vegetation and for the mineral substrate u_b is the only parameter required to define the reflectance spectrum of their combination.

4. Discussion

The simple model presented in the previous sections leads to several important insights into the behavior of the irradiance reflectance spectra of minerals and vegetation in the underwater environment. The first significant result is that we are now able to estimate the ratio of the reflectance of materials immersed in water to their dry state. Figure 9 shows the ratio of irradiance reflectance for limestone and for beach sand that can be computed using Equations (14) and (17) from our model.

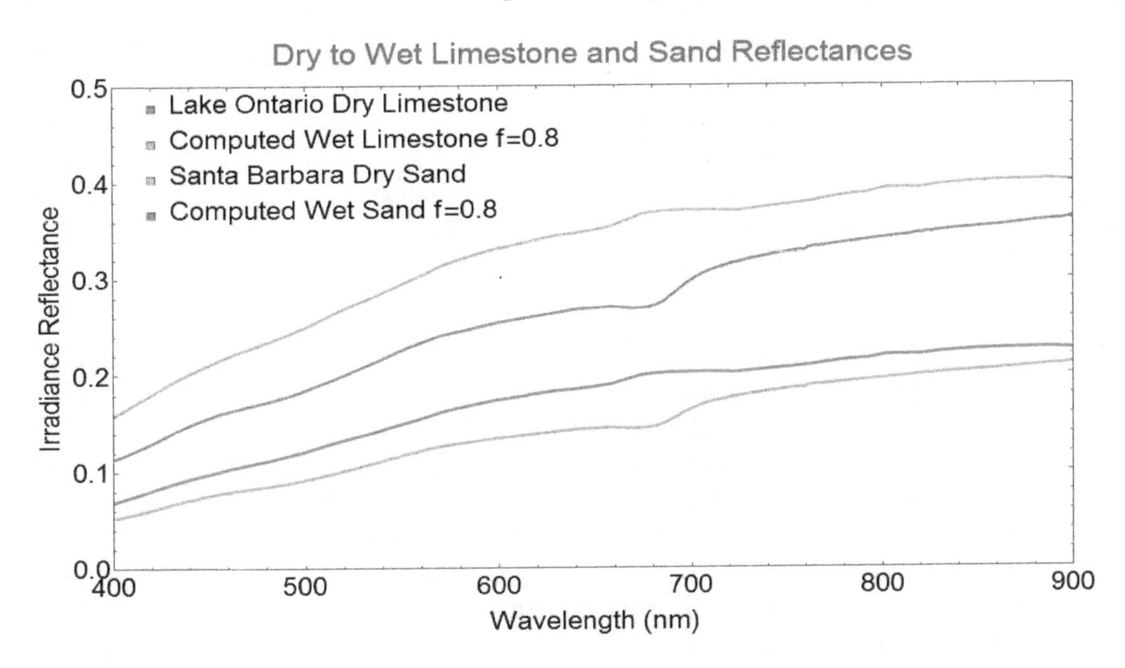

Figure 9. Computed irradiance reflectance for wet limestone and beach sand. The spectral signature for dry limestone comes from the shore of Lake Ontario. The signature of dry sand comes from a beach in Santa Barbara.

The ability to transfer reflectance spectra measured in air to their in-water equivalent is of great importance in practice as there are many comprehensive sources of spectral signatures measured in air while very few data are available under water due to the obvious difficulties in measurement. These underwater reflectance spectra are the backbone of all near shore shallow water hyperspectral surveys and the accuracy of any depth or bottom cover composition depends directly on their estimates.

Using the algebraic radiative transfer model, we have shown that we can estimate directly from reflectance measurement the relative absorption spectrum $<d>\, a(\lambda)$. Using these spectra, we have managed to obtain a simple and accurate fitting function whose structure is nevertheless based on fundamental physical considerations in far wing line broadening of the absorption from the lowest energy electronic transition in the material. Note that, strictly speaking, this fitting function is only appropriate for dielectric materials since the presence of the conduction bands in metals is not accounted for. The existence of this simple function valid over the range of wavelength of relevance to underwater hyperspectral measurements opens the possibility of identifying the material by a direct fit to the absorption parameters obtained from an inversion of the irradiance reflectance measurements of the water column using the algebraic radiative transfer model. The variance of the ν and λ_o parameters seen in Table 4 which control the spectral shape of the material may be indicative of a fundamental difficulty in obtaining directly bottom-type identification from the measured airborne hyperspectral reflectance. Addressing the scope and precise nature of this problem will be the subject of a further study.

We have also obtained a similarly simple four-parameter fitting function to the reflectance spectra of vegetation. The first parameter is the mean size of the vegetative cells $<d>$, the second parameter is the chlorophyll-a absorption saturation factor u_{cp}, the third parameter is the mass fraction of cellulose contained in a cell f_{cl}, and the fourth parameter is the bulk chlorophyll-a mass concentration ρ_{chl}. The backscattering term is controlled by the relative index of cellulose in water and the mean size of the cells while the absorption is the weighed sum of the absorption of water, cellulose and chlorophyll-a. The cell absorption is composed the absorption of pure water, cellulose and chlorophyll-a. The shape of the absorption of chlorophyll-a is controlled by the package effect through the absorption saturation parameter u_{cp} and its magnitude is controlled by the chlorophyll-a bulk density ρ_{chl}. The parameters of the model are interrelated. Relationships such as the one given in Equation (27) open up the possibility of obtaining estimates of parameters such as the ratio of the total volume of the chloroplasts to the total cell volume which could be used as an indicator of cell health.

We have shown that the reflectance of a mix of vegetation and minerals is not just a simple relative area coverage problem. When vegetation grows on top of a mineral substrate there results a combined spectrum which depends in a highly non-linear fashion on the product of thickness of the vegetation times its backscattering coefficient u_b. The overall effect for the reflectance spectra $R_m(\lambda)$ of pixels which are partially covered in vegetation is a combination of this non-linear mixing and area coverage factor.

$$R_m(\lambda) = \left[R_b(\lambda)(1 - f_{vp}) + f_{vp}\, R_t(\lambda)\right], \tag{41}$$

f_{vp} is the fractional per pixel vegetation cover. Given that the vegetation is generally expected to be of the same type and in substantially the same state of health over areas larger than a pixel it will in many cases be possible to separate the area coverage factor f_{vp} from the backscattering thickness factor u_b. This new information has the potential to increase significantly the level of knowledge about the ecologically relevant status and distribution of the near shore underwater vegetation.

In summary, the model we proposed here helps limit the number of parameters that need to be fitted for an analysis of the marine environment with hyperspectral irradiance reflectance spectra. This is an important factor because of the restricted wavelength band available when working in the underwater environment. The added complexity of the overlying water column absorption and scattering spectrum renders extremely difficult and unstable any inversion directly based on fitting linear combinations of bottom reflectance spectra. The low reflectance values and the low signal-to-noise ratios as depth increases severely affect the detectable level of spectral variation. The spectral angle is often near or within the noise band so the only hope for reasonable depth and type of bottom estimates and identifications are to use general parameters in low numbers. This is the case with our model. Furthermore, all the parameters in the model have a physical basis and are amenable to being further constrained in their fitting range by any information available from other sources such as the size and shape of chloroplasts, size of mineral grains and size of the spacing for near-surface fragmentation and porosity of rocks and sand.

We are currently using as a default reference the specific absorption spectrum of Bricaud et al. [9]. Given the extensive work in relation to coral reefs on the end member spectra and their variability [21] it may be possible in future to derive reference spectra better suited to modeling vegetation that also include a better balanced and more comprehensive mix of accessory pigments. As we have seen, a substantial part of the vegetation reflectance signature differences and spectral variability may be explained by the choloroplast absorption saturation effect. The remaining differences could, therefore, be less significant than appear at first glance, thus potentially reducing the number of distinct spectral absorption compositions required to model the end member signatures set. The other potential contribution of the approach we have taken of modeling the reflectance signatures by a radiative transfer model is that the variability in the spectral signatures clearly outlined for instance by Hochberg et al. [21] can be related explicitly to several parameters of interest in the study of corals such as the thickness of the thin translucent organic cover z_b over the mineral substrate (Equation (20)), the size of the scattering features of the cells, and their chloroplast pigment concentration.

We must remember that bathymetry in unknown waters is one of the main drivers for restricting the number of variables to optimize to obtain an estimate of the water column depth. The standard approach of using a combination of linear mixes is problematic when there is no ground truth or a priori knowledge of the bottom to restrict the space of end members for bottom reflectance. However, there is a way to use the best features of our model and the linear mixing method. We first solve for depth using the generic bottom reflectance model proposed here. Once the depth is estimated, we can use the measured water surface irradiance reflectance to derive the bottom spectrum that would produce that measured surface irradiance reflectance. Given this bottom spectrum we can then use the standard linear mixing method to determine the bottom vegetation and mineral types that compose it, thus extracting valuable information about the bottom type. Given this new information we can recompute the depth and correct for any error in the original approximate model, therefore maximizing the benefits of both approaches. This mixed method will be the subject of future investigations.

Author Contributions: G.F., J.-P.A. and M.L. conceptualized the study together. G.F. developed the model and wrote up the work. J.-P.A. led and was responsible for all the trials that gathered the hyperspectral data from airborne and ground-based detectors. He also analyzed the results to extract the irradiance reflectance spectra. M.L. validated the reflectance spectra using a precision polarized bi-directional reflectance function measurement system he designed and developed.

Acknowledgments: The totality of the funding for this work and its publication costs was from Defence Research and Development Canada (DRDC). The authors would like to thank Dorte Krause-Jensen and Birgit Olesen for kindly using their expertise and taking the time to identify the vegetation species in the sample used in Section 3.3. The authors would also like to thanks both reviewers for very detailed and constructive comments which helped significantly improve this paper.

Appendix A

The index of refraction equations for water, calcite, quartz and cellulose used in this paper are given below. Calcite and crystalline quartz are birefringent materials with two orientation dependent indices of refraction, the extraordinary index for propagation along the direction of the optical axis and the ordinary index for propagation orthogonal to the optical axis. These were measured by Gosh [8].

$$n_{co}^2 - 1 = 0.73358749 + \frac{0.96464345\,\lambda^2}{\lambda^2 - 0.0194325203} + \frac{1.8283145\,\lambda^2}{\lambda^2 - 120.}, \tag{A1}$$

$$n_{ce}^2 - 1 = 0.35859695 + \frac{0.82427830\,\lambda^2}{\lambda^2 - 0.0106689543} + \frac{0.14429128\,\lambda^2}{\lambda^2 - 120.}, \tag{A2}$$

$$n_{qo}^2 - 1 = 0.28604141 + \frac{1.07044083\,\lambda^2}{\lambda^2 - 0.0100585997} + \frac{1.10202242\,\lambda^2}{\lambda^2 - 100.}, \tag{A3}$$

$$n_{qe}^2 - 1 = 0.28851804 + \frac{1.09509924\lambda^2}{\lambda^2 - 0.0102101864} + \frac{1.15662475\,\lambda^2}{\lambda^2 - 100.}, \tag{A4}$$

$$n_{cl}^2 - 1 = \frac{1.124\,\lambda^2}{\lambda^2 - 0.011087}, \tag{A5}$$

$$n_{wq} = 1.31405 - 2.02 \times 10^{-6} T_c^2 + \frac{0.01586 - 4.23 \times 10^{-6} T_c}{\lambda} - \frac{0.004382}{\lambda^2} + \frac{0.0011455}{\lambda^3}, \tag{A6}$$

The wavelength in the expressions above is in microns. n_{co} and n_{ce} are, respectively, the ordinary and extraordinary index of calcite. n_{qo} and n_{qe} are the ordinary and extraordinary index of crystalline quartz. n_{cl} is the index of pure solid cellulose measured by Sultanova et al. [9]. n_{wq} is the index of pure water measured by Quan and Fry [10] which is strictly only valid to its full accuracy between 0.4 to 0.7

microns. In that last expression, T_c is the temperature in degrees centigrade. However, we need for our approach an expression for the index of water that is valid in the near IR. Schriebener et al. [11] have proposed such an expression valid from 0.2 to 2.5 microns. We have verified that it does match with the available data and the Quan and Fry formula over its range of applicability.

$$\frac{(n_{ws}^2-1)}{(n_{ws}^2+2)}\left(\frac{1}{\overline{\rho}}\right) = a_0 + a_1\overline{\rho} + a_2\overline{T} + a_3\overline{\lambda}\,\overline{T} + \frac{a_4}{\overline{\lambda}^2} + \frac{a_5}{\left(\overline{\lambda}^2 - \overline{\lambda_{uv}}^2\right)} + \frac{a_6}{\left(\overline{\lambda}^2 - \overline{\lambda_{ir}}^2\right)} + a_7\overline{\rho}^2 \tag{A7}$$

In this expression we have:

$$\overline{T} = \frac{T_k}{273.15}$$

$$\overline{\lambda} = \frac{\lambda}{0.589}$$

$$\overline{\rho} = \frac{\rho}{1\text{ g cm}^3}$$

Table A1. Coefficients of the water index of refraction Formula (A7).

Coefficient	Coefficient
$a_o = 0.244257733$	$a_4 = 1.58920570 \times 10^{-3}$
$a_1 = 9.74634476 \times 10^{-3}$	$a_5 = 2.45934259 \times 10^{-3}$
$a_2 = -3.73234996 \times 10^{-3}$	$a_6 = 0.900704920$
$a_3 = 2.68678472 \times 10^{-4}$	$a_7 = -1.66626219 \times 10^{-2}$
$\overline{\lambda_{uv}} = 0.2292020$	$\overline{\lambda_{ir}} = 5.432937$

Formula (A7) is the one we use in this paper because of the extended range we require. We have verified that it matches to one part in a thousand the index formula given by Quan and Fry and that it tracks closely the available experimental data on water index in the near IR and UV.

If we assume that the orientation of the optical axis of the calcite and quartz grains is random, we need to compute the resulting average index as follows. The ordinary index n_o is the same no matter the angular orientation of the incoming ray with the optical axis of the crystal. The extraordinary index n_e varies as a function of the angles with respect to the optical axis $n_e(\varphi, \theta)$. The shape of the variation is this spheroid defined by:

$$\frac{k_x^2}{n_e^2} + \frac{k_y^2}{n_e^2} + \frac{k_z^2}{n_o^2} = \frac{\omega^2}{c^2} \tag{A8}$$

The light-wave propagation vector is $\boldsymbol{k}$ and its angular frequency is ω with the speed of light being given by c. The optical axis is along the z direction. Transforming to cylindrical coordinates the spheroid is symmetrical about the angle φ and elliptical in θ. If we assume that the distribution of the optical axis is random we can derive an expression for the mean extraordinary index.

$$n_e(\theta) = \sqrt{n_o^2 \cos\theta^2 + n_e^2 \sin\theta^2}\,, \tag{A9}$$

$$<n_e> = \frac{\int_0^{\pi/2} n_e(\theta)\sin\theta\, d\theta}{\int_0^{\pi/2} \sin\theta\, d\theta}\,, \tag{A10}$$

The result of the integral is:

$$<n_e> = \frac{n_o}{2}\left\{1 + \frac{n_e^2}{n_o^2}\frac{1}{\sqrt{1 - n_e^2/n_o^2}} + \ln\left[\frac{n_o}{n_e} + \sqrt{1 - n_e^2/n_o^2}\right]\right\}, \tag{A11}$$

The final result for the mean index for random birefringent crystal orientation is:

$$n_e(\theta) = <n> = \frac{n_o + n_e}{2}, \quad \text{(A12)}$$

We use these formulae to compute the mean index of both calcite and quartz crystals.

Appendix B

To compute the absorption saturation effect, first we need formulas for the absorption efficiency Q_a of the chloroplasts. The original model assumed that the chloroplasts were spherical and that their absorption efficiency can be modeled using the anomalous diffraction theory which is applicable since there is almost no difference in the real part of the index of refraction for the chloroplasts and the surrounding cell medium.

For spherical chloroplasts, the absorption efficiency is given by:

$$Q_{a-sph}(z) = 2\left[\frac{1}{2} + \frac{e^{-z}}{z} + \frac{(e^{-z}-1)}{z^2}\right], \quad \text{(A13)}$$

$$z = a_o^*(\lambda)u\,,$$

$$u = \rho_{cp} d_{cp}\,,$$

$a_o^*(\lambda)$ is the specific mass absorption coefficient of chlorophyll-a at low concentration in units of $m^2 \cdot gr^{-1}$. ρ_{cp} is the chlorophyll-a mass density inside the chloroplast in $gr \cdot m^{-3}$ and d_{cp} is the diameter of the chloroplast in meters.

Note that in the limit of small z we have:

$$Q_{a-sph}(z) = \frac{2z}{3}, \quad \text{(A14)}$$

Since in the limit of small concentrations the chloroplast absorption will be unsaturated and equal to the limiting absorption, we can write the absorption saturation gain function for a spherical chloroplast as:

$$G_{a-sph}(z) = \left(\frac{3}{2\,z}\right) Q_{a-sph}(z)\,, \quad \text{(A15)}$$

The saturated absorption spectrum for spherical chloroplasts can, therefore, be computed as:

$$a_v^*(\lambda) = \left(\frac{3}{2\,z}\right) Q_{a-sph}(z) a_o^*(\lambda)\,, \quad \text{(A16)}$$

It is interesting to estimate what the effect of chloroplast shape maybe in the estimate of this packaging effect. Chloroplasts are often disk shaped and we will use the form for the absorption efficiency for randomly oriented disks.

$$Q_{a-dsk}(z_d) = 2 - E_3(z_d)\,, \quad \text{(A17)}$$

With,

$$z_d = a_o^*(\lambda)\rho_{cp}\tau_{cp} = a_o^*(\lambda)\, u_{cp}, \quad \text{(A18)}$$

τ_{cp} is the thickness of the disk. We also define f as the ratio of the thickness τ_{cp} of the disk to its diameter d_{cp}. $E_3(z)$ is the exponential integral function of order 3 which is defined as:

$$E_n(z) = \int_1^\infty \frac{e^{-z\,t}}{t^n} dt\,,$$

In the limit of small z we have:

$$Q_{a-dsk}(z_d) = 2\,z_d\,, \tag{A19}$$

The gain function for randomly oriented disks becomes:

$$G_{a-dsk}(z_d) = \left(\frac{1}{2\,z_d}\right) Q_{a-dsk}(z_d)\,, \tag{A20}$$

Finally the saturated absorption spectrum for disks can be computed as:

$$a_v^*(\lambda) = \left(\frac{1}{2\,z_d}\right) Q_{a-dsk}(z_d) a_o^*(\lambda)\,, \tag{A21}$$

The exact formulas given above can be approximated to within a 10% relative error by the following simple exponential forms.

$$a_v^*(\lambda) = \left(\frac{3}{2\,z}\right)\left(1 - e^{-2z/3}\right) a_o^*(\lambda)\,, \tag{A22}$$

$$a_v^*(\lambda) = \left(\frac{1}{2\,z_d}\right)\left(1 - e^{-2z_d}\right) a_o^*(\lambda)\,, \tag{A23}$$

References

1. Albert, A.; Mobley, C. An analytical model for subsurface irradiance and remote sensing reflectance in deep and shallow case-2 waters. *Opt. Express* **2003**, *11*, 2873–2890. [CrossRef] [PubMed]
2. Jonasz, M.; Fournier, G. *Light Scattering by Particles in Water: Theoretical and Experimental Foundations*; Academic Press: New York, NY, USA, 2007; pp. 39–42, 77–80, ISBN 10: 0-12-388751-8.
3. Szudy, J.; Bayliss, W.E. Uniform Frank-Condon treatment of pressure broadening of spectral lines. *J. Quant. Spectrosc. Radiat. Transf.* **1975**, *15*, 641–668. [CrossRef]
4. Wooten, F. *Optical Properties of Solids*; Academic Press: New York, NY, USA, 1972; pp. 42–52, ISBN 9781483220765.
5. Aas, E. Two-stream irradiance model for deep waters. *Appl. Opt.* **1987**, *26*, 2095–2101. [PubMed]
6. Jonasz, M.; Fournier, G. *Light Scattering by Particles in Water: Theoretical and Experimental Foundations*; Academic Press: New York, NY, USA, 2007; pp. 119–120, ISBN 10: 0-12-388751-8.
7. Fournier, G.; Neukermans, G. An Analytical Model for Light Backscattering by Coccoliths and Coccospheres of Emiliania Huxleyi. *Opt. Express* **2017**, *25*, 14996–15009. [CrossRef] [PubMed]
8. Morel, A.; Bricaud, A. Theoretical results concerning light absorption in a discrete medium, and application to specific absorption of phytoplankton. *Deep Sea Res.* **1981**, *28*, 1375–1393. [CrossRef]
9. Bricaud, A.; Morel, A.; Babin, M.; Allali, K.; Claustre, H. Variations of light absorption by suspended particles with the chlorophyll-a a concentration in oceanic (Case 1) waters: Analysis and implications for bio-optical models. *J. Geophy. Res.* **1998**, *103*, 31033–31044. [CrossRef]
10. Ciotti, A.M.; Lewis, M.R.; Cullen, J.J. Assessment of the relationships between dominant cell size in natural phytoplankton communities and the spectral shape of the absorption coefficient. *Limnol. Oceanogr.* **2002**, *47*, 404–417. [CrossRef]
11. Ghosh, G. Dispersion-equation coefficients for the refractive index and birefringence of calcite and quartz crystals. *Opt. Commun.* **1999**, *163*, 95–102. [CrossRef]
12. Sultanova, N.; Kasarova, S.; Nikolov, I. Dispersion properties of optical polymers. *Acta Phys. Pol. A* **2009**, *116*, 585–587. [CrossRef]
13. Quan, X.; Fry, E.S. Empirical equation for the index of refraction of seawater. *Appl. Opt.* **1995**, *34*, 3477–3480. [CrossRef] [PubMed]
14. Schiebener, P.; Straub, J.; Levelt Sengers, J.M.H.; Gallagher, J.S. Refractive Index of Water and Steam as Function of Wavelength, Temperature and Density. *J. Phys. Chem. Ref. Data* **1990**, *19*, 677–717. [CrossRef]
15. Baranoski, G.V.G. Modeling the interaction of infrared radiation (750 to 2500 nm) with bifacial and unifacial plant leaves. *Remote Sens. Environ.* **2006**, *100*, 335–347. [CrossRef]

16. Gitelson, A. The peak near 700 nm on radiance spectra of algae and water: Relationship of its magnitude and position with chlorophyll-a concentration. *Int. J. Remote Sens.* **1992**, *13*, 3367–3373. [CrossRef]
17. Pope, R.M.; Fry, E.S. Absorption spectrum (380–700 nm) of pure water. II. Integrating cavity measurements. *Appl. Opt.* **1997**, *36*, 8710–8723.
18. Kou, L.; Labrie, D.; Chýlek, P. Refractive indices of water and ice the 0.65 to 2.5 m spectral range. *Appl. Opt.* **1993**, *32*, 3531–3540. [CrossRef] [PubMed]
19. Jacquemoud, S.; Ustin, S.L.; Verdebout, J.; Schmuck, J.; Andreoli, G.; Hosgood, B. Estimating leaf biochemistry using the PROSPECT leaf optical properties model. *Remote Sens. Environ.* **1996**, *56*, 194–202. [CrossRef]
20. Adolfo, A.; Martin, P.; UV-Visible NIR Microspectroscopy of Nanocrystalline cellulose. *CRAIC Technol.* **2013**. Available online: http://www.warsash.com.au/news/articles/craic-application-paper.pdf (accessed on 10 May 2018).
21. Hochberg, E.J.; Atkinson, M.J.; Andrefouet, S. Spectral reflectance of coral reef bottom-types worldwide and implications for coral reef remote sensing. *Remote Sens. Environ.* **2003**, *85*, 159–173. [CrossRef]

Permissions

All chapters in this book were first published in MDPI; hereby published with permission under the Creative Commons Attribution License or equivalent. Every chapter published in this book has been scrutinized by our experts. Their significance has been extensively debated. The topics covered herein carry significant findings which will fuel the growth of the discipline. They may even be implemented as practical applications or may be referred to as a beginning point for another development.

The contributors of this book come from diverse backgrounds, making this book a truly international effort. This book will bring forth new frontiers with its revolutionizing research information and detailed analysis of the nascent developments around the world.

We would like to thank all the contributing authors for lending their expertise to make the book truly unique. They have played a crucial role in the development of this book. Without their invaluable contributions this book wouldn't have been possible. They have made vital efforts to compile up to date information on the varied aspects of this subject to make this book a valuable addition to the collection of many professionals and students.

This book was conceptualized with the vision of imparting up-to-date information and advanced data in this field. To ensure the same, a matchless editorial board was set up. Every individual on the board went through rigorous rounds of assessment to prove their worth. After which they invested a large part of their time researching and compiling the most relevant data for our readers.

The editorial board has been involved in producing this book since its inception. They have spent rigorous hours researching and exploring the diverse topics which have resulted in the successful publishing of this book. They have passed on their knowledge of decades through this book. To expedite this challenging task, the publisher supported the team at every step. A small team of assistant editors was also appointed to further simplify the editing procedure and attain best results for the readers.

Apart from the editorial board, the designing team has also invested a significant amount of their time in understanding the subject and creating the most relevant covers. They scrutinized every image to scout for the most suitable representation of the subject and create an appropriate cover for the book.

The publishing team has been an ardent support to the editorial, designing and production team. Their endless efforts to recruit the best for this project, has resulted in the accomplishment of this book. They are a veteran in the field of academics and their pool of knowledge is as vast as their experience in printing. Their expertise and guidance has proved useful at every step. Their uncompromising quality standards have made this book an exceptional effort. Their encouragement from time to time has been an inspiration for everyone.

The publisher and the editorial board hope that this book will prove to be a valuable piece of knowledge for researchers, students, practitioners and scholars across the globe.

List of Contributors

Xiaogang Xing
State Key Laboratory of Satellite Ocean Environment Dynamics, Second Institute of Oceanography, Ministry of Natural Resources, Hangzhou 310012, China
Pilot National Laboratory for Marine Science and Technology (Qingdao), Qingdao 266000, China

Emmanuel Boss
School of Marine Sciences, University of Maine, Orono, ME 04469, USA

Fei Chai
State Key Laboratory of Satellite Ocean Environment Dynamics, Second Institute of Oceanography, Ministry of Natural Resources, Hangzhou 310012, China
School of Marine Sciences, University of Maine, Orono, ME 04469, USA

Jie Zhang
Laboratoire d'Océanographie de Villefranche, Sorbonne Université, 06238 Villefranche-sur-Mer, France

Christian Bräuer-Burchardt, Christoph Munkelt, Ingo Gebhart, Matthias Heinze, Stefan Heist and Peter Kühmstedt
Fraunhofer Institute for Applied Optics and Precision Engineering IOF, Albert-Einstein-Str. 7, 07745 Jena, Germany

Gunther Notni
Fraunhofer Institute for Applied Optics and Precision Engineering IOF, Albert-Einstein-Str. 7, 07745 Jena, Germany
Department of Mechanical Engineering, Ilmenau University of Technology, Gustav-Kirchho-Platz 2, 98693 Ilmenau, Germany

Bingqiang Sun and Ping Yang
Department of Atmospheric Sciences, Texas A&M University, College Station, TX 77843, USA

George W. Kattawar
Department of Physics and Astronomy and Institute for Quantum Science and Engineering, Texas A&M University, College Station, TX 77843, USA

Xiaodong Zhang
Department of Earth System Science and Policy, University of North Dakota, Grand Forks, ND 58202, USA

Daniel Koestner, Dariusz Stramski and Rick A. Reynolds
Marine Physical Laboratory, Scripps Institution of Oceanography, University of California San Diego, La Jolla, CA 92093-0238, USA

Knut Stamnes, Nan Chen, Yongzhen Fan, Wei Li and Zhenyi Lin
Stevens Institute of Technology, Hoboken, NJ 07030, USA

Børge Hamre and Jakob Stamnes
University of Bergen, 5020 Bergen, Norway

Snorre Stamnes
NASA LARC, Hampton, VA 23681, USA

Jacopo Agagliate and David McKee
Department of Physics, University of Strathclyde, 107 Rottenrow, Glasgow G4 ONG, UK

Rüdiger Röttgers and Kerstin Heymann
Helmholtz-Zentrum Geesthacht, Max-Planck-Straße 1, 21502 Geesthacht, Germany

John D. Hedley
Numerical Optics Ltd., Tiverton EX16 8AA, UK

Maryam Mirhakak and Heidi M. Dierssen
Department of Marine Sciences, University of Connecticut, Groton, CT 06340, USA

Adam Wentworth
Materials Science and Engineering Department, University of Connecticut, Storrs, CT 06269, USA

Guoqing Wang and Zhongping Lee
School for the Environment, University of Massachusetts Boston, Boston, MA 02125, USA

Colleen B. Mouw
Graduate School of Oceanography, University of Rhode Island, Narragansett, RI 02882, USA

Lucile Duforêt-Gaurier, David Dessailly and Hubert Loisel
Univ. Littoral Cote d'Opale, Univ. Lille, CNRS, UMR 8187, LOG, Laboratoire d'Océanologie et de Géosciences, F 62930 Wimereux, France

William Moutier
Royal Meteorological Institute of Belgium, 1180 Brussels, Belgium

Georges Fournier, Jean-Pierre Ardouin and Martin Levesque
DRDC Valcartier Research Centre, Québec, QC G3J1X5, Canada

Index